W9-BLV-028

FOURIER SERIES and ORTHOGONAL FUNCTIONS

Harry F. Davis

Professor of Applied Mathematics
University of Waterloo, Ontario

Dover Publications, Inc.
NEW YORK

Published in Canada by General Publishing Company, Ltd., 30 Lesmill Road, Don Mills, Toronto, Ontario.
Published in the United Kingdom by Constable and Company, Ltd.

This Dover edition, first published in 1989, is an unabridged, slightly corrected republication of the work originally published in 1963 by Allyn and Bacon, Inc., Boston, Massachusetts.

Manufactured in the United States of America
Dover Publications, Inc., 31 East 2nd Street, Mineola, N.Y. 11501

Library of Congress Cataloging-in-Publication Data

Davis, Harry F.
Fourier series and orthogonal functions / Harry F. Davis.
p. cm.
Reprint. Originally published : Boston : Allyn and Bacon, 1963.
Includes index.
ISBN 0-486-65973-9
1. Mathematical physics. 2. Fourier series. 3. Functions, Orthogonal. I. Title.
QA401.D32 1989 88-34072
530.1′5—dc19 CIP

TO MYRNA

PREFACE

This is an introductory text on the theory of Fourier series and orthogonal functions, and applications of the Fourier method to the solution of boundary-value problems. I have tried to present these topics in a clear manner, always assuming that the reader has no prior knowledge of any of them and wishes to get an over-all picture before studying specialized topics.

An attempt is made to give the reader an intuitive understanding of the theory and the way the theory is used in solving problems. Considerable emphasis is given to the physical interpretation of the theory. At the same time, the theory is developed in a logically careful manner. Convergence theorems for Fourier series are proved in detail. The basic properties of Legendre polynomials, spherical harmonics, and Bessel functions are derived rigorously.

In writing this book, a central idea has been to blend modern theory with topics of down-to-earth practicality. I have tried to carry out this idea genuinely, rather than give it the usual perfunctory lip service. I hope thereby to provide a book that better fits a modern curriculum and gives a foundation for later study in functional analysis, abstract harmonic analysis, and quantum mechanics.

This book is intended as much for mathematicians as for engineers and physicists. Several topics are included here which are not usually found in undergraduate texts. The most important of these are summability theory, generalized functions, and spherical harmonics. Numerical methods have not been completely overlooked.

Most of the book can be read by anyone familiar with partial derivatives, multiple integrals, vectors, and elementary differential equations. Prior knowledge of partial differential equations or advanced vector analysis is not required.

It is not necessary that the topics be taken up in the order presented. After reading the first chapter, the reader can proceed to any one of the later chapters; the index should be consulted in order to locate the definitions of unfamiliar terms.

The reader can, if he likes, overlook many details in reading this book. The author is aware that undergraduate engineers often have little interest in detailed proofs. Nevertheless, in writing this text the author has sincerely tried, within his limited capabilities (and what he presumes are the limitations of undergraduate readers) to present the best that modern scholarship has to offer. Emphasis is on general mathematical ideas used in engineering and physics *and the relationships between them.* In the author's opinion, this calls for mathematical rigor, balanced by adequate motivation for the more difficult ideas.

The author makes few claims to novelty in the presentation of this material. Several minor discussions are included which, so far as the author is aware, have never appeared in print before, simply because they are usually deemed too advanced for undergraduates but too elementary for graduate students. The last section in Chapter 6 is the only one in which we assume more than the basic prerequisites; a reader who has not studied modern algebra should make no attempt to study this section. The Appendix is intended solely for beginning graduate students in pure mathematics; it is doubtful whether anyone else would find it interesting.

There are about 570 exercises. Answers to most of them will be found at the end of the book. The reader who lacks time to solve exercises should nevertheless consider them as an integral part of the text; he should read them and also the comments in the Answers and Notes section. The more difficult exercises are starred. The exercises have been devised to encourage the reader to review what has been read, apply the theory to specific problems, and give him a continuing sense of achievement as he progresses through the book.

I wish to thank Professor Andrew Browder of Brown University and Professor David J. Benney of the Massachusetts Institute of Technology, for their comments and criticism, always pertinent and valuable; Professors M. A. McKiernan and D. G. Wertheim for their interest and assistance; Mrs. Alma Fielding, the only person I recognize as a better mathematical typist than myself; my students H. P.

McSlide and M. M. McSnoyd, who worked through all the exercises and wrote most of the Answers and Notes section; the editorial staff of Allyn and Bacon, for tolerating my eccentricities and for their superb cooperation; Colonel W. R. Sawyer, truly a scholar, soldier, and gentleman, who has inspired a generation of Officer Cadets and has profoundly influenced the writing of this book; and above all others Myrna MacPhie, who was present at the conference which led to this book, encouraged me in every stage of preparing it, and is now my wife.

H. F. D.

CONTENTS

CHAPTER 1

LINEAR SPACES

1.1 • FUNCTIONS

We begin by stating certain conventions that will be used throughout this book.

The words *number* and *scalar*, when not otherwise qualified, may be regarded as synonyms for *real number*. On the other hand, if they are consistently interpreted to mean *complex number*, most of the statements made are still valid.

Similarly, the beginner who wishes to do so can take the word *function* to mean *real-valued function*, unless indicated otherwise, and in this case complex conjugates can be ignored. Simply read $f(x)$ for $\overline{f(x)}$ and α^2 for $|\alpha|^2$, since for real functions $\overline{f(x)} = f(x)$ and for real numbers $|\alpha|^2 = \alpha^2$.

The term *domain of a function* is used for brevity instead of *domain of definition of a function*. The reader must distinguish between functions for which no domain is specified, and those for which a domain *is* specified. If no domain is specified, and no indication given to the contrary, the reader should assume that $f(x)$ is defined for *every real number* x. To carry matters a little further, suppose we ask the reader to consider a function f with the property $f(-x) = -f(x)$. Then the reader may consider the function $\sin x$, or any other function he chooses (having this property) but not csc x, since csc x is not defined when x is an integral multiple of π.

If the domain is specified, then the reader must regard the function as defined *only* within this domain. This will be clarified later by examples.

A distinguishing characteristic of advanced mathematics is that a function is regarded as a single object, just as points, numbers, and vectors are regarded as single objects. We usually emphasize this by using a single letter to denote a function. For example, if we specify that $f(x) = x^2$ for every x, then we have specified a function f. The single letter f denotes the function; $f(x)$ refers to a *number*, namely the value the function corresponds with x. In this example, if $x = 3$, then $f(x) = 9$.

We shall not be pedantic in this matter, however. To avoid circumlocution, we will sometimes refer to "the function x^2" or "the function $\sin x$" where we should more properly say "the function f defined by the requirement that for every real number x the value $f(x)$ is x^2" or "the function g where $g(x) = \sin x$ for every x."

Throughout this book *function* always means *single-valued function.* For any x in the domain of a function f, $f(x)$ is a single number which is not ambiguous; we do not discuss multiple-valued functions.

If two functions f and g have the same domain, we say they are *equal* and write $f = g$ if and only if $f(x) = g(x)$ for every x in their domain. Functions having different domains of definition are never regarded as equal.

The *sum* of two functions is defined only when the functions have the same domain. If f and g are functions with the same domain of definition D, their sum $f + g$ is, by definition, the function having domain D whose values are the sum of the corresponding values of f and g. That is, the values of $f + g$ are given by

$$(f + g)(x) = f(x) + g(x)$$

for every x in D.

The same qualification applies to the *product* of two functions, which is defined by

$$(fg)(x) = [f(x)][g(x)].$$

This is sometimes called the *pointwise product* to distinguish it from the convolution product that will be introduced in Section 2.6.

For any number C, the function Cf is similarly defined by

$$(Cf)(x) = Cf(x)$$

and has the same domain as f. Any function Cf is called a *scalar multiple* of f.

We often find it useful to distinguish between the *number* zero and a *function* that is identically equal to zero. Such a function is called a *zero function* and will conventionally be denoted θ. That is, we have $\theta(x) = 0$ for every x in the domain. This notation is ambiguous unless the domain of definition is specified, since we regard two functions having different domains as distinctly different functions, even if both of them is a zero function.

Some of the functions given as examples below are obtained by piecing together expressions with which the reader is already familiar. Nevertheless, the reader must regard each of these as a single function and not a "patchwork" of different functions.

Many students have the incorrect idea that, to specify a function, one must have a simple "formula" for determining its values. We will show by an example that this need not be the case. (See Example 5.)

EXAMPLE 1: Let f be the function having domain $0 \leqq x \leqq 1$, defined by $f(x) = x^2$. The graph of this function is the heavy curve in Figure 1.1.

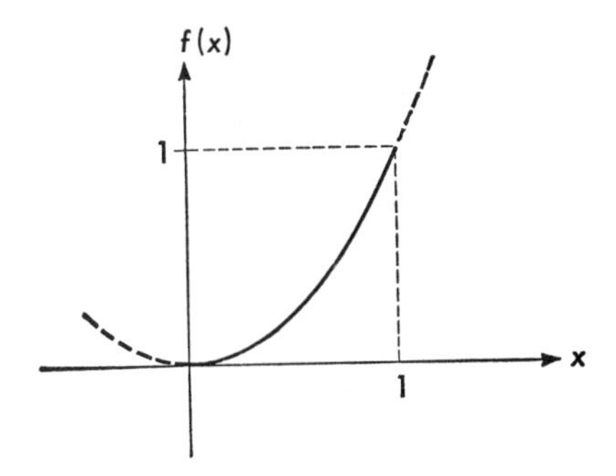

Figure 1.1

The natural tendency of the student is to extend the graph as the dotted curve indicates. This is a mistake we wish the reader to avoid. If you are told the domain is $0 \leqq x \leqq 1$, then you must forget about your old friend the function x^2 discussed in elementary calculus, for this is not the same function. This function is not defined for values of x other than those in this interval.

Observe that $f(0) = 0$, $f(\frac{1}{3}) = \frac{1}{9}$, and $f(1) = 1$. Notice that $f(-1)$ and $f(1.2)$ are not defined. If you were asked to solve the equation $f(x) = 1 - (3x/2)$, you would obtain $x = \frac{1}{2}$ but you would reject the possibility $x = -2$.

EXAMPLE 2: Now consider the function f defined for all x by

$$f(x) = \begin{cases} 0 & x < 0 \\ x^2 & 0 \leqq x \leqq 1 \\ 1 & x > 1 \end{cases}$$

Part of the graph of this function is pictured in Figure 1.2.

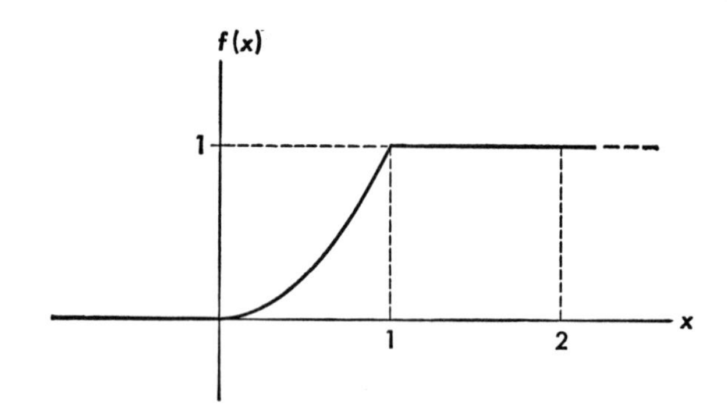

Figure 1.2

Here you can fill in the dotted parts indicated; indeed, you should imagine the graph extending infinitely far to the left and to the right. This function is *not* equal to the function defined in Example 1; its domain of definition is not the same.

Despite the unusual appearance of this function, it is continuous for all values of x, and even has a derivative which is defined for every x except $x = 1$. It is not difficult to evaluate integrals involving such a function; for example, to find $\int_{-1}^{3} f(x)\,dx$ we proceed as follows:

$$\begin{aligned} \int_{-1}^{3} f(x)\,dx &= \int_{-1}^{0} f(x)\,dx + \int_{0}^{1} f(x)\,dx + \int_{1}^{3} f(x)\,dx \\ &= \int_{-1}^{0} 0\,dx + \int_{0}^{1} x^2\,dx + \int_{1}^{3} 1\,dx \\ &= 0 + \tfrac{1}{3} + 2 = 2\tfrac{1}{3}. \end{aligned}$$

Since the domain of the function defined in this example is not the same as that defined in Example 1, the *sum* of these two functions is not defined.

EXAMPLE 3: Let f be defined by

$$f(x) = x - n \qquad \text{whenever } n \leqq x < n + 1 \qquad (n = 0, 1, 2, 3, \cdots).$$

The domain is the set of all nonnegative real numbers (i.e., the positive reals and also zero). This function is not defined to the left of $x = 0$. (It is immaterial that the function *could* be defined for such values of x, by modifying the definition.)

Notice that the graph of such a function, when sketched as in Figure 1.3, does not convey all the necessary information. There is

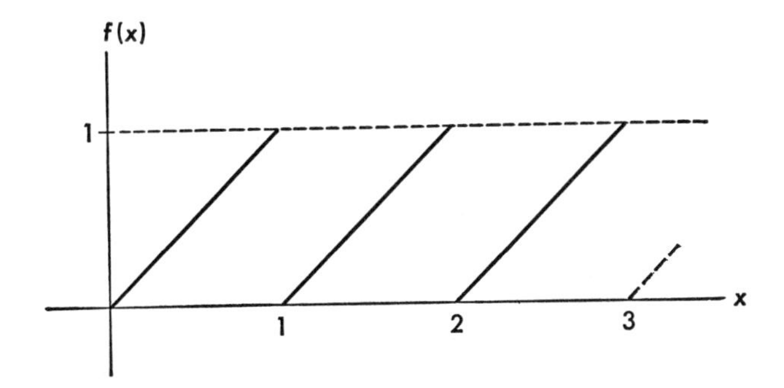

Figure 1.3

no way to tell from the figure whether $f(2) = 1$ or $f(2) = 0$. (We see from the definition that $f(2) = 0$.) This function is discontinuous at $x = 1, 2, 3, \cdots$, but for all other values of x in its domain it is not only continuous but is differentiable, with $f'(x) = 1$.

EXAMPLE 4: Consider a taut string of length L, stretched along the x axis between $x = 0$ and $x = L$. Suppose the string is constrained to move in the xy plane and small vibrations are set up in the string, which is photographed by a high-speed camera at a certain instant of time. Consider the photograph as the graph of a function f. The domain of this function is the interval $0 \leqq x \leqq L$. It would not make sense to speak of $f(x)$ for x not in this interval (Figure 1.4).

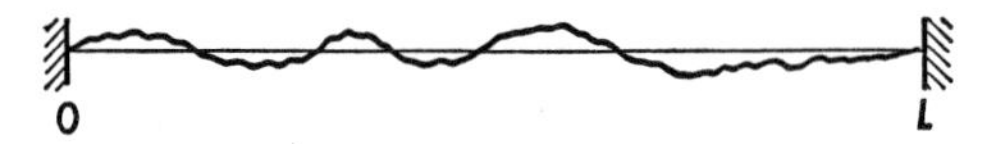

Figure 1.4

This example is intended to dispel any idea the reader may have that it is unimportant in physics or engineering to be concerned with the domain of a function, or with functions defined only in finite intervals.

EXAMPLE 5: For each positive integer n, let $f(n)$ denote the number of distinct prime factors of n. (A prime number is a positive integer $p \geqq 2$ that is not divisible by any positive integer other than itself and 1; for example, 2, 3, 5, 7, 11, 13, 17, $\cdots$.) Then $f(21) = 2$, since there are two prime factors of 21, namely 3 and 7. The domain of this function is the set of all positive integers. It would not make sense to speak of $f(x)$ for $x = \frac{3}{5}$ or $x = \pi$.

EXAMPLE 6: Let f be defined by

$$f(1) = 3, f(2) = 0, f(3) = -1, f(4) = 7$$

and let this constitute the complete definition of f. Then the domain of f consists of the first four positive integers, and $f(x)$ makes sense only when $x = 1, 2, 3,$ or 4. In such an instance, we completely specify f by writing down its values in order: 3, 0, -1, 7.

More generally, if we know that f has as its domain the first n positive integers, we can specify f by writing its values in order: $f(1), f(2), \cdots, f(n)$. Conversely, any sequence of numbers can be considered to determine such a function. Thus $(3, 2, -7, 5, 8, 8)$ represents a function f defined for the first six positive integers; in this case, $f(4) = 5$ and $f(1) = 3$. An infinite sequence of numbers represents a function whose domain is the entire set of positive integers.

If an infinite sequence is constructed in an obvious way, it is common to represent it by giving a few typical terms, followed by three dots meaning "use your own imagination." Thus it is presumed that anyone can tell from $(2, 4, 6, 8, \cdots)$ that the 23rd entry is 46, since the associated function is "obviously" $f(n) = 2n$, and $f(23) = 46$.

EXAMPLE 7: What is the sum of the sequences $(2, 3, 7, 9)$ and $(5, 2, -1, -4)$?

It may appear at first that this question has no relevance to the material in this section, since we have not discussed the addition of sequences. (Pause to consider this for a moment.)

In the preceding example, we saw that each of these sequences can be considered as a function defined for $x = 1, 2, 3$, and 4. Denoting the first sequence by f, we have $f(1) = 2$, $f(2) = 3$, $f(3) = 7$, and $f(4) = 9$. Denoting the second sequence by g, we have $g(1) = 5$, etc. The sum of functions having a common domain has been defined earlier; in this case, the domain is the set consisting of the first four positive integers, and according to the definition, the answer is $(7, 5, 6, 5)$.

EXAMPLE 8: Find

(a) $(3, 2, 2, 7) + 2(3, 1, 1, 1)$

and (b) $(1, 2, 3, 4, \cdots) + (1, -1, 1, -1, 1, \cdots)$.

Solution: (a): $(9, 4, 4, 9)$, and (b): $(2, 1, 4, 3, 6, \cdots)$ for reasons illustrated in Examples 6 and 7. Notice that the idea of a scalar multiple of a function enters into (a), as well as the idea of adding two functions.

Comment on the Exercises

Every section of this book is followed by a set of exercises. These exercises are intended to help to develop understanding; they are not tests. Many students feel that it is efficient to attempt to solve the exercises *before* thoroughly mastering the theoretical material, and the author tends to agree with them.

The starred exercises are more difficult than the others; some of them are intended for readers having more than the minimal prerequisites stated in the Preface.

Students using this book for self-study are invited to correspond with the author.

• EXERCISES

1. For the function f defined in Example 2 (Figure 1.2),

(a) Determine $f(-2)$.

(b) Determine $f(\frac{1}{2})$.

(c) For what values of x is $f(x) = x$?

(d) Find all "solutions" of $f(x) = x/2$.

(e) Determine $\int_0^2 f(x)\,dx$.

2. Considered as functions of a real variable, what is the domain of definition of each of the following functions?
(a) $1/x$.
(b) $(1 + x^2)^{-1}$.
(c) $\dfrac{x}{(x + 1)(x - 2)}$.
(d) e^{-1/x^2}.

3. In most books on analysis (including this one) the *limit convention* is used. That is, the value of a function at a point where a denominator vanishes is understood to be the limit, provided this limit is finite and is not ambiguous. After you have decided in your own mind what this means, determine the domain of definition of each of the following functions, when this convention is adopted:
(a) Those in Exercise 2.
(b) $x/|x|$.
(c) $\dfrac{\sin x}{x}$.
(d) $\dfrac{\sin (n + \frac{1}{2})x}{2\pi \sin \dfrac{x}{2}}$.
(e) $\dfrac{x}{\sin x}$.

4. See if you can determine the function defining the following sequence, and hence determine what the next value in the sequence should be:

$$0, 1, 1, 1, 1, 2, 1, 1, 1, 2, 1, 2, 1, 2, 2, 1, 1, 2, 1, 2, 2, 2, 1, 2, 1, 2, 1, 2, 1, \cdots.$$

***5.** Can you give an explicit formula for the function defining the sequence in Exercise 4? (Don't spend too much time on this.)

6. Give a precise definition of $f - g$, where f and g are functions. Is the difference of two functions always defined?

***7.** (For students familiar with matrix theory.) What is the domain of definition of the function known as the "determinant"?

8. According to the definition given in the text, is the function $x + \dfrac{1}{x}$ the sum of the functions x and $\dfrac{1}{x}$?

1.2 • VECTORS

In the preceding section we pointed out that a sequence of numbers can be regarded as a function. In this section we will point out that points and vectors can also be regarded as functions. Then we will begin the more difficult task of showing that it is sometimes useful to regard functions as vectors or points.

We recall that it is possible to introduce a Cartesian coordinate system into Euclidean space and thereby represent each point in the space by a sequence of three real numbers (x, y, z). In the preceding section, we saw that any sequence of numbers can be regarded as a function. Therefore, points can be regarded as functions.

Any vector in Euclidean space can be represented, relative to a Cartesian coordinate system, by a sequence of three real numbers, its scalar components along the three axes. Therefore vectors can also be regarded as functions. In this connection, it will be noted that the rule for addition of sequences, given in the preceding section, is the familiar rule for addition of vectors in component form, since the sum of two vectors has components that are the sums of the corresponding components of the two vectors. The rule for finding a scalar multiple of a sequence is also the same as the rule for finding a scalar multiple of a vector, when the vector is written in component form.

Now let us explore the idea of regarding functions as points or vectors.

It is useful to think of any real sequence of length n as representative of a point in an n-dimensional Euclidean space. Alternatively, we may regard it as the sequence of components of a vector in such a space. We ignore the difficulties inherent in "visualizing" such spaces, and we assume no prior knowledge, on the part of the reader, with the geometry of spaces of dimension greater than three. Moreover, we recognize that it would be offensive to students of synthetic geometry to make no distinction between Euclidean spaces, defined axiomatically, and coordinate sequences. Therefore we will use the term *Cartesian n-space* throughout this book.

Definition. For any positive integer n, the collection of all real sequences of length n is called *Cartesian n-space* and will be denoted R^n.

Since the geometrical interpretation of R^2 is the Euclidean plane, and that of R^3 is Euclidean space (the "ordinary" space of solid geometry), we will shamelessly refer to any real sequence $(x_1, x_2, \cdots, x_n)$ as a *point* or a *vector* in R^n. When we do not wish to conjure prejudices, we will simply say *element* of R^n.

It will be noted that we draw no sharp distinction between a point and a vector. This is because we imagine that any nonzero vector can be represented by a directed line segment extending from the origin (i.e., its "tail" is at the origin), and this provides an obvious correspondence between points and vectors (the point at the "head" of the arrow representing the vector). The zero vector corresponds to the point at the origin.

There is no need to explain the simpler operations with vectors in R^n. For example, we do not need to explain how to add two such vectors, for this follows from the definition of the sum of two functions, as noted in the preceding section. Other operations will be defined in the next chapter.

The definition is extended in an obvious manner to complex sequences of length n. The collection of all such sequences is called *complex Cartesian n-space.*

We ask the reader who feels there is something "impractical" about spaces of dimension greater than three to withhold judgment until later.

Before proceeding, the reader may find the following two examples instructive. The first one shows that it is useful to regard functions as vectors even in the two-dimensional case. (The precise definition of "dimension" will be given in Section 1.4. Until then we will continue to use it rather loosely.)

EXAMPLE 1: Any function of the form

$$g(t) = A \sin (kt + \phi) \tag{1}$$

where A, k, and ϕ are real constants is said to be *sinusoidal.* We assume that both A and k are not negative in the following discussion; this involves no loss in generality because of the leeway in choosing ϕ.

Notice that both $\sin t$ and $\cos t$ are sinusoidal. To obtain $\cos t$, take $A = 1$, $k = 1$, and $\phi = \pi/2$.

We call A the *amplitude*, ϕ the *phase angle* or simply the *phase*, k the *angular frequency* or angular velocity, and $k/2\pi$ the *frequency*. If $A = 0$ then k and ϕ are not uniquely determined; g is then the zero function. If A is positive, k and ϕ are uniquely determined by the requirement that k be positive and ϕ be in some specified range, say $0 \leqq \phi < 2\pi$.

Any sinusoidal function can be written in the form

$$g(t) = B \sin kt + C \cos kt \tag{2}$$

and conversely, any function of this form is sinusoidal. The proof is left as an exercise. The motivation for the terminology just introduced will also be left to the Exercises; many readers will already be familiar with these terms.

When written in this form, it is easy to see that the sum of two sinusoidal functions of angular frequency k is again a sinusoidal function of angular frequency k. To add two such functions, we simply add the corresponding B's and C's. This is precisely the same as the rule for addition of vectors in R^2. Indeed, for each fixed positive k, we can identify the class of all sinusoidal functions having angular frequency k with the set of all vectors in the Euclidean plane.

This is illustrated in Figure 1.5. If the function is represented by (B, C), and the axes chosen as indicated, then the amplitude A is

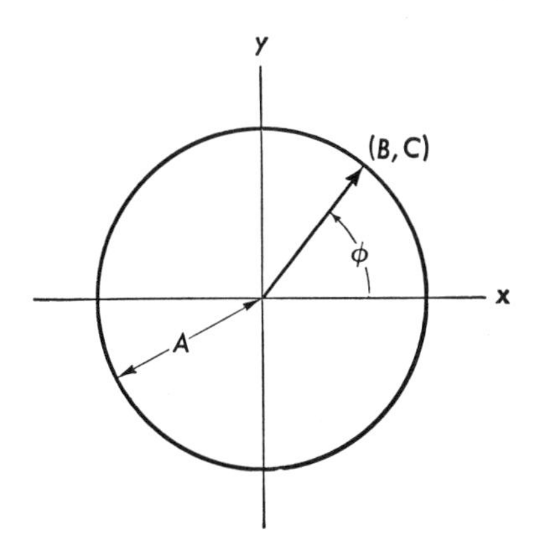

Figure 1.5

represented by the magnitude of the vector, and the phase ϕ is the angle between the vector and the positive x direction, measured as shown.

The easiest way to find the sum of two or more sinusoidal functions having the same angular frequency k, when they are presented in the form of (1), is of course to rewrite all of them in the form of (2). Then it is an easy task to sum them and, if desired, write the sum in the form of (1), an easy exercise in trigonometry. To guard against making sign errors and errors in determining the phase angle of the sum, it is almost a necessity to draw a rough diagram in which each function (and ultimately the sum) is represented by a vector in the plane. This is essentially the procedure used by electrical engineers, except that (for reasons which need not concern us yet) they take the plane to be an Argand diagram (and hence the vectors to be complex numbers).

This example will be considered in greater detail in the Exercises. It has been emphasized mainly because of the importance of sinusoidal functions in our later work. This is one of the simplest instances in which it is useful to think of functions as vectors.

EXAMPLE 2: Let us consider an nth order differential equation of the form

$$\frac{d^n y}{dx^n} + a_1(x)\frac{d^{n-1}y}{dx^{n-1}} + \cdots + a_{n-1}(x)\frac{dy}{dx} + a_n(x)y = 0 \tag{3}$$

where the coefficients $a_1, a_2, \ldots, a_n$ are either constants or functions defined and continuous for every x. The *general solution* of such an equation can always be written in the form

$$y(x) = C_1u_1(x) + C_2u_2(x) + \cdots + C_nu_n(x) \tag{4}$$

where the C's are scalars and the u's are linearly independent. (The meaning of the term "linearly independent" will be reviewed in Section 1.4.)

The reader must clearly keep in mind the meaning of the term "general solution." There are actually infinitely many solutions, since the C's can be chosen arbitrarily. Corresponding to each particular solution of (3) there is a uniquely determined sequence $(C_1, C_2, \cdots, C_n)$. Therefore each solution of (3) can be represented by a vector in Cartesian n-space.

Now let us compare these two examples. In Example 1 each function could be represented in a rather natural way by a vector (B,C), but it will be noted that if we had written (2) in the form $g(t) = B \cos kt + C \sin kt$, interchanging the order in which we wrote $\cos kt$ and $\sin kt$, the vector (B,C) would in general represent a different function. Thus there are at least two equally "natural" ways to represent such functions by vectors.

This problem becomes more acute when we study Example 2. For instance, if $n = 4$ there are $4!$ different ways in which we can arrange the u's in order, and therefore 24 equally "natural" ways to represent particular solutions by vectors. Actually the problem is even worse than this would indicate.

For example, every solution of

$$\frac{d^3y}{dx^3} + 4\frac{dy}{dx} = 0 \tag{5}$$

can be written in the form $C_1 + C_2 \cos 2x + C_3 \sin 2x$. But the reader can readily see that it is also possible to write every solution in the form $C_1 \sin^2 x + C_2 \cos^2 x + C_3 \sin x \cos x$. If the first form is used, the vector $(1,1,1)$ represents the function $1 + \cos 2x + \sin 2x$, but if the second form is used this vector represents the function $1 + \sin x \cos x$.

What these examples show is that there exist some classes of functions for which vector representation is possible. They also show that there is no single manner for accomplishing this, which is rather an unsatisfactory state of affairs. In the next section, we will remedy this by considering each such class of functions as a "space" itself, and each function of the class as itself a "vector." This will remove the artificiality of relating functions with sequences of numbers, and yet we will be able to preserve many cherished geometrical ideas that will be useful in our later work.

• EXERCISES

1. The sequence $(1, 3, 7, 2, 9)$ is a point in R^5. Let the corresponding function on the set consisting of the first five positive integers be denoted g. What is $g(4)$?

2. A vector in Euclidean 3-space has magnitude 6 and makes equal angles with all three axes of a Cartesian coordinate system. Determine $f(1)$, $f(2)$, and $f(3)$ for the corresponding function f.

3. By analogy with the three-dimensional case, determine the midpoint of the line segment in R^4 joining $(1, 2, 6, 0)$ with $(1, 4, 8, 2)$.

4. **(a)** If (1) and (2) represent the same function, write B and C in terms of A and ϕ.

(b) Similarly, write A and ϕ in terms of B and C.

5. Write $\sin x - \cos x$ in the form $A \sin (kx + \phi)$.

6. Determine what is wrong with the following argument. Suppose we are given $f(x) = -\sin x + \cos x$. Then $B = -1$ and $C = 1$, so $A = \sqrt{2}$, and $\phi = \tan^{-1}(-1) = -\pi/4$. Therefore $f(x) = \sqrt{2} \sin (x - \pi/4)$.

7. Let $f(x) = 3 \sin 2x - 4 \cos 2x$.
   (a) What is the amplitude of this function?
   (b) What is its angular frequency?
   (c) What is its phase angle in degrees?

8. Let $f(t) = 5 \sin (2t + \pi/2) - 4 \sin (2t - \pi/4)$. Write $f(t)$ in the form $A \sin (kt + \phi)$.

9. Imagine a particle moving about the circumference of the circle $x^2 + y^2 = A^2$ counterclockwise with uniform angular velocity k radians/second. Suppose that, at $t = 0$, its angular displacement is ϕ.
   (a) Determine its angular displacement θ as a function of t.
   (b) Determine its projection on the y axis (i.e., its y coordinate) as a function of t.
   (c) Hence, explain the reason for the terminology introduced in Example 1.

10. If $k = 0$, is it necessarily true that (1) can be written in the form of (2)? Are B and C then uniquely determined?

11. (This exercise is mainly for engineering students.)
   (a) Explain how to use a sliderule to write a function (1) in the form of (2).
   (b) Show how to obtain A and ϕ if given a function in the form of (2). [Hint: Do *not* use the expression for A derived in Exercise 4(b), since this involves squaring and adding.]

12. Which of the following functions are sinusoidal?
   (a) $\sin x + \cos x$.
   (b) $\sin x + \sin 2x$.
   (c) $\sin x \cdot \cos x$.
   (d) $\sin^2 x$.
   (e) $1 - 2 \cos^2 x$.

1.3 • LINEAR SPACES

By a *linear class of functions* or a *linear space* we mean a class of functions, all having the same domain, with the properties (a) if two functions belong to the class, their sum does also, and (b) if f is a member of the class, every scalar multiple of f is also.

For example, the collection of all continuous functions is a linear space, since the sum of two continuous functions is continuous, and any scalar multiple of a continuous function is continuous.

The terms *vector space* and *function space* are sometimes used by other authors. We caution the reader that most books on modern algebra use the term "vector space" in a more general sense, and a few books call *any* class of functions a "function space."

The elements of a linear space are sometimes called *vectors*. The function identically equal to zero is then called the *zero vector* and any other function is said to be *nonzero*.

If every element of a linear space A is also an element of a linear space B, we say that A is a *subspace* of B. According to this definition, every linear space is a subspace of itself. The zero vector alone constitutes a subspace, called the *trivial subspace*.

EXAMPLE 1: For any set D, let us call the collection of *all* functions having domain D the *function space supported by D*. Every function f having domain D is an element of this space. If D consists of exactly n points, the function space supported by D is our old friend Cartesian n-space, since the points can be numbered from 1 to n and each function considered to have domain $1, 2, \cdots, n$. In the general case, one thinks of each function f as a vector having a component $f(x)$ for each x in D.

EXAMPLE 2: Let us consider the special case of Example 1 obtained by taking D to be the real line. This space is too "large" to be very useful or theoretically interesting, since it contains every function, whether continuous or not, and there is no suitable way for defining what we would mean by the "magnitude" of a vector in this space. Nevertheless, there are a few interesting remarks that can be made about this linear space.

If $f(x) = f(-x)$ for every x, f is said to be an *even* function. If $f(x) = -f(-x)$, then f is called an *odd* function. One reason for this terminology is that if n is a positive integer, x^n is even whenever n is even, and is odd whenever n is odd. (Also see Exercise 12.)

It is easy to show (see Exercise 4) that every function can be written in one and only one way as the sum of an even function and an odd function. Geometrically, we visualize this as shown in Figure 1.6.

The class of all even functions is a linear space, and so also is the class of all odd functions. The only function that is *both* even and odd is the zero function θ. Both of these linear spaces are subspaces of the linear space of all functions. We represent them schematically

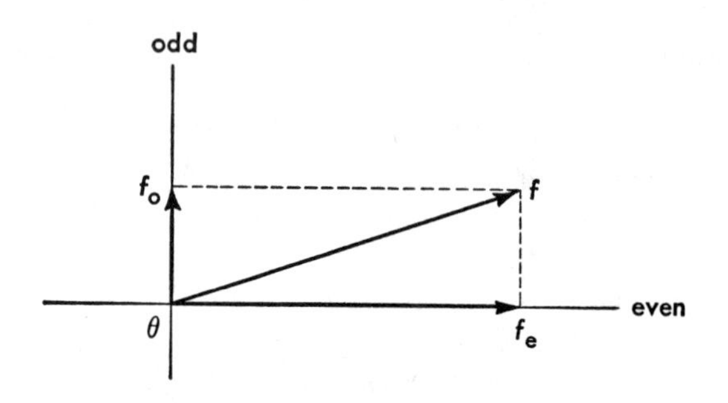

Figure 1.6

by lines in the figure. The zero function is the point of intersection of these lines. Any nonzero function f can be represented by a nonzero vector extending from the origin, which can be written uniquely as a vector sum of vectors contained in the two subspaces mentioned.

Schematic diagrams such as this should not be taken too literally. For example, it might appear from this figure that the linear space of all even functions is one-dimensional, whereas in fact it is infinite-dimensional.

Example 3: A function f is said to be *periodic* if for some fixed positive number t we have $f(x + t) = f(x)$ for all x. The number t is said to be a *period* of f. For each fixed t, the class of all functions of period t is a linear space.

Example 4: The class of all functions defined and continuous in the interval $a \leqq x \leqq b$ is a linear space. This space is denoted $C[a,b]$ throughout this book. If f is of class $C[a,b]$, the *norm* of f is the number $||f||$ defined by

$$||f|| = \left[\int_a^b |f(x)|^2\, dx\right]^{1/2}.$$

(More properly, this is the *mean square norm;* another type of norm will be discussed later.) We will see later on that the norm of f can be interpreted as the "length" of f when we think of f as a vector.

Example 5: The class of all infinite sequences $x = (x_1, x_2, x_3, \cdots)$ for which $\sum_{n=1}^{\infty} |x_n|^2 < \infty$ is a linear space (see Exercise 23). This space is called *Hilbert coordinate space.* The magnitude of a vector x in Hilbert coordinate space is defined to be the number

$$\|x\| = \left[\sum_{n=1}^{\infty} |x_n|^2\right]^{1/2}.$$

(For further details see Sec. 2.3.)

Many other examples will arise in later sections of this book.

• EXERCISES

1. Determine which of the following classes of functions are linear spaces. The domain is understood to be the real line.
(a) All periodic functions with period 2π.
(b) All periodic functions.
(c) All f having the property $f(0) = 0$.
(d) All sinusoidal functions.
(e) All positive functions, i.e., $f(x) > 0$ for all x.
(f) All functions possessing at most a finite number of discontinuities.
(g) All functions possessing exactly one discontinuity.
(h) All functions with the property $f(0) = f(1)$.
***(i)** (For more advanced students.) All functions possessing at most a finite number of relative maxima and minima in each interval of finite length.

2. Give a geometrical interpretation for the various subspaces of R^2.

3. Give a geometrical interpretation for the various subspaces of R^3.

4. If f is a function, the *even part* of f is the function f_e defined by $f_e(x) = [f(x) + f(-x)]/2$, and the *odd part* is defined by $f_o(x) = [f(x) - f(-x)]/2$.
(a) Show that f_e is an even function and f_o is an odd function.

(b) Show that the only way a function f can be written as the sum of an even function and an odd function is by writing $f = f_e + f_o$.

5. Find the even and odd parts of e^x.

6. If a differentiable function is even, is its derivative also even?

7. If f is an odd continuous function, what is $\int_{-c}^{c} f(x)\,dx$?

8. If $f(x)$ is a function, what is the "physical" interpretation of $f(x - vt)$, if v is positive and t represents *time*?

9. Every real number t for which $f(x + t) = f(x)$ for all x is called a *period* of f. Show that if t_1 and t_2 are periods of f, $t_1 - t_2$ is also a period of f.

10. What is the least positive period common to all of the functions $\sin 2x$, $\sin 3x$, $\sin 4x$, $\cdots$?

11. What is the least positive period common to all the functions $\sin^2 x$, $\sin^2 2x$, $\sin^2 3x$, $\cdots$?

12. Prove that, if $A_0 + A_1x + A_2x^2 + \cdots$ is a power series converging to $f(x)$ for all x, and if f is an even function, then $A_k = 0$ for odd k.

13. Classify the following functions as even or odd:

(a) $\sin 4x$.	**(d)** $\cos^2 x$.	**(g)** $e^x - e^{-x}$.
(b) $\cos 4x$.	**(e)** ex^2.	**(h)** $f(x) - f(-x)$.
(c) $\sin^2 x$.	**(f)** $\lvert x \rvert$.	**(i)** $f(x) + f(-x)$.

14. If two functions are considered "evenly equivalent" whenever their even parts are equal, show that every function is "evenly equivalent" to an even function.

15. If m and n are positive integers, what is the minimum positive period of $\sin nx + \sin mx$?

16. Let $f(x) = 1$ when x is rational, and $f(x) = 0$ otherwise.
    (a) Is this function periodic?
    (b) If so, does this function have a least positive period?

17. Give an example of a function with least positive period 4 whose square has least positive period 2.

18. Is it necessarily true that, if f and g both have minimal positive period T, their sum also has minimal positive period T?

*19. Prove in detail that $\sin x + \sin \sqrt{2}x$ is not a periodic function.

***20.** Prove in detail that a continuous nonconstant periodic function has a least positive period.

***21.** Prove that the sum of two continuous periodic functions is periodic if and only if they have some positive period in common, or else show by an example that the statement is false.

22. Show that any function f can be written uniquely as the sum of a constant function and a function that is zero at $x = 0$.

23. Show that Hilbert coordinate space satisfies the two properties required of any linear space.

1.4 • FINITE-DIMENSIONAL LINEAR SPACES

If all the functions in a linear class of functions are real-valued, we speak of a *real linear space.* If they are allowed to be complex-valued (and the scalars are also permitted to be complex) we speak of a *complex linear space.* Unless specified otherwise, we permit either alternative when we say "linear space."

If $f_1, f_2, \cdots, f_k$ are functions all having the same domain, and $\alpha_1, \alpha_2, \cdots, \alpha_k$ are scalars, any expression of the form $\alpha_1 f_1 + \alpha_2 f_2 + \cdots + \alpha_k f_k$ is called a *linear combination* of the functions.

At no time in this book will the term "linear combination" refer to an infinite series. If we should happen to be given an infinite sequence of functions, it will be understood that a linear combination of functions in the sequence never involves more than a finite number of them.

A single function f_1 is a linear combination of $f_1, f_2, \cdots, f_k$. Simply take $\alpha_1 = 1$, $\alpha_2 = \cdots = \alpha_k = 0$. The zero function is also a linear combination, since we can take $\alpha_i = 0$ $(i = 1, 2, \cdots, k)$. This is called the *trivial* linear combination. If for at least one i, $\alpha_i \neq 0$, the linear combination is called *nontrivial.*

This is not to say, however, that a nontrivial linear combination may not also be the zero function! For example, if we let $f_1(x) = \sin^2 x$, $f_2(x) = \cos^2 x$, and $f_3(x) = 1$, then $f_1 + f_2 - f_3$ is a nontrivial linear combination of these three functions, but nevertheless this is the zero function. This is possible of course because these three functions are related through the identity $\sin^2 x + \cos^2 x = 1$.

Strictly speaking, one should distinguish between a formal linear combination of functions, and the function defined by this linear combination. For example, as we have just seen, $f_1 + f_2 - f_3$ may be the same function as $0f_1 + 0f_2 + 0f_3$. Different linear combinations may represent the same function. To avoid circumlocution, we will not make this distinction; the meaning will always be clear from the context.

A finite set of functions $f_1, f_2, \cdots, f_k$ all having the same domain is said to be *linearly dependent* if there is a nontrivial linear combination of the functions equal to the zero function. The set is *linearly independent* if the only linear combination $\alpha_1 f_1 + \alpha_2 f_2 + \cdots + \alpha_k f_k$ that is identically zero is the linear combination in which $\alpha_1 = \alpha_2 = \cdots = \alpha_k = 0$.

The following theorem provides a clue to why the word "dependent" is appropriate, for it states essentially that in a linearly dependent set one of the functions "depends" on the others.

A sequence $f_1, f_2, \cdots, f_k$ of functions is linearly dependent if and only if one of the functions is a linear combination of the others.

The proof is left to the student as a very simple exercise. It is important to note that "one" means "at least one," not "exactly one." This is a matter of simple logic. If a farmer says "one of my chickens has whooping cough," and in fact two of his chickens have whooping cough, his assertion is nevertheless true.

An infinite sequence of functions is said to be linearly independent if and only if every finite set of functions in the sequence is linearly independent. Otherwise, it is said to be linearly dependent. The theorem above is valid for infinite sequences as well as for finite sequences.

The following *example* may be instructive. We consider the infinite sequence $1, x, x^2, x^3, \cdots$. Every linear combination of terms in this sequence is a polynomial. Again we emphasize that we do not consider an infinite series to be a linear combination (unless, of course, all but a finite number of the terms are equal to zero). The largest integer n for which the coefficient of x^n is not zero is called the *degree* of the polynomial. Thus $1 + x^3$, $2 - x^2 + 4x^7$, and 8 are polynomials of degree 3, 7, and 0 respectively. This puts the zero polynomial in the anomalous position of having no degree (having no degree is not the same as having zero degree). A standard theorem in algebra states that a polynomial of degree n has at most n zeros, and therefore cannot be identically zero. The only polynomial identically zero is the zero polynomial for which all coefficients are zero. This shows that the infinite sequence $1, x, x^2, x^3, \cdots$ is linearly independent.

The next example provides motivation for the definitions which follow. An equation of the form

$$\frac{d^n y}{dx^n} + a_1 \frac{d^{n-1} y}{dx^{n-1}} + \cdots + a_{n-1} \frac{dy}{dx} + a_n y = 0 \tag{1}$$

where $a_1, a_2, \cdots, a_n$ are continuous functions is called a *homogeneous linear equation of order* n. The class of all solutions of (1) is called its *solution space*. It is easy to see that this is a linear space, since the sum of two solutions is also a solution, and any scalar multiple of a solution is a solution.

The zero function is obviously a solution of (1), called the *trivial solution*. Are there necessarily any nontrivial solutions? A standard existence theorem for differential equations states that there are plenty of nontrivial solutions (unless $n = 0$, a trivial case). The theorem states that *there exist* n *solutions* $u_1, u_2, \cdots, u_n$, *forming a linearly independent set, and every solution of* (1) *is a linear combination* $C_1 u_1 + C_2 u_2 + \cdots + C_n u_n$.

This theorem is sometimes loosely worded "there is a general solution of (1) containing n arbitrary constants."

If the coefficients $a_1, a_2, \cdots, a_n$ are defined and continuous only in an interval $a \leqq x \leqq b$, the above theorem is still valid, but all solutions must be understood to have this interval as their domain of definition.

Because of this theorem, we call the solution space of (1) an *n-dimensional linear space* and the functions $u_1, u_2, \cdots, u_n$ a *basis* for this space.

There is nothing unique about the basis. For example, every solution of

$$\frac{d^3 y}{dx^3} + 4 \frac{dy}{dx} = 0 \tag{2}$$

can be written in the form $C_1 + C_2 \cos 2x + C_3 \sin 2x$, so the functions 1, $\cos 2x$, $\sin 2x$ provide a basis for the solution space. It is also possible to write every solution in the form $D_1 \sin^2 x + D_2 \cos^2 x + D_3 \sin x \cos x$, so another basis is provided by $\sin^2 x$, $\cos^2 x$, $\sin x \cos x$.

We formalize these remarks in the following definitions. A collection of elements of a linear space is called a *spanning set* for the space, or is said to *span* the space, if every element of the linear space is a linear combination of elements in this collection.

A collection of elements of a linear space is called a *basis* for the linear space if it spans the space and is linearly independent. (Note that an infinite collection of elements is by definition linearly independent if and only if every finite sequence drawn from the collection is linearly independent.)

These definitions are far more general than needed in any kind of practical work, and their full significance will not be discussed in this book. It can be proved by transfinite methods beyond the scope of this book that every nontrivial linear space has a basis. This statement sounds very important, but it has no known practical value. If a space has a *finite* basis, then we are interested in finding a basis; that is what we do when we solve an equation like (1). If a space has no finite basis, then we are usually not interested in finding any basis, as we shall see later.

If a space has a finite basis, then it is said to be *finite-dimensional*. If it has an infinite basis, it is *infinite-dimensional*. A linear space is infinite-dimensional if and only if for every positive integer n, no matter how large, it is possible to find a linearly independent sequence of n elements all belonging to the space. A trivial linear space consisting of only one element (the zero element, a function identically zero) is said to be *zero-dimensional*. A trivial linear space has no basis and hence is neither finite-dimensional nor infinite-dimensional. The following theorem shows, among other things, that a finite-dimensional linear space cannot also be infinite-dimensional.

The Fundamental Theorem of Linear Algebra

If a linear space V has a basis of n elements, for a positive integer n, then

(i) *every basis for V has exactly n elements,*

(ii) *every linearly independent subset of V has at most n elements, and is a basis for V if and only if it has exactly n elements,*

(iii) *every subset of V that spans V must have at least n elements, and is a basis if and only if it has exactly n elements,*

(iv) *if the elements of a fixed basis in V are taken in a definite order, every element of V is represented by a unique coordinate sequence.*

To explain what (iv) means, let the basis be $u_1, u_2, \cdots, u_n$ and let f be an element of V. Then $f = C_1u_1 + C_2u_2 + \cdots + C_nu_n$, and the sequence $(C_1, C_2, \cdots, C_n)$ represents f. No other sequence can represent f, for if $f = D_1u_1 + D_2u_2 + \cdots + D_nu_n$, then on subtracting we obtain a linear combination $(C_1 - D_1)u_1 + (C_2 - D_2)u_2 + \cdots + (C_n - D_n)u_n$ that is identically zero, implying that each coefficient $C_i - D_i$ equals zero and therefore $C_i = D_i$ $(i = 1, 2, \cdots, n)$.

The sequence $(C_1, C_2, \cdots, C_n)$ is the *coordinate sequence* representing f relative to the basis.

A linear space having a basis of n elements is said to be *n-dimensional.* We have already given one example: the solution space of a homogeneous linear differential equation of order n. Another example is Cartesian n-space. The above theorem justifies the definition of "n-dimensional" by stating that a space cannot be n-dimensional and also m-dimensional, for positive integers n and m, unless $n = m$.

The proof of the above theorem is completely elementary and can be found in any book on linear algebra.

• EXERCISES

1. Give two different methods for proving that $\sin x$ and $\cos x$ span a two-dimensional linear space. (In other words, give two proofs that $\sin x$ and $\cos x$ form a linearly independent set.)

2. Prove that $1, x, x^2, x^3, \cdots$ is a linearly independent sequence by a method different from that used in the text. [Hint: If a function is identically zero, all of its derivatives are also identically zero.]

3. If u and v are differentiable, their *Wronskian* is the function $[u,v]$ defined by

$$[u,v](x) = u(x)v'(x) - v(x)u'(x).$$

(a) Show that, if $[u,v]$ is nonzero for at least one value of x, the functions u, v must be linearly independent.

***(b)** Show by an example that it is possible for two linearly independent functions to have a Wronskian that is identically zero.

***(c)** Show that if $[u,v]$ is identically zero, the functions u and v are linearly dependent in any interval in which one or the other of the functions is nonzero.

4. Find a basis for the solution space of $d^ny/dx^n = 0$.

5. Are the functions x and $|x|$ linearly independent?

6. As elements of $C[0,1]$, are x and $|x|$ linearly independent?

7. Let V be the three-dimensional linear space spanned by 1, x, and x^2. Find a basis for V relative to which the coordinate sequence of any element f is the same as the sequence $f(-1), f(0), f(1)$.

8. An n-dimensional space has basis $1, x, x^2, \cdots, x^{n-1}$. What is the coordinate sequence of $(1 + x)^{n-1}$ relative to this basis?

9. A two-dimensional space is spanned by $\cos^2 x$ and $\sin^2 x$. What is the coordinate sequence (in this case, coordinate "pair") relative to this basis of the function identically equal to unity?

10. What is the dimension of the linear space spanned by the four functions $1, x, x^2$, and $1 - x$?

11. Is there any difference between the linear space spanned by $\sin x$ and $\cos x$ and that spanned by $\sin x + \cos x$ and $\sin x - \cos x$?

12. Can any linearly independent set contain the function θ? (Recall that, throughout this book, θ denotes the zero function, unless indicated otherwise.)

13. Is the function $\sin 2x$ an element of the linear space spanned by $\sin x$ and $\cos x$?

14. Let H_n denote the linear class of functions spanned by $1, \cos x, \cos 2x, \cdots, \cos nx, \sin x, \sin 2x, \cdots, \sin nx$.
    (a) What is the dimension of H_n?
    (b) Is H_n a subspace of H_{n+1}?
    (c) For what values of n is $\sin^2 x$ an element of H_n?
    (d) For what values of n is $\cos^3 x$ a function of class H_n?
    ***(e)** Is the function $\dfrac{\sin (n + \frac{1}{2})x}{\sin \frac{1}{2}x}$ (defined at $x = 0$ by the limit convention) a function of class H_n?

15. Given n linearly independent functions $u_1, u_2, \cdots, u_n$, with a common domain D. A set of n points $x_1, x_2, \cdots, x_n$ in D is called a *D-set* for these functions if it possesses the following property: given any sequence of n scalars $C_1, C_2, \cdots, C_n$, there is one and only one linear combination of these functions whose value at x_i is C_i for each $i = 1, 2, \cdots, n$. Prove in detail that every linearly independent sequence of functions with a common domain $u_1, u_2, \cdots, u_n$ has a D-set.

1.5 • INFINITE-DIMENSIONAL LINEAR SPACES

If the idea of regarding functions as vectors is to be useful, we must be able to do more than add them and multiply them by scalars. We should want to be able to discuss them geometrically, to define a notion of "magnitude" of a vector, and perhaps to know what is meant by the "angle" between two vectors.

From a theoretical viewpoint this is always possible, but from a practical viewpoint it is seldom feasible unless we restrict our considerations to very special classes of functions. Thus, we will sometimes consider only functions that are continuous; sometimes we will demand more, say that they be differentiable as well; sometimes less, as for example that they be piecewise continuous. We will almost always require that the functions under consideration be integrable. In this section, we will briefly explain what we mean by these terms.

A function f is said to be *bounded* in an interval $a \leqq x \leqq b$ if there exists a number M such that $|f(x)| \leqq M$ for every x in the interval. It is said to be *integrable* over the interval if it is bounded in the interval and the integral $\int_a^b f(x)\, dx$ exists according to the usual definition found in any elementary calculus text.

The reader may be familiar with so-called *improper* integrals. For example, $\int_0^1 dx/\sqrt{x}$ does not exist according to the usual definition, since $1/\sqrt{x}$ is not bounded in the interval $0 \leqq x \leqq 1$, but for any positive ϵ the function is bounded in the interval $\epsilon \leqq x \leqq 1$ and the integral $\int_\epsilon^1 dx/\sqrt{x}$ exists. Moreover, this integral tends to a finite limit as ϵ tends to zero. Therefore $\int_0^1 dx/\sqrt{x}$ exists as an "improper integral." We will seldom have any occasion to use improper integrals. We mention this matter only because we wish to caution the reader that some of the assertions we make are not valid for improper integrals. For instance, if a function is integrable over an interval, then its square is also. This would not be valid for improper integrals; $1/\sqrt{x}$ has an improper integral over $0 \leqq x \leqq 1$, but its square $1/x$ does not.

More advanced books introduce another definition of integration, due to Lebesgue, which permits the integration of a larger class of functions. A student of mathematics may study the Lebesgue integral for several months before he is sufficiently familiar with it to handle it with confidence. In the Answers and Notes section we will include a few comments intended for more advanced readers familiar with the Lebesgue theory. (Other readers should ignore them.)

When we say that f is integrable over $a \leqq x \leqq b$ we do not necessarily mean that it would be an easy task to compute the value of $\int_a^b f(x)\, dx$.

A function f is said to be *piecewise continuous* in an interval $a \leqq x \leqq b$ if there exist n points $a = x_1 < x_2 < x_3 < \cdots < x_n = b$ such that f is continuous in each interval $x_i < x < x_{i+1}$ and has finite one-sided limits $f(x_i+)$ and $f(x_{i+1}-)$ at the endpoints of each such interval ($i = 1, 2, \cdots, n - 1$).

If f is piecewise continuous in an interval $a \leqq x \leqq b$ then it is necessarily bounded in that interval. Indeed, it is integrable over the interval; the class of all piecewise continuous functions is a subspace of the class of all integrable functions. However, not every integrable function is piecewise continuous. For example, consider the function

$$f(x) = \begin{cases} 1 & \text{if } x = 1/n,\ n = 1, 2, 3, \cdots \\ 0 & \text{otherwise.} \end{cases}$$

This function has an infinite number of discontinuities in the interval $0 \leqq x \leqq 1$ (see Figure 1.7) and therefore is not piecewise continuous

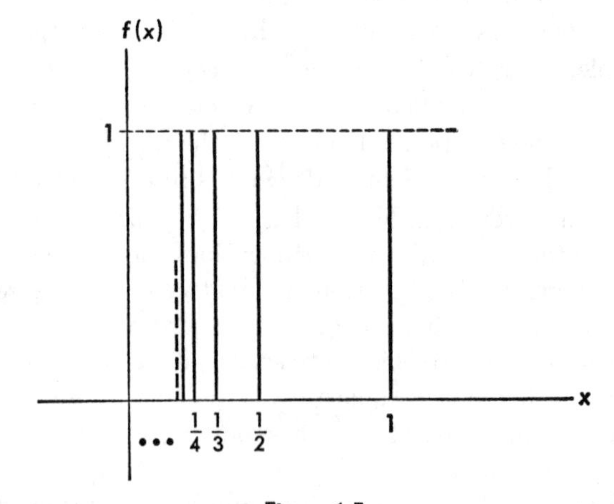

Figure 1.7

in this interval. Nevertheless, $\int_0^1 f(x)\,dx$ exists and equals zero, as the reader can verify if he cares to take time to review the precise definition of the definite integral given in elementary calculus texts.

The class of all functions defined and *continuous* throughout $a \leqq x \leqq b$ forms an even smaller linear space. This space is denoted $C[a,b]$ throughout this book; it is "smaller" only in the sense that it is a subspace of the class of all functions that are piecewise continuous in this interval.

It is important to keep in mind that f is of class $C[a,b]$ provided only that it is continuous in the open interval $a < x < b$ and also $f(a) = f(a+)$ and $f(b) = f(b-)$. The behavior of the function outside the interval $a \leqq x \leqq b$ is quite immaterial; in working with $C[a,b]$ we ignore completely the possibility that $f(x)$ can be defined for x outside the interval (Section 1.1). For example, $f(x) = e^{-1/x}$ (defined at $x = 0$ by the limit convention) is of class $C[0,1]$ although $f(0-)$ does not exist.

A similar comment applies to the derivative of a function whose domain is a finite interval. For any x_0 in the domain, $f'(x_0)$ is defined in the usual manner as the limit (if it exists) as x tends to x_0 of the difference quotient $[f(x_0) - f(x)]/[x_0 - x]$, with the stipulation that x be restricted to the domain. Thus, when $|x|$ is used to define a function of class $C[-1,1]$, it is not differentiable at $x = 0$; but considered as a function of class $C[0,1]$ it is differentiable (even at $x = 0$).

Keeping these remarks in mind, we denote by $C^n[a,b]$ the class of all functions possessing continuous derivatives of order n in the interval $a \leqq x \leqq b$. For each positive integer n, $C^{n+1}[a,b]$ is a subspace of $C^n[a,b]$. Moreover, for each n, it is easy to construct a function of class $C^n[a,b]$ that is not of class $C^{n+1}[a,b]$. Thus we have an infinite sequence of linear spaces, $C^1[a,b]$, $C^2[a,b]$, $C^3[a,b]$, $\cdots$, with each space a subspace of those preceding it and truly less inclusive than those preceding it. The imagination boggles in any attempt to "visualize" this geometrically; one might suppose that such a sequence would have to terminate ultimately, yet there are many functions of class $C^n[a,b]$ for *every* positive integer n.

In particular, each function in the sequence $1, x, x^2, x^3, \cdots$, considered as a function having domain $a \leqq x \leqq b$, is of class $C^n[a,b]$, independently of n. Since this is a linearly independent sequence of infinite length, it follows that each of the linear spaces $C^n[a,b]$ is infinite-dimensional. Therefore *a fortiori* all of the linear spaces discussed in this section are infinite-dimensional (provided we exclude the degenerate interval $a = b$).

In all of the linear spaces discussed in this section, it is possible to define the "length" of a "vector" f to be the number $\|f\| = [\int_a^b |f(x)|^2\,dx]^{1/2}$. The "distance" between two vectors (interpreted geometrically to mean the distance between two points having position vectors f and g) is the number

$$\|f - g\| = \left[\int_a^b |f(x) - g(x)|^2\,dx\right]^{1/2}.$$

With these definitions, which will be discussed in much greater detail in the next chapter, many of the simpler geometrical theorems remain valid. For example, if we are given three vectors (= points) f, g, and h, the distance between f and h cannot exceed the sum of the distances between f and g and between g and h, which is the familiar triangle inequality.

Of course, the reader is not expected as yet to appreciate the *usefulness* of these definitions; the reader having a clear understanding of this has no need to read this book. At this point we only hope that there breathes no reader with soul so dead that he is not both interested and excited at the prospect of studying the geometry of infinite-dimensional spaces. The next chapter is devoted mainly to this topic.

At this point we wish to emphasize that a certain amount of browsing may be helpful. It is by no means necessary or desirable that the reader of this book (or any other textbook) study the topics in the order in which they are presented. If you choose to do some skipping, *make use of the index* to locate the definitions of unfamiliar terms. The reader should not allow a section that he finds particularly difficult to be a barrier to further study.

• EXERCISES

1. Let $f(x) = |x - n|$ whenever $n - 1 \leqq x < n + 1$ $(n = 0, 2, 4, 6, \cdots)$.

(a) What is the domain of f?

(b) Draw a rough sketch of the graph of f.

(c) For what values of x is f continuous?

(d) For what values of x is f differentiable?

(e) Compute $\int_0^k f(x)\,dx$ for arbitrary positive k.

(f) If $g(x) = x$ for all x in the domain of f, for what values of x does $f + g$ have the value 4?

2. Which of the following functions are integrable over $0 \leqq x \leqq L$? (L is an arbitrary positive number.)

(a) $f(x) = x^2$.

(b) $f(x) = 1/x^2$.

(c) $f(x) = \sin x$.

(d) $f(x) = \begin{cases} 0 & 0 \leqq x < L/2 \\ 1 & L/2 \leqq x \leqq L. \end{cases}$

(e) $f(x) = x \sin(1/x)$, defined at $x = 0$ by the limit convention.

(f) $f(x) = \dfrac{1}{x}\sin\dfrac{1}{x}$ when $x \neq 0$; $f(0) = 0$.

(g) $f(x) = e^{-1/x}$, defined at $x = 0$ by the limit convention (as x tends to $0+$).

***(h)** $f(x) = \begin{cases} 1 & \text{when } x \text{ is rational} \\ 0 & \text{when } x \text{ is irrational.} \end{cases}$

(i) $f(x) = \left|x - \dfrac{L}{2}\right|.$

(j) $f(x) = \displaystyle\int_0^x \left|x - \frac{L}{2}\right| dx.$

***3.** Which function in the above list is bounded but not integrable?

4. Which functions in the above list are piecewise continuous in the interval $0 \leqq x \leqq L$?

5. Which are continuous in the interval?

6. Which functions in the above list are of class $C[0,L]$ but not of class $C^1[0,L]$?

7. Which are of class $C^1[0,L]$ but not of class $C^2[0,L]$?

8. Which are of class $C^n[0,L]$ for every positive integer n?

9. Is the function $\sin(1/x)$ continuous in the open interval $0 < x < 1$?

10. Is the function f defined by $f(x) = \sin(1/x)$ when $x \neq 0$, and $f(0) = 0$, piecewise continuous in the closed interval $0 \leqq x \leqq 1$? (After all, this function has only one discontinuity in this interval!)

11. **(a)** The function $\sin x$ can be expanded in a power series. Is it a linear combination of the functions $1, x, x^2, x^3, \cdots$?

(b) A linear space has basis $1, x, x^2, x^3, \cdots$. Is e^x an element of this linear space?

CHAPTER 2

ORTHOGONAL FUNCTIONS

This chapter is entirely theoretical. It is not necessary to master this chapter before studying Chapter 3; some readers may prefer to study these chapters in reverse order. Readers using this book for self-study should not be discouraged if they feel forced to skip large portions of this chapter.

The central idea of this chapter is the following. We are given a sequence of functions $u_1(x)$, $u_2(x)$, $u_3(x)$, $\cdots$ which have the *orthogonality property:*

$$\int_a^b u_m(x)\overline{u_n(x)}\,dx = 0$$

whenever $n \neq m$. We desire to expand an "arbitrary" function $f(x)$ in an infinite series

$$f(x) = C_1u_1(x) + C_2u_2(x) + C_3u_3(x) + \cdots.$$

Superficially, this seems quite easy. To find C_n for any fixed n, we multiply both sides of this equation by $\overline{u_n(x)}$ and integrate over the interval (a,b),

$$\int_a^b f(x)\overline{u_n(x)}\,dx = C_1\int_a^b u_1(x)\overline{u_n(x)}\,dx + C_2\int_a^b u_2(x)\overline{u_n(x)}\,dx + \cdots.$$

Because of the orthogonality property, every term on the right side vanishes except the nth, and we obtain

$$\int_a^b f(x)\overline{u_n(x)}\,dx = C_n\int_a^b |u_n(x)|^2\,dx$$

which we can solve for C_n.

In this chapter, we investigate this idea from a rigorous viewpoint. We will place special emphasis on geometrical ideas; in particular, we will explain why the term "orthogonality" is appropriate here. We will also introduce the notion of a boundary-value problem. However, we defer consideration of physical applications to later chapters.

2.1 • INNER PRODUCTS

An *inner product* in a linear space is a function of pairs of elements of the space, that is, to each pair f and g in the linear space there is a number $(f|g)$, satisfying the following axioms:

(1) $(f|f) \geqq 0$ for every f, and $(f|f) = 0$ if and only if $f = \theta$

(2) $(f|g) = \overline{(g|f)}$ for every pair f and g

(3) $(\alpha f + \beta g|h) = \alpha(f|h) + \beta(g|h)$

(4) $(f|\alpha g + \beta h) = \bar{\alpha}(f|g) + \bar{\beta}(f|h)$

where α and β are arbitrary numbers.

The reason for preferring this "abstract" definition over others that could be given is that it says nothing whatever about how the scalar product is constructed, whether by a sum or an integral. Any theorems depending only on this general definition will apply to any scalar product, no matter how it is constructed or what the nature of the linear space may be. This is especially important in applications to physics and engineering, because the inner products arising are not always as simple as those given as examples later in this section.

By (1), $(f|f)$ is never negative, so it has a nonnegative square root, called the *norm* of f and denoted $\|f\|$:

$$\|f\| = (f|f)^{1/2}. \tag{5}$$

Think of $\|f\|$ as the "length" or "magnitude" of f.

We now state and prove two of the most important inequalities in mathematics. The first is the *Schwarz inequality:*

$$|(f|g)| \leqq \|f\|\|g\|. \tag{6}$$

Proof: If $(f|g) = 0$, this inequality is obviously valid, since the right side of (6) cannot be negative. Let us therefore assume that $(f|g) \neq 0$. By (1) we have, for every scalar a,

$$(f + ag|f + ag) \geqq 0.$$

Expanding this, using (3) and (4), we obtain

$$(f|f) + a(g|f) + \bar{a}(f|g) + |a|^2(g|g) \geqq 0.$$

Now let b be an arbitrary real number and take $a = b(f|g)$. On substituting, we obtain

$$(f|f) + 2b|(f|g)|^2 + b^2|(f|g)|^2(g|g) \geqq 0.$$

The left side is a quadratic expression in b, having real coefficients. The inequality shows that it cannot have two distinct real roots b_1 and b_2, since if it did there would be values of b for which the expression would be negative. Therefore the discriminant (the part of the quadratic formula inside the square root) is not positive (or the quadratic formula would give two roots):

$$|(f|g)|^4 - (f|f)|(f|g)|^2(g|g) \leqq 0.$$

Since $(f|g) \neq 0$, we can divide both sides by $|(f|g)|^2$ to obtain $|(f|g)|^2 - (f|f)(g|g) \leqq 0$ or $|(f|g)|^2 \leqq \|f\|^2\|g\|^2$, and the Schwarz inequality follows by taking positive square roots.

If f and g are elements of a *real* linear space, and neither has zero norm, the angle between f and g is defined by

$$\cos \theta = \frac{(f|g)}{\|f\|\,\|g\|}.$$

This makes sense by virtue of the Schwarz inequality, which guarantees that $(f|g)/\|f\|\,\|g\|$ will be in the range from -1 to 1. As a consequence of this definition, we can write

$$(f|g) = \|f\|\,\|g\| \cos \theta.$$

This gives us a geometrical interpretation of the inner product: it is the product of the lengths (magnitudes) of f and g, times the cosine of the angle between them. The reader may recognize this as a generalization of the "dot product" (sometimes called the *scalar product*) of vectors in space.

Notice, however, that this relation cannot be valid in a *complex* linear space (unless we permit θ to be complex) because, in general, $(f|g)$ need not be a real number.

Next we prove the *Minkowski inequality:*

$$\|f + g\| \leqq \|f\| + \|g\|. \tag{7}$$

In the proof, we make use of the Schwarz inequality, the fact that the sum of a complex number z with its conjugate $\bar{z}$ is a real number, and the inequality $z + \bar{z} \leqq 2|z|$.

Proof: $\|f+g\|^2 = (f+g|f+g) = (f|f) + (g|f) + (f|g) + (g|g) \leqq \|f\|^2 + 2|(f|g)| + \|g\|^2 \leqq \|f\|^2 + 2\|f\|\,\|g\| + \|g\|^2 = (\|f\| + \|g\|)^2$. The Minkowski inequality now follows by taking positive square roots.

Replacing f by $f-h$ and g by $h-g$, we obtain the *triangle inequality:*

$$\|f-g\| \leqq \|f-h\| + \|h-g\|. \tag{8}$$

Here is a somewhat less important inequality:

$$|\|f\| - \|g\|| \leqq \|f-g\|. \tag{9}$$

Proof: $\|f\| = \|f-g+g\| \leqq \|f-g\| + \|g\|$, hence $\|f\| - \|g\| \leqq \|f-g\|$. Similarly, $\|g\| - \|f\| \leqq \|g-f\| = \|f-g\|$.

This inequality shows that, if the norm of $f-g$ is sufficiently small, the norm of f must be nearly equal to the norm of g.

Here is an interesting expression:

$$\begin{aligned}(f|g) &= \tfrac{1}{4}[(f+g|f+g) - (f-g|f-g) \\ &\qquad + i(f+ig|f+ig) - i(f-ig|f-ig)] \\ &= \tfrac{1}{4}[\|f+g\|^2 - \|f-g\|^2 + i\|f+ig\|^2 - i\|f-ig\|^2]\end{aligned} \tag{10}$$

showing that the inner product can be written entirely in terms of the norm.

We complete our derivations by showing that $\|af\| = |a|\|f\|$. *Proof:* $\|af\|^2 = (af|af) = a\bar{a}(f|f) = |a|^2\|f\|^2$, and take positive square roots.

Much later in this book we will discuss norms that are defined directly, without using any inner product and quite unrelated to any inner product. It is therefore useful to list here the three fundamental properties required of any norm.

$$\|f\| \geqq 0, \text{ and } \|f\| = 0 \text{ if and only if } f = \theta. \tag{11}$$

$$\|af\| = |a|\|f\|. \tag{12}$$

$$\|f+g\| \leqq \|f\| + \|g\|. \tag{13}$$

We see that these are precisely the familiar properties of the modulus (absolute value) of a complex number. If z_1 and z_2 are complex numbers, $|z_1 - z_2|$ is the distance between the points z_1 and z_2 in the complex plane. Similarly, we think of $\|f-g\|$ as the distance between two elements f and g in a linear space. The reason (8) is called the triangle inequality is that in three-dimensional space it reduces to the familiar triangle inequality of geometry: the length of one side of a triangle does not exceed the sum of the lengths of the other two sides.

Sometimes it is convenient to drop the requirement that $(f|g) = 0$ only when $f = \theta$. When we do this, we call $(f|g)$ a *pseudo-inner product* and the corresponding norm $\|f\|$ a *pseudo-norm.* We then say that two elements f and g are *equal almost everywhere* if and only if $\|f - g\| = 0$. In particular (taking $g = \theta$) we say that f is *equal to zero almost everywhere* if and only if $\|f\| = 0$. We will see later that this is a most appropriate expression; in the applications we discuss, a function f having the property $\|f\| = 0$ will for all practical purposes be the same as the function identically equal to zero. We will not use this terminology very often, however.

EXAMPLE 1: In real Cartesian 3-space, the inner product of two elements

$$x = (x_1,x_2,x_3) \quad \text{and} \quad y = (y_1,y_2,y_3)$$

is given by

$$(x|y) = x_1y_1 + x_2y_2 + x_3y_3.$$

The reader will recognize this as the usual "scalar product" or "dot product" of two vectors, given in terms of their components. The usual notation is $x \cdot y$ but for various reasons it is more convenient for our purposes to use the notation $(x|y)$. Some books use (x,y) instead.

EXAMPLE 2: In complex Cartesian n-space, the inner product of two elements

$$x = (x_1, x_2, \cdots, x_n), \quad y = (y_1, y_2, \cdots, y_n)$$

is customarily defined by

$$(x|y) = x_1\bar{y}_1 + x_2\bar{y}_2 + \cdots + x_n\bar{y}_n.$$

The presence of the complex conjugates ensures that $(x|x)$, the inner product of an element with itself, will be a real number, and in fact a nonnegative real number, which is needed in order to satisfy axiom (1).

The norm of an element is [by (5)]

$$\|x\| = [|x_1|^2 + |x_2|^2 + \cdots + |x_n|^2]^{1/2}$$

which is the natural generalization, to n dimensions, of the length of a vector.

The definition is the same for *real* Cartesian n-space; in this case, the complex conjugates can be omitted.

EXAMPLE 3: In the linear space of all continuous functions of period 2π, an inner product is defined by

$$(f|g) = \int_0^{2\pi} f(x)\overline{g(x)}\, dx.$$

In many books the similarity between this and Example 2 is emphasized; one thinks of f and g as vectors with components $f(x)$ and $g(x)$, and by replacing the sum in Example 2 by an integral we obtain this definition. This is, however, placing the emphasis in the wrong place; the connection between the two examples is simply that both are special cases of the more general definition given at the beginning of this section.

The reader will notice that it is simpler to write $(f|g)$ than to write $\int_0^{2\pi} f(x)\overline{g(x)}\, dx$. Even if this were the only inner product to be discussed in this book (which is not the case), we would still prefer to write $(f|g)$. Skeptical readers should convince themselves of the economy of this notation by writing out the *proof* of the Schwarz inequality using the integral notation.

EXAMPLE 4: Derive the inequality

$$\int_a^b |h(x)|\, dx \leqq \left[(b-a)\int_a^b |h(x)|^2\, dx\right]^{1/2}$$

valid whenever h is a (bounded) integrable function.

Solution: Let $f(x) = |h(x)|$ and let $g(x) = 1$ for all x. Writing out the Schwarz inequality for the special case of the inner product defined by

$$(f|g) = \int_a^b f(x)\overline{g(x)}\, dx,$$

we obtain, on the left side, $|\int_a^b f(x)\overline{g(x)}\, dx| = \int_a^b |h(x)|\, dx$, and on the right side

$$\|f\|\,\|g\| = \left[\int_a^b |h(x)|^2\, dx\right]^{1/2}\left[\int_a^b dx\right]^{1/2} = \left[(b-a)\int_a^b |h(x)|^2\, dx\right]^{1/2}.$$

We see that the desired inequality is simply the Schwarz inequality, for this very special case.

EXAMPLE 5: The inner product defined in the preceding example is called the *inner product of the functions f and g with respect to the interval (a,b)*. This can be generalized slightly in the following way. Let r be a fixed function, defined and continuous in the interval $a \leqq x \leqq b$, having the property $r(x) > 0$ whenever $a < x < b$. Define the inner product by

$$(f|g) = \int_a^b f(x)\overline{g(x)}r(x)\, dx.$$

This is called the inner product with respect to the *weight function* r.

EXAMPLE 6: In a similar manner, one can generalize the definition given in Example 2. We let $m_1, m_2, \cdots, m_n$ be positive "weights" and define the inner product in Cartesian n-space by

$$(x|y) = m_1 x_1 \overline{y}_1 + m_2 x_2 \overline{y}_2 + \cdots + m_n x_n \overline{y}_n.$$

In some applications, the numbers m_k are the masses of certain particles.

Unless stated otherwise, we will always use the definition given in Example 2, which is the "usual" inner product in Cartesian n-space.

EXAMPLE 7: Consider the linear space of all functions defined and piecewise continuous in the interval $a \leqq x \leqq b$. (For the precise definition of piecewise continuity, see Section 1.5.) It is possible for $\int_a^b |f(x)|^2\, dx$ to be equal to zero, even though $f(x)$ is not identically zero in the interval. Therefore, we can have $\|f\| = 0$ even though $f \neq \theta$, and it follows that the inner product defined as in Example 4 is actually a pseudo-inner product. However, if $\|f\| = 0$ for a piecewise continuous function f, $f(x)$ will equal zero for every x in the interval with at most a finite number of exceptions. To construct a specific example of a function that would be said to equal zero *almost everywhere* in the interval, let $f(x) = 0$ except at one point $x = x_0$, and let $f(x_0) = 1$.

EXAMPLE 8: If f and g represent current and voltage respectively, their inner product $(f|g)$, taken over an interval of periodicity, is proportional to the rate at which power is being supplied; $\|f\|$ and $\|g\|$ are proportional to the root-mean-square values of current and voltage, and the Schwarz inequality makes an obvious statement about the power factor. We omit further details; the interested reader is referred to the book *Mathematics of Circuit Analysis* by E. A. Guillemin (Wiley, 1949).

• EXERCISES

1. Find the angle between (1,2,2,0) and (0,0,3,4) in Cartesian 4-space.

2. Find the inner product of $(1,i,i,2)$ and $(-1,i,i,3)$ in complex Cartesian 4-space. (This may seem like a silly exercise, but three students out of four get the wrong answer.)

3. Show that the following four vectors are mutually orthogonal:

$$(1,1,1,1),\quad (1,i,-1,-i),\quad (1,-1,1,-1),\quad (1,-i,-1,i).$$

4. Find the magnitudes of the four vectors in Exercise 3.

5. Prove that, if nonzero vectors x and y are orthogonal, neither can be a scalar multiple of the other.

Find the inner product $(f|g)$, relative to the interval $-\pi \leqq x \leqq \pi$ in each of the following instances:

6. $f(x) = \sin x$, $g(x) = \cos x$.

7. $f(x) = \sin 2x$, $g(x) = \sin(-2x)$.

8. $f(x) = e^{ix}$, $g(x) = e^{-ix}$.

9. $f(x) = x$, $g(x) = e^{ix}$.

10. f and g are integrable, f is odd and g is even.

11. Prove that (9) follows from (7) and (8) and hence need not be listed as a separate axiom.

12. **(a)** Find the norm relative to the interval $-1 \leqq x \leqq 1$ of the function

$$f(x) = \begin{cases} -1 & x < 0 \\ 1 & x \geqq 0. \end{cases}$$

(b) Find the norm $\|f - g\|$ relative to the interval $-1 \leqq x \leqq 1$ if f is defined in part (a) and g is defined by

$$g(x) = \begin{cases} -1 & x \leqq 0 \\ 1 & x > 0. \end{cases}$$

13. Prove in detail that if equality holds in the Schwarz inequality one of the functions is a scalar multiple of the other.

2.2 • ORTHOGONAL FUNCTIONS AND VECTORS

This section can be omitted without loss in continuity.

For simplicity, we will consider only sine polynomials, i.e., functions of the form

$$(1)\qquad f(x) = B_1 \sin x + B_2 \sin 2x + \cdots + B_n \sin nx.$$

Since this sum has only a finite number of terms, there is no question about its convergence as there would be with an infinite series. The sum function f is continuous; indeed, it possesses derivatives of all orders, and can be integrated over any interval of finite length. Since each term has period 2π, it is clear that f is periodic and has period 2π.

The problem we propose is the following. Suppose someone arbitrarily selects values for the n coefficients $B_1, B_2, \cdots, B_n$, evaluates $f(x)$ for a number of values of x, and presents us with a carefully drawn graph of the function f. Is there any way that we can determine from this graph the values of the coefficients $B_1, B_2, \cdots, B_n$? We propose to call this the "sine polynomial game," and hope it will enjoy great popularity the world over. Aside from its recreational value, we have a serious purpose; it is to derive a formula for the coefficients B_k in terms of the function f.

The procedure is very simple. We make use of an inner product, defined by

$$(2) \qquad (f|g) = \int_0^\pi f(x)g(x)\,dx.$$

Since the functions we are considering are real-valued, we can omit the complex conjugation indicated in the preceding section. Since functions of form (1) are *odd* there is no need to use the full interval of periodicity, so we integrate only over the interval $0 \leqq x \leqq \pi$. (Don't misunderstand this; if all the functions were *even*, we would also need only half the period.)

By calculation we obtain the "orthogonality relations"

$$(3) \qquad (\sin px|\sin mx) = \int_0^\pi \sin px \sin mx\,dx = \begin{cases} 0 & p \neq m \\ \pi/2 & p = m. \end{cases}$$

It follows that, if we take the inner product of both sides of (1) with $\sin mx$, every term except one will vanish on the right side, and the nonvanishing term will be $\pi B_m/2$. The left side is $(f|\sin mx) = \int_0^\pi f(x) \sin mx\,dx$, and so we obtain

$$(4) \qquad B_m = \frac{2}{\pi}\int_0^\pi f(x) \sin mx\,dx.$$

In principle, this permits us to play the sine polynomial game. Since f is presented graphically, we should have to find each B_m by graphical integration. This is not easy. A naive procedure would be to form tables of values $f(x) \sin mx$, for $m = 1, 2, \cdots, n$, and determine the areas by approximation procedures, using (say) Simpson's rule. Less naive graphical methods have been devised; see Willers,

Practical Analysis, Dover (1948),* pages 334–355, for the tedious details.

However, if we make a gentleman's agreement that no sine polynomial of order greater than some fixed n will be used, i.e., if we know in advance the value of n in (1), then the following procedure (of theoretical value) can be used. To fix ideas, we take $n = 3$.

We know the unknown function is of the form

$$f(x) = B_1 \sin x + B_2 \sin 2x + B_3 \sin 3x. \tag{5}$$

Rough plots of the three functions $\sin x$, $\sin 2x$, and $\sin 3x$ in the interval $0 \leqq x \leqq \pi$ are shown in Figure 2.1. We divide the interval up into $n + 1 = 4$ parts and form a table of the values of these three functions at the endpoints of the subintervals.

x	0	$\pi/4$	$\pi/2$	$3\pi/4$	π
$\sin x$	0	$\sqrt{2}/2$	1	$\sqrt{2}/2$	0
$\sin 2x$	0	1	0	-1	0
$\sin 3x$	0	$\sqrt{2}/2$	-1	$\sqrt{2}/2$	0

The first and last columns are not really needed. Let us ignore them, and look at the middle three entries in each row.

By now the reader will suspect what we have up our sleeve: vectors! The three central entries in each row are the components of a vector in Cartesian 3-space. In an earlier section, we thought of all functions of the form $B \sin x + C \cos x$ as vectors in a plane; now we think of all functions of the form (5) as vectors in space.

Let $u = (u_1,u_2,u_3)$ and $v = (v_1,v_2,v_3)$ be elements of Cartesian 3-space. Their *inner product* or *scalar product* (called "dot product" in elementary texts), denoted $u \cdot v$ or $(u|v)$, is defined to be the number

$$u \cdot v = u_1v_1 + u_2v_2 + u_3v_3. \tag{6}$$

An equivalent definition is

$$u \cdot v = \|u\| \, \|v\| \cos \theta \tag{7}$$

where $\|u\|$ and $\|v\|$ denote the magnitudes of u and v respectively, and if u and v have nonzero magnitude $\cos \theta$ is the cosine of the angle between the two vectors. A detailed discussion of the geometrical significance and elementary applications of the inner product will be found in Davis, *Introduction to Vector Analysis* (Allyn and Bacon, 1961), together with a proof of the equivalence of (6) and (7). All we are interested in here is the definition (4) and the observation that $u \cdot v = 0$ if and only if (a) either u or v (or both) is the zero vector, or (b) u and v are perpendicular nonzero vectors. Whether nonzero

*Currently out of print.

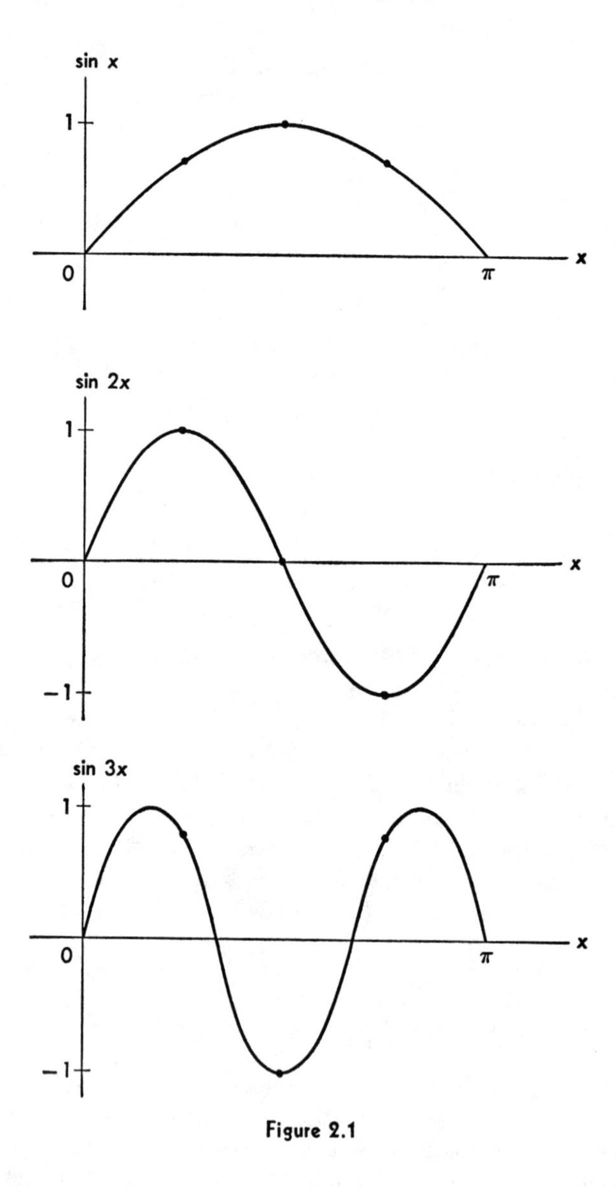

Figure 2.1

or not, we say u and v are orthogonal if $u \cdot v = 0$. (Orthogonality is more general than perpendicularity; perpendicularity suggests the vectors are nonzero.)

Now let us define three vectors using rows from the preceding table:

$$u = (\sqrt{2}/2,\ 1,\ \sqrt{2}/2) \tag{8}$$

$$v = (1,\ 0,\ -1) \tag{9}$$

$$w = (\sqrt{2}/2,\ -1,\ \sqrt{2}/2) \tag{10}$$

and let p be the sequence representing the unknown function f, i.e.,

$$p = (f(\pi/4),\ f(\pi/2),\ f(3\pi/4)). \tag{11}$$

From (3) we see that

$$p = B_1 u + B_2 v + B_3 w. \tag{12}$$

By direct calculation, we see that u, v, and w are mutually orthogonal:

$$u \cdot v = 0, \quad u \cdot w = 0, \quad v \cdot w = 0 \tag{13}$$

and also

$$u \cdot u = 2, \quad v \cdot v = 2, \quad w \cdot w = 2. \tag{14}$$

By taking inner products of both sides of (12) with u, v, and w in turn, we obtain formulas analogous to (4):

$$B_1 = \frac{(p \cdot u)}{(u \cdot u)}, \quad B_2 = \frac{(p \cdot v)}{(v \cdot v)}, \quad B_3 = \frac{(p \cdot w)}{(w \cdot w)} \tag{15}$$

$$B_1 = \tfrac{1}{2}(p \cdot u), \quad B_2 = \tfrac{1}{2}(p \cdot v), \quad B_3 = \tfrac{1}{2}(p \cdot w). \tag{16}$$

Suppose then that we are presented with a graph looking like Figure 2.2.

We see at once that the "even harmonic" $\sin 2x$ is missing, since

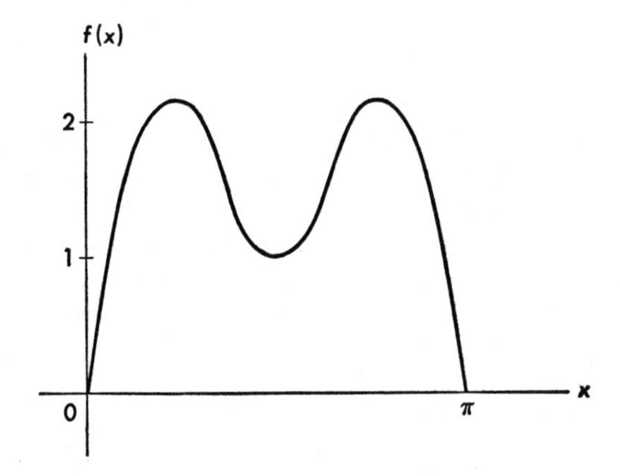

Figure 2.2

the presence of a nonzero $B_2 \sin 2x$ term would destroy the symmetry about $\pi/2$. Suppose the values we read from the graph are

$$f(\pi/4) = 2.13, \quad f(\pi/2) = 1.00, \quad f(3\pi/4) = 2.13;$$

then $B_1 = \frac{1}{2}[2.13\sqrt{2}/2 + 1 + 2.13\sqrt{2}/2] = 2$ approximately, $B_2 = 0$, and $B_3 = 1$ approximately. We suspect this is the function $f(x) = 2 \sin x + \sin 3x$.

Lest the reader think this game has no practical significance, we note that the special case just discussed is related to the normal modes of oscillation of three equal masses attached to a taut string supported at both ends; the masses are represented by heavy dots in Figure 2.1. Moreover, it is related to certain numerical procedures that we will discuss in Section 3.11.

The reader may have observed the connection between the procedure described above and the theory of D-sets (Section 1.4, Exercise 15). In this case, the D-set consists of three points, namely $\pi/4$, $\pi/2$, and $3\pi/4$. The remarkable feature of this D-set is that the vectors thereby associated with the functions $\sin x$, $\sin 2x$, and $\sin 3x$ satisfy the orthogonality relations (13), which considerably simplifies the determination of the coefficients B_1, B_2, and B_3.

A similar procedure can be used to determine the n coefficients in (1), if we are given the values of f at the n points $m\pi/(n + 1)$ $(m = 1, 2, \cdots, n)$. Further discussion will be reserved for Section 3.11; the important point is that this is possible only if we know in advance the value of n. If, in playing this "game," we have no agreement in advance on the maximum value of n that is permissible, this procedure via D-sets is not likely to work. It will generally not work for me if my opponent chooses a value of n that is larger than the number of points in my particular D-set.

On the other hand, (4) provides a valid formula for computing the coefficients no matter how large n may be (provided n is finite). This suggests the possibility that we generalize this "game" to permit (1) to be an *infinite series.*

But if we permit the function to be given by an infinite series, problems of convergence immediately arise and it is not a priori clear that (4) is valid. These problems will be discussed in the next section, in some generality.

• EXERCISES

1. Find a function of the form $f(x) = B_1 \sin x + B_2 \sin 2x + B_3 \sin 3x$, such that $f(\pi/4) = 3$, $f(\pi/2) = 0$, and $f(3\pi/4) = 1$.

2. Using the identity

(17) $$1 + z + z^2 + \cdots + z^n = \frac{1 - z^{n+1}}{1 - z},$$

derive the identity

(18)
$$1 + \cos\theta + \cdots + \cos n\theta = \frac{1 - \cos\theta + \cos n\theta - \cos(n+1)\theta}{2 - 2\cos\theta}.$$

[Hint: Let $z = e^{i\theta}$ in (17) and equate the real parts of both sides.]

3. Use (18) to derive the identity

(19) $$\tfrac{1}{2} + \cos\theta + \cdots + \cos n\theta = \frac{\sin(n + \frac{1}{2})\theta}{2\sin\frac{1}{2}\theta}.$$

4. Show that the identity

(20) $$\sum_{k=0}^{n} \cos[km\pi/(n+1)] = 1$$

is valid whenever m is an odd integer.

5. Show that, if m is an even integer (but not zero or an even multiple of $n + 1$), we have

(21) $$\sum_{k=0}^{n} \cos[km\pi/(n+1)] = 0.$$

6. Show that, if p and q are integers, the sum

(22) $$\sum_{k=0}^{n} \sin px_k \sin qx_k \qquad [x_k = k\pi/(n+1)]$$

either equals zero or equals $(n + 1)/2$. Discuss the relevance of this to the "sine-polynomial game."

7. Derive the following chain of equalities:

(23) $$\sin\theta + \sin 2\theta + \cdots + \sin n\theta$$
$$= \frac{\sin\theta - \sin(n+1)\theta + \sin n\theta}{2 - 2\cos\theta}$$
$$= \frac{\cos\frac{\theta}{2} - \cos(n + \frac{1}{2})\theta}{2\sin\frac{1}{2}\theta} = \frac{\sin\left(\frac{n+1}{2}\right)\cdot\sin\frac{n}{2}\theta}{\sin\frac{1}{2}\theta}.$$

8. Show that

(24) $$\sin\theta + \sin 3\theta + \cdots + \sin(2n-1)\theta = \frac{\sin^2 n\theta}{\sin\theta}.$$

9. Find $\sum_{k=0}^{n} \cos^2 px_k$ where p is an integer and $x_k = k\pi/(n+1)$. (Distinguish various possible cases.)

10. Invent a game using cosine polynomials, and (using the identities derived above) write down the appropriate formulas for determining the coefficients.

2.3 • ORTHOGONAL SEQUENCES

Sequences of functions having the orthogonality property mentioned at the beginning of this chapter are called *orthogonal sequences*. These sequences are the main subject of this book.

The introduction to orthogonal sequences given in this section is somewhat abstract. It is included here because there is a growing feeling among mathematicians, especially pure mathematicians, that the student should be introduced as early as possible to this more general approach, rather than be taught it later after studying many special cases. Moreover, there is a considerable economy of notation in this approach which some feel makes the subject even easier.

Not all students will agree with this, nor will all instructors be inclined to begin this section so early in the course. Therefore this book has been written in such a manner that one can proceed directly to Chapter 3 at this point. Very few references will be made to this section later in the book. However, those readers who do master this section before proceeding will undoubtedly study later topics with a greater depth of understanding.

Throughout this section, V denotes a linear space equipped with an inner product. In most of the applications made later, V will be a linear class of functions f for which $f(x)$ is defined in some interval $a \leqq x \leqq b$; one of the main examples is $C[a,b]$, which was defined in Section 1.5. However, some of the theorems to follow are valid for ordinary vectors in space, and insofar as possible the reader should visualize the elements of V as vectors. The reader who finds it difficult to think of a function as a "vector" should reread Sections 1.1 and 1.5.

Here is another *example* of a linear space to which these theorems apply. Let V denote the class of all real-valued functions defined and continuous on the surface of a sphere with center at the origin. A function f of class V has values $f(\phi,\theta)$, where ϕ and θ are the spherical coordinates of a point on the surface of the sphere. If the radius of the sphere is $r = 1$, the element of area on the surface of the sphere is $dA = r \sin \phi \, d\phi \, d\theta = \sin \phi \, d\phi \, d\theta$. We define the inner product of two members f and g of V by

$$(f|g) = \iint f(\phi,\theta)g(\phi,\theta) \, dA \tag{1}$$

or, if we wish to be more explicit,

$$(f|g) = \int_0^{2\pi} \int_0^{\pi} f(\phi,\theta)g(\phi,\theta) \sin \phi \, d\phi \, d\theta. \tag{2}$$

The reader will notice that this does not differ from the definition of the inner product of functions of class $C[a,b]$, except that here we are integrating over the surface of a sphere rather than over an interval $a \leqq x \leqq b$.

It cannot be too strongly emphasized, however, that the following discussion is not restricted to the example just given. The following theorems are valid quite generally for any (real or complex) linear space V.

If f and g are of class V, we say f and g are *orthogonal* elements of V whenever $(f|g) = 0$. We have already noted that this is related to the geometric notion of orthogonality; we think of f and g as mutually orthogonal vectors.

Caution: If f and g are orthogonal elements of $C[a,b]$, say, it must *not* be assumed that the *graphs* of f and g intersect at right angles. The term "orthogonal" as used here is not to be confused with the use of the same word in the term "orthogonal trajectories." It is not profitable for the reader to try to relate orthogonality of two functions f and g to their graphs.

The zero element, which we denote θ throughout, is orthogonal to every element of V. We have $(\theta|f) = 0$ for every element f.

Theorem 1 (Pythagorean Theorem). *If* $(f|g) = 0$, *then* $\|f + g\|^2 = \|f\|^2 + \|g\|^2$. (See Figure 2.3).

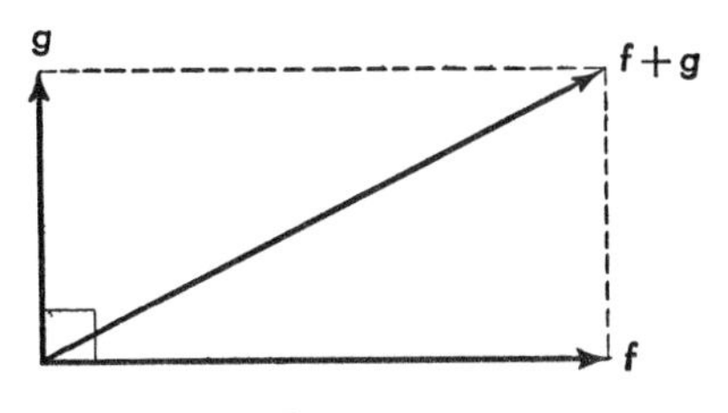

Figure 2.3

Proof: $\|f + g\|^2 = (f + g|f + g) = (f|f) + (g|f) + (f|g) + (g|g) = \|f\|^2 + \|g\|^2$, since $(f|g) = 0$ and hence also $(g|f) = 0$.

Of paramount importance in our later work is the notion of a *projection*, which we now introduce in its simplest form. Projections are of fundamental significance in the theory of Fourier series and in many other branches of mathematics. On a philosophical level, one can wax poetic on the subject of projections; that they provide the truly essential link between mathematics and physical reality, that every physical measurement is related to a projection, and so on. We will spare the reader these philosophical ideas, since a knowledge of quantum mechanics is required before one can begin to judge their validity. The general definition of a projection will be found in Section 6.8.

Definition 1. Let ϕ be an element of V having *unit norm*, i.e., $\|\phi\| = 1$. For any f of class V the *projection of f in the direction of ϕ* is denoted proj $(f:\phi)$ and is defined by

$$\text{proj }(f:\phi) = (f|\phi)\phi. \tag{3}$$

The geometrical idea is shown in Figure 2.4, where the vector OP represents proj $(f:\phi)$.

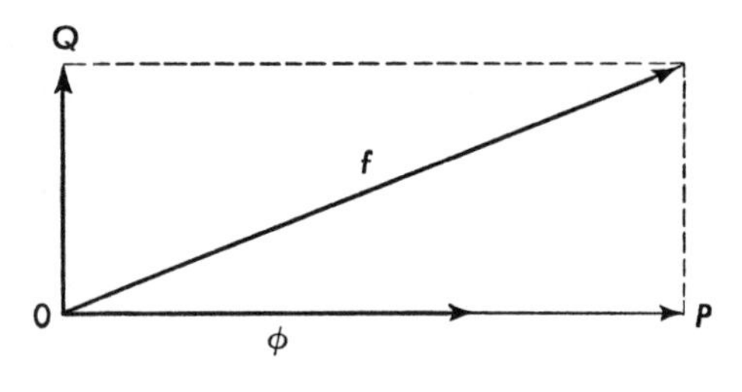

Figure 2.4

Notice that proj $(f:\phi)$ is an element of V; it is not simply a scalar. In vector algebra it is called the *vector component* of f in the direction of ϕ, i.e., the scalar component of f in the direction of a unit vector, times the unit vector.

If ϕ does not have unit norm, the projection of f in the direction of ϕ is defined by

$$\text{proj }(f:\phi) = \frac{(f|\phi)}{\|\phi\|}\,\phi \tag{4}$$

provided that ϕ is not the zero element of V. When ϕ has unit norm, (4) reduces to (3) since $\|\phi\| = 1$ in this case. In the following discussion, we will assume that ϕ has unit norm.

Theorem 2. *The projection of f in the direction of ϕ is equal to f if and only if f is a scalar multiple of ϕ.*

Proof: If proj $(f{:}\phi) = f$, then by (3) we have $f = (f|\phi)\phi$, i.e., f is a scalar multiple of ϕ. Conversely, if $f = \alpha\phi$ for some scalar α, then proj $(f{:}\phi) = (f|\phi)\phi = (\alpha\phi|\phi)\phi = \alpha(\phi|\phi)\phi = \alpha\phi = f$.

In Figure 2.4 the vector OQ represents $f -$ proj $(f{:}\phi)$, and is clearly orthogonal to ϕ. The next theorem shows quite generally that, if we subtract from f its projection in the direction of ϕ, we always obtain an element orthogonal to ϕ.

Theorem 3. $f -$ proj $(f{:}\phi)$ *is orthogonal to* ϕ.

Proof:

$$(f - \text{proj}\ (f{:}\phi)|\phi) = (f - (f|\phi)\phi|\phi) = (f|\phi) - (f|\phi)(\phi|\phi) = (f|\phi) - (f|\phi) = 0.$$

As noted in Theorem 2, $f -$ proj $(f{:}\phi) = \theta$ if and only if f is a scalar multiple of ϕ.

We shall now generalize Definition 1. At this point, the reader should review the definition of *subspace* given in Section 1.3.

Let us assume we are given n mutually orthogonal elements of V, all having unit norm: $\phi_1, \phi_2, \cdots, \phi_n$, where $(\phi_j|\phi_k) = 0$ whenever $j \neq k$ and $\|\phi_j\| = 1$ for every j. These n elements of V span a subspace of V which we denote H_n.

Definition 2. The projection of f into the subspace H_n is defined by

$$(5)\ \text{proj}\ (f{:}\phi_1, \phi_2, \cdots, \phi_n) = (f|\phi_1)\phi_1 + (f|\phi_2)\phi_2 + \cdots + (f|\phi_n)\phi_n.$$

In the special case $n = 2$, this is illustrated in Figure 2.5 by the vector OP. In this figure, H_2 is represented by a plane; OP is in the plane and OQ is perpendicular to it. Notice that OQ represents $f -$ proj $(f{:}\phi_1,\phi_2)$.

Theorem 4. *The projection of f into H_n is equal to f itself if and only if f is in H_n.*

Proof: If $f =$ proj $(f{:}\phi_1, \phi_2, \cdots, \phi_n)$, then by the definition (5) f is a linear combination of $\phi_1, \phi_2, \cdots, \phi_n$, and therefore is an element of

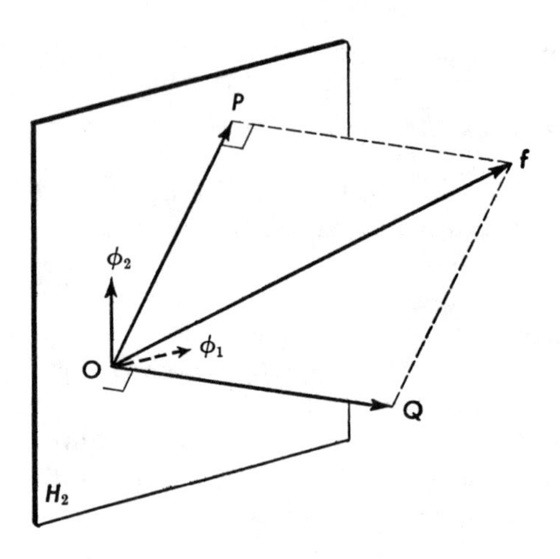

Figure 2.5

H_n. Conversely, if f is in H_n, then by definition $f = \sum_{k=1}^{n} \alpha_k\phi_k$ (where the α_k's are scalars) and $(f|\phi_j) = \left(\sum_{k=1}^{n} \alpha_k\phi_k|\phi_j\right) = \sum_{k=1}^{n} \alpha_k(\phi_k|\phi_j) = \alpha_j$ (notice that, by the orthogonality, every term except one vanishes). Hence,

$$f = \sum_{k=1}^{n} \alpha_k\phi_k = \sum_{k=1}^{n} (f|\phi_k)\phi_k$$

which equals proj $(f:\phi_1, \phi_2, \cdots, \phi_n)$ by definition.

Theorem 5. $f -$ proj $(f:\phi_1, \phi_2, \cdots, \phi_n)$ *is orthogonal to every element of* H_n.

Proof: By a calculation similar to that in the proof of Theorem 3, it is easy to show that $f -$ proj $(f:\phi_1, \phi_2, \cdots, \phi_n)$ is orthogonal to every one of the elements $\phi_1, \phi_2, \cdots, \phi_n$. If g is an element of H_n, then $g = \sum_{k=1}^{n} \alpha_k\phi_k$ for some choice of scalars $\alpha_1, \alpha_2, \cdots, \alpha_n$ (by the definition of H_n); letting $h = f -$ proj $(f:\phi_1, \phi_2, \cdots, \phi_n)$ we have $(h|g) = \left(h|\sum_{k=1}^{n} \alpha_k\phi_k\right) = \sum_{k=1}^{n} \bar{\alpha}_k(h|\phi_k) = 0$ since $(h|\phi_k) = 0$ for every k.

PROBLEM: Let V be the linear class of all continuous real-valued functions of period 2π, with inner product defined by $(f|g) =$

$\int_0^{2\pi} f(x)g(x)\,dx$. Derive an expression for proj $(f{:}\phi_1,\phi_2)$ if $\phi_1(x) = \sin nx$ and $\phi_2(x) = \cos nx$, assuming n is a positive integer.

Solution: We have $(\phi_1|\phi_2) = \int_0^{2\pi} \sin nx \cos nx\,dx = 0$, $\|\phi_1\|^2 = (\phi_1|\phi_1) = \int_0^{2\pi} \sin^2 nx\,dx = \pi$ and a similar calculation gives $\|\phi_2\|^2 = \pi$. Therefore ϕ_1 and ϕ_2 are *orthogonal* but they do not have unit norm. We *normalize* these elements to obtain $\dfrac{1}{\sqrt{\pi}} \sin nx$, $\dfrac{1}{\sqrt{\pi}} \cos nx$. Using Definition 2, we obtain

$$\begin{aligned}
\text{proj } (f{:}\phi_1,\phi_2) &= \left[\int_0^{2\pi} f(t)\,\frac{\sin nt}{\sqrt{\pi}}\,dt\right]\frac{\sin nx}{\sqrt{\pi}} \\
&\quad + \left[\int_0^{2\pi} f(t)\,\frac{\cos nt}{\sqrt{\pi}}\,dt\right]\frac{\cos nx}{\sqrt{\pi}} \\
&= \frac{1}{\pi}\int_0^{2\pi} f(t)[\sin nt \sin nx + \cos nt \cos nx]\,dt \\
&= \frac{1}{\pi}\int_0^{2\pi} f(t) \cos n(x - t)\,dt
\end{aligned}$$

which is the desired expression.

Definition 3. A sequence $\phi_1, \phi_2, \phi_3, \cdots$ of elements of V is said to be *orthogonal* if $(\phi_j|\phi_k) = 0$ whenever $i \neq j$ and $\|\phi_k\| \neq 0$ for every k. It is *orthonormal* if it is orthogonal and $\|\phi_k\| = 1$ for every k.

This definition applies both to finite and infinite sequences.

Theorem 6. *Every orthogonal sequence is linearly independent.*

Proof: Suppose that $\phi_1, \phi_2, \phi_3, \cdots$ is orthogonal. It is necessary to prove that any finite linear combination of elements of this sequence that equals the zero element must have all coefficients equal to zero. For example, suppose we have $\sum_{k=1}^{n} \alpha_k\phi_k = \theta$. Then $\left(\sum_{k=1}^{n} \alpha_k\phi_k \middle| \phi_j\right) = (\theta|\phi_j) = 0$ for every j. We have (on expanding the sum and using the linearity of the inner product) $\sum_{k=1}^{n} \alpha_k(\phi_k|\phi_j) = 0$, and since the sequence is orthogonal this reduces to $\alpha_j(\phi_j|\phi_j) = 0$, for every j. Since $(\phi_j|\phi_j) \neq 0$ (why?) it follows that $\alpha_j = 0$ for every j. The same procedure can be used to show that any finite linear combination that equals θ has all coefficients equal to zero, which is what we wanted to prove.

The converse of Theorem 6 is *not* valid. A linearly independent sequence may not be orthogonal. However, there is a process, which we shall now describe, whereby we can use a linearly independent sequence to *construct* an *orthonormal* sequence. This is conventionally called the *Gram-Schmidt orthogonalization process*, although the word "orthonormalization" would be more appropriate.

Theorem 7. *Given any linearly independent sequence of elements of V,*

$$u_1, u_2, u_3, \cdots, \tag{6}$$

there exists an orthonormal sequence of elements of V,

$$\phi_1, \phi_2, \phi_3, \cdots \tag{7}$$

with the property that the first k elements of (7) *is a basis for the subspace spanned by the first k elements of* (6), *for each k.*

These sequences can be either finite or infinite. If V is n-dimensional, then neither sequence can have more than n elements (Section 1.4) and the two sequences will have the same number of elements. If V is not finite-dimensional (for instance, if V is $C[a,b]$) then infinite linearly independent sequences exist; if (6) is an infinite sequence then (7) will be an infinite sequence also.

It follows from this theorem that ϕ_1 must be a scalar multiple of u_1 (take $k = 1$ in the theorem), that ϕ_2 must be a linear combination of u_1 and u_2 (take $k = 2$), and in general ϕ_k is a linear combination of $u_1, u_2, \cdots, u_k$.

Indeed, this is the way we construct the sequence (7). The proof of Theorem 7 follows from the method of construction, which we now outline.

We construct ϕ_1 first, by simply *normalizing* u_1. That is, we let $\phi_1 = \alpha^{-1}u_1$ where $\alpha = \|u_1\|$. This is permissible, since $\alpha \neq 0$ ($\alpha = 0$ would contradict the linear independence of (6)), and $\|\phi_1\| = \|\alpha^{-1}u_1\| = |\alpha^{-1}|\|u_1\| = \alpha^{-1}\alpha = 1$, as desired.

Now consider $u_2 - \text{proj}\,(u_2:\phi_1)$. Since u_2 is not a scalar multiple of u_1 (by the hypothesis of linear independence) it cannot be a scalar multiple of ϕ_1 either, so $u_2 - \text{proj}\,(u_2:\phi_1)$ cannot be the zero element (see the remark following Theorem 3). By Theorem 3, this element is orthogonal to ϕ_1. We normalize this element, and call the resulting element ϕ_2:

$$\phi_2 = \frac{u_2 - \text{proj}\,(u_2:\phi_1)}{\|u_2 - \text{proj}\,(u_2:\phi_1)\|}.$$

We continue in this manner. We let ϕ_k be the vector obtained by normalizing $u_k - \text{proj}\,(u_k{:}\phi_1, \cdots, \phi_{k-1})$. By Theorem 5, this vector is orthogonal to $\phi_1, \phi_2, \cdots, \phi_{k-1}$, and (by the hypothesis of linear independence) cannot be the zero element θ. Therefore at each stage we obtain a nonzero element $u_k - \text{proj}\,(u_k{:}\phi_1, \cdots, \phi_{k-1})$ that is normalized to obtain ϕ_k.

If (6) is a finite sequence, the process stops after we have run through all the u's. If (6) is an infinite sequence, the process never terminates, and in principle yields an infinite sequence of ϕ's. We leave further details to the reader.

If V is an n-dimensional space, with basis $u_1, u_2, \cdots, u_n$, this orthogonalization process provides a way for constructing a basis $\phi_1, \phi_2, \cdots, \phi_n$ that is *orthonormal.*

The following theorem is hardly more than a rewording of Theorem 4; there is no need to repeat the proof.

Theorem 8. *If $\phi_1, \phi_2, \cdots, \phi_n$ is an orthonormal basis for an n-dimensional linear space V, then every f in V is of the form $f = \sum_{k=1}^{n} \alpha_k\phi_k$ where $\alpha_k = (f|\phi_k)$.*

In other words, the coordinate sequence representing f, i.e., the sequence $(\alpha_1, \alpha_2, \cdots, \alpha_n)$, is especially easy to find if the basis is orthonormal.

Theorem 8 is actually a special case of the following more general theorem, which we will prove in detail; it is valid whether V is finite-dimensional or not.

Theorem 9. *If $\phi_1, \phi_2, \cdots, \phi_n$ is an orthonormal basis for an n-dimensional subspace H_n of a linear space V, and if f is an element of V, then that element h of H_n for which $\|f - h\|$ is a minimum is $h = \sum_{k=1}^{n} \alpha_k\phi_k$ where $\alpha_k = (f|\phi_k)$. In other words, the element of H_n that most closely approximates f is the element $h = \text{proj}\,(f{:}\phi_1, \phi_2, \cdots, \phi_n)$.*

If the reader will pause to look at Figure 2.5 and review the remarks made earlier, he will see that this has a geometric interpretation: the shortest distance from f to H_n is the perpendicular distance.

Proof: Let h be a general element of H_n, i.e., an element of the form $\sum_{k=1}^{n} \beta_k\phi_k$. Let $\alpha_k = (f|\phi_k)$. We desire to prove that for $\|f - h\|$ to be a minimum we must take β_k to be equal to α_k for every $k = 1, 2, \cdots, n$.

With sufficient patience, this can be proved by standard methods of second-year calculus; $\|f - h\|$ is a function of the n variables $\beta_1, \beta_2, \cdots, \beta_n$, and if we set the partial derivatives of this function equal to zero we obtain the desired result. Here, however, is an easier method, involving no calculus.

We have

$$\begin{aligned}\|f-h\|^2 &= (f-h|f-h) = (f|f) - (h|f) - (f|h) + (h|h)\\ &= \|f\|^2 - \sum_{k=1}^{n} \beta_k(\phi_k|f) - \sum_{k=1}^{n} \bar{\beta}_k(f|\phi_k) + \sum_{k=1}^{n} \beta_k\bar{\beta}_k\\ &= \|f\|^2 - \sum_{k=1}^{n} \beta_k\bar{\alpha}_k - \sum_{k=1}^{n} \bar{\beta}_k\alpha_k + \sum_{k=1}^{n} |\beta_k|^2\\ &= \|f\|^2 + \sum_{k=1}^{n} (\beta_k - \alpha_k)(\bar{\beta}_k - \bar{\alpha}_k) - \sum_{k=1}^{n} |\alpha_k|^2\\ &= \|f\|^2 - \sum_{k=1}^{n} |\alpha_k|^2 + \sum_{k=1}^{n} |\beta_k - \alpha_k|^2.\end{aligned}$$

Keep in mind that the only "variables" here are the β's. We see at once that this will be minimized by taking $\beta_k = \alpha_k$ for every k, so that every term in the last sum will be zero. If we do this we have

$$h = \sum_{k=1}^{n} \alpha_k\phi_k = \sum_{k=1}^{n} (f|\phi_k)\phi_k = \text{proj}\,(f:\phi_1, \phi_2, \cdots, \phi_n)$$

as asserted.

Corollary 1. *Under the above hypotheses,*

$$\|f - \sum_{k=1}^{n} \alpha_k\phi_k\|^2 = \|f\|^2 - \sum_{k=1}^{n} |\alpha_k|^2 \tag{8}$$

if $\alpha_k = (f|\phi_k)$.

We recall that $\|f - h\|$ is geometrically interpreted as the distance between f and h. We can say that the *smaller* $\|f - h\|$ is the *better* is the "approximation" that h provides for f. This explains the use of the term "approximates" in Theorem 9.

In particular, if f and h are elements of $C[a,b]$, the problem of minimizing $\|f - h\|$ is equivalent to the problem of minimizing $\int_a^b |f(x) - h(x)|^2\,dx$. Thus we obtain the following as a special case:

Corollary 2. *For a given function f, and for given functions $\phi_1, \phi_2, \cdots, \phi_n$ satisfying the condition that*

$$\int_a^b \phi_j(x)\overline{\phi_k(x)}\,dx \tag{9}$$

is zero when $j \neq k$ and is unity whenever $j = k$, the integral

$$\int_a^b |f(x) - \alpha_1\phi_1(x) - \alpha_2\phi_2(x) - \cdots - \alpha_n\phi_n(x)|^2\, dx \tag{10}$$

assumes its minimum value if we choose the α's so that

$$\alpha_k = \int_a^b f(x)\overline{\phi_k(x)}\, dx \qquad (k = 1, 2, \cdots, n). \tag{11}$$

Of course, if we are dealing with real-valued functions, the complex conjugations can be omitted.

Readers familiar with the *method of least squares* will recognize this as a variety of least-squares approximation. The *geometric* notion of finding that point in H_n which is nearest f corresponds to the *analytic* notion of finding that function of class H_n approximating f most closely in the least-squares sense.

Many of the functions occurring in physics, as for instance those representing pressure or temperature, are idealizations on a macroscopic level of a much more complicated microscopic situation. The value of such a function at a point is actually an average value over a neighborhood of the point. (What we observe as temperature is a statistical average related to the motions of many molecules.) We are therefore interested in approximations based on an averaging process, such as a least-squares approximation, rather than some of the other types of approximation that are possible. The reader is cautioned, however, that other kinds of "approximation" will be discussed later, and the last sentence of Theorem 9 is not generally valid for these kinds of approximation.

The Condition of Finality

This subsection can be omitted without loss in continuity.

Suppose we are given a linearly independent sequence of functions of class V,

$$u_1,\ u_2,\ u_3,\ \cdots \tag{12}$$

and suppose we wish to minimize the value of

$$\|f - \alpha_1 u_1 - \alpha_2 u_2 - \cdots - \alpha_n u_n\|. \tag{13}$$

Since we do not assume the sequence to be orthonormal, Theorem 9 does not apply.

We pose the following question. Can we find the values of the α's step-by-step in the following manner? First minimize $\|f - \alpha_1 u_1\|$, then (using the value of α_1 thus obtained) minimize $\|f - \alpha_1 u_1 - \alpha_2 u_2\|$, then (using the values of α_1 and α_2 already obtained) minimize $\|f - \alpha_1 u_1 - \alpha_2 u_2 - \alpha_3 u_3\|$, and so on.

If the values of the α's can be determined in this manner for every f of class V, we say the sequence (12) satisfies the *condition of finality* in the determination of coefficients. What this means, then, is that if we find values $\alpha_1, \alpha_2, \cdots, \alpha_k$ that minimize

$$\|f - \alpha_1 u_1 - \alpha_2 u_2 - \cdots - \alpha_k u_k\|,$$

then we do not need to revise these values when we attempt to minimize $\|f - \alpha_1 u_1 - \alpha_2 u_2 - \cdots - \alpha_{k+1} u_{k+1}\|$; all we need to do is determine the value of α_{k+1}. At each step, the value of an α is determined *once and for all.*

On the other hand, if it is necessary at each step to revise the coefficients of $u_1, u_2, \cdots, u_k$, as well as determine the coefficient of u_{k+1}, then the sequence does *not* satisfy the condition of finality.

Theorem 10. *A sequence (12) satisfies the condition of finality if and only if it is orthogonal.*

Proof: If the sequence is orthogonal, it can be normalized to give an orthonormal sequence, and Theorem 9 applies. In terms of the original sequence, the coefficients needed to minimize (13) are

$$\alpha_k = \frac{(f|u_k)}{(u_k|u_k)} \tag{14}$$

and they are determined once and for all.

If the sequence satisfies the condition of finality, we prove that it must be orthogonal in the following manner. For each positive integer n, let H_n denote the n-dimensional subspace of V spanned by $u_1, u_2, \cdots, u_n$. We have already noted that the element h_n in H_n making $\|f - h_n\|$ a minimum is that for which $f - h_n$ is orthogonal to every element of H_n. Similarly, the element h_{n+1} in H_{n+1} making $\|f - h_{n+1}\|$ a minimum is that for which $f - h_{n+1}$ is orthogonal to every element of H_{n+1}. If the condition of finality is satisfied, h_{n+1} and h_n differ only by a scalar multiple of u_{n+1}, and for some choices of f this will be a *nonzero* scalar multiple. Since both $f - h_n$ and $f - h_{n+1}$ are orthogonal to every element of H_n, and in particular to $u_1, u_2, \cdots, u_n$, it follows that their difference is also, and therefore u_{n+1} is orthogonal to $u_1, u_2, \cdots, u_n$. Since this is true for every n, it follows that $u_1, u_2, u_3, \cdots$ is an orthogonal sequence, which was to be proved.

Because of Theorem 10, we shall restrict ourselves almost entirely to orthogonal sequences in the sequel.

Infinite Series

Suppose we have an *orthonormal* sequence of elements of V,

$$\phi_1, \phi_2, \phi_3, \cdots, \tag{15}$$

and we are given an element f in V. We form the coefficients $\alpha_k = (f|\phi_k)$ and consider the series

$$\alpha_1\phi_1 + \alpha_2\phi_2 + \alpha_3\phi_3 + \cdots. \tag{16}$$

We assume that (15) is an infinite sequence, which means that V cannot be finite-dimensional and therefore Theorem 8 does not apply. We seek some theorem that will *replace* Theorem 8 when V is infinite-dimensional.

According to Corollary 1 of Theorem 9, we have, for each fixed n,

$$\sum_{k=1}^{n} |\alpha_k|^2 \leqq \|f\|^2 \tag{17}$$

since the left side of equation (8) is never negative. The left side of (17) is bounded above by $\|f\|^2$, independently of n, and as n increases the left side of (17) cannot decrease. Therefore the sequence of partial sums of the infinite series $\sum_{k=1}^{\infty} |\alpha_k|^2$ is convergent, and the limit of this sequence cannot exceed $\|f\|^2$. From this we conclude that

$$\sum_{k=1}^{\infty} |\alpha_k|^2 \leqq \|f\|^2 \tag{18}$$

which is known as *Bessel's inequality*, and also that

$$\lim_{k\to\infty} \alpha_k = 0 \tag{19}$$

since the terms of a convergent infinite sequence must tend to zero, and $|\alpha_k|^2$ tends to zero if and only if α_k tends to zero. We summarize this explicitly in the following:

Riemann's Lemma. *For any orthonormal sequence* $\phi_1, \phi_2, \phi_3, \cdots$ *and any* f *in* V, $\lim_{k\to\infty} (f|\phi_k) = 0$.

Another form of this lemma will be found in Section 6.7.

An infinite series

$$\sum_{k=1}^{\infty} \alpha_k u_k \tag{20}$$

where the α's are scalars and the u's are elements of V (not necessarily forming an orthonormal sequence) is said to *converge in the mean* or to *converge in the least-squares sense* if there is an element g in V with the property that

$$\left\|g - \sum_{k=1}^{n} \alpha_k u_k\right\| \tag{21}$$

tends to zero with increasing n.

The obvious question now arises: is it necessarily true that (16) converges in the mean to the function f (the same function used in constructing the coefficients α_k)? The answer is *no*. It is possible for (16) to fail to converge in the mean, or for it to converge in the mean to a function g that is not equal to f. We will now proceed to explore this matter further.

Approximating Bases

A linearly independent sequence of elements of a linear space V,

$$u_1, u_2, u_3, \cdots \tag{22}$$

is called an *approximating basis* for V if, for every element f of V, there are linear combinations of elements of (22) approximating f as closely as we like. In other words, for every f in V and every positive number ϵ there exists some linear combination $C_1u_1 + C_2u_2 + \cdots + C_nu_n$ such that

$$\|f - C_1u_1 - C_2u_2 - \cdots - C_nu_n\| < \epsilon.$$

This does *not* imply that there is any infinite series $C_1u_1 + C_2u_2 + C_3u_3 + \cdots$ which converges in the mean to f.

EXAMPLE: We will see later that $1, x, x^2, x^3, \cdots$ is an approximating basis for $C[-1,1]$ (indeed, it is an approximating basis for every space $C[a,b]$). Let $f(x) = |x|$. Clearly f is an element of $C[-1,1]$, and (this is not obvious) it is possible to find polynomials that approximate $|x|$ in the least-squares sense as closely as we please in the interval $[-1,1]$. However, there is no series $C_0 + C_1x + C_2x^2 + C_3x^3 + \cdots$ converging in the mean to $|x|$ in the interval $[-1,1]$. The difficulty that arises here is this: if we have a polynomial of degree n that very closely approximates $|x|$, and we wish to improve the approximation by finding a polynomial of higher degree, it is usually necessary to change the coefficients already determined in addition to adding extra terms in higher powers of x.

This has already been discussed in connection with the condition of finality. From that discussion, we see at once that if (22) is an *orthogonal* approximating basis, then there *must* be an infinite series which converges in the mean to f. For simplicity, we consider only *orthonormal* approximating bases, since one can always normalize an orthogonal set to obtain an orthonormal set.

Here, then, is the theorem we have been seeking, which replaces Theorem 8 when V is infinite-dimensional.

Theorem 11. *If $\phi_1, \phi_2, \phi_3, \cdots$ is an orthonormal approximating basis in V, and f is an element of V, then the series*

$$\alpha_1\phi_1 + \alpha_2\phi_2 + \alpha_3\phi_3 + \cdots \tag{23}$$

where $\alpha_k = (f|\phi_k)$, converges in the mean to f.

Proof: By the definition of approximating basis, for any prescribed positive ϵ there is a linear combination $C_1\phi_1 + C_2\phi_2 + \cdots + C_n\phi_n$ such that $\|f - C_1\phi_1 - C_2\phi_2 - \cdots - C_n\phi_n\| < \epsilon$. By Theorem 9, the *best* such approximation is $\alpha_1\phi_1 + \alpha_2\phi_2 + \cdots + \alpha_n\phi_n$; that is, $\|f - \alpha_1\phi_1 - \alpha_2\phi_2 - \cdots - \alpha_n\phi_n\|$ must be less than ϵ also. Since ϵ may be taken arbitrarily small, it follows (by definition) that (23) converges in the mean to f.

In this instance, the left side of equation (8) must tend to zero with increasing n, so the right side does also, and we obtain *Parseval's equality:*

$$\|f\|^2 = \sum_{k=1}^{\infty} |\alpha_k|^2 \tag{24}$$

valid under the hypotheses stated in Theorem 11. This is an improvement on (18). The improvement is possible because of the extra assumption that the orthonormal sequence is also an approximating basis.

Many examples of approximating bases will be given later. Notice the essential distinction between a *basis* and an *approximating basis.* If (22) is an approximating basis, all we can expect is that an arbitrary element of V can be *approximated* by linear combinations; in the definition of basis we demanded that every element of V be exactly *equal* to a linear combination. Later on we will construct various approximating bases to spaces such as $C[-1,1]$; no *basis* for $C[-1,1]$ has ever been explicitly constructed. (The *existence* of a basis for $C[-1,1]$ can be proved by transcendental methods beyond the scope of this book; such methods are nonconstructive in nature.)

Corollary. *If $\phi_1, \phi_2, \phi_3, \cdots$ is an orthonormal approximating basis in V, and f is an element of V such that $(f|\phi_k) = 0$ for $k = 1, 2, 3, \cdots$, then $f = \theta$.*

It follows at once from Parseval's equality that under these hypotheses we must have $\|f\| = 0$, and therefore $f = \theta$. As noted earlier, $f = \theta$ must sometimes be interpreted to mean $f(x) = 0$ *almost everywhere* rather than $f(x) = 0$ *identically* throughout its domain (see Section 2.1).

Terminology

This subsection can be omitted without loss in continuity.

Some authors use the term *closed sequence* instead of *approximating basis*, and use the term *complete sequence* to mean a sequence $\phi_1, \phi_2, \phi_3, \cdots$ such that $(f|\phi_k) = 0$ for every k implies $f = \theta$. It follows that every closed sequence is complete, but not every complete sequence need be closed. In some linear spaces every complete sequence is also closed, and the distinction vanishes. (This happens to be the case with the Hilbert coordinate space, discussed at the end of this section, but we will not prove it.) The reader is cautioned that other authors use the terms *closed* and *complete* with just the reverse meaning; what we call an "approximating basis" is called a "complete sequence," etc. The reader should carefully check to see which convention is being used. We use neither of these terms in this book.

The definition of *approximating basis* can be given in a different way. If W is a collection of elements of V, we say that W is *dense* in V if, for every element f in V and every positive ϵ, there is an element g in W such that $\|f - g\| < \epsilon$. If we are given a linearly independent sequence $u_1, u_2, u_3, \cdots$ of elements of V, and we let W denote the linear space spanned by these elements, then the sequence is said to be a *basis* if V and W are identical, and an *approximating basis* if W is dense in V. (Keep in mind that the elements of W are the *finite* linear combinations that can be formed using $u_1, u_2, u_3, \cdots$.)

Hilbert Coordinate Space

This subsection can be omitted without loss in continuity. It is included at the request of several instructors who wish to use it as a point of digression for more specialized lectures.

Definition. *Hilbert coordinate space*, denoted R^∞, is the class of all infinite sequences of scalars, $\alpha_1, \alpha_2, \alpha_3, \cdots$, with the property that $\sum_{k=1}^{\infty} |\alpha_k|^2$ is convergent. The sum of two such sequences,

$$(\alpha_1, \alpha_2, \alpha_3, \cdots) + (\beta_1, \beta_2, \beta_3, \cdots)$$

is defined to be

$$(\alpha_1 + \beta_1, \beta_2 + \beta_2, \alpha_3 + \beta_3, \cdots)$$

and the product of a sequence $(\alpha_1, \alpha_2, \alpha_3, \cdots)$ by a scalar k is defined to be

$$(k\alpha_1, k\alpha_2, k\alpha_3, \cdots).$$

The inner product of two sequences $\alpha = (\alpha_1, \alpha_2, \alpha_3, \cdots)$ and $\beta = (\beta_1, \beta_2, \beta_3, \cdots)$ is defined by

$$(\alpha|\beta) = \sum_{k=1}^{\infty} \alpha_k \bar{\beta}_k \tag{25}$$

with the norm defined via the inner product in the usual way:

$$\|\alpha\| = (\alpha|\alpha)^{1/2} = \left[\sum_{k=1}^{\infty} |\alpha_k|^2\right]^{1/2}. \tag{26}$$

We omit the routine verification that R^∞ is a linear space and that for α and β of class R^∞ the series in (25) defining $(\alpha|\beta)$ must converge. (See Exercise 23, Sec. 1.3.)

Now suppose that V is a linear space, that $\phi_1, \phi_2, \phi_3, \cdots$ is an orthonormal approximating basis in V, and f is an element of V. The *coordinate sequence* representing f, relative to the approximating basis, is the sequence $(\alpha_1, \alpha_2, \alpha_3, \cdots)$ where $\alpha_k = (f|\phi_k)$. According to Bessel's inequality, the coordinate sequence is an element of R^∞. Moreover, it is impossible for two distinct functions f and g to be represented by the same coordinate sequence, since if $(f|\phi_k) = (g|\phi_k)$ for every k, then $(f - g|\phi_k) = 0$ for every k and by the Corollary to Theorem 11, $f = g$.

Thus, different elements of V are represented by different elements of R^∞. It is not hard to see that the coordinate sequence representing the sum of two functions of class V is the sum of their coordinate sequences, and the coordinate sequence for the product of a function f with a scalar k is the product of k with the coordinate sequence representing f. According to Parseval's identity, the norm of the coordinate sequence representing f equals the norm of f. It is not difficult to prove that if α and β are coordinate sequences representing f and g respectively, their inner product $(\alpha|\beta)$ equals the inner product $(f|g)$.

The obvious question arises: does every sequence $\alpha_1, \alpha_2, \alpha_3, \cdots$ in R^∞ correspond to an element of V? In other words, if we have a fixed orthonormal approximating basis $\phi_1, \phi_2, \phi_3, \cdots$ in V, and scalars $\alpha_1, \alpha_2, \alpha_3, \cdots$ for which we know $\sum_{k=1}^{\infty} |\alpha_k|^2$ is convergent, does the infinite series $\sum_{k=1}^{\infty} \alpha_k\phi_k$ necessarily converge in the mean to an element of V? The answer is *no*, at least for the linear spaces we discuss in this book. This is because we restrict attention to relatively "nice" functions, and some sequences in R^∞ correspond to functions that are not integrable in the usual elementary sense.

• EXERCISES

1. Show that $\sin x$, $\sin 2x$, $\sin 3x$, $\cdots$ is an orthogonal sequence in $C[0,2\pi]$.

2. Is the sequence described in Problem 1 an orthogonal sequence in $C[0,\pi]$?

3. Is $1, x, x^2, x^3, \cdots$ an orthogonal sequence in $C[0,1]$? In $C[-1,1]$?

4. Prove the *parallelogram equality* (the proof is not too much different from that of Theorem 1):

$$\|f + g\|^2 + \|f - g\|^2 = 2\|f\|^2 + 2\|g\|^2.$$

5. What is the geometric interpretation of the equality of Exercise 4?

6. Let f and g be elements of $C[0,1]$ defined by $f(x) = 1$ and $g(x) = x$. Using (4), find the projection of f in the direction of g.

7. Prove Theorem 5. That is, fill in the details omitted in the proof given in the text.

8. Using the Gram-Schmidt process, find an orthonormal basis for the four-dimensional subspace of $C[-1,1]$ spanned by 1, x, x^2, and x^3.

9. The first four *Legendre polynomials* are $P_0(x) = 1$, $P_1(x) = x$, $P_2(x) = \frac{1}{2}(3x^2 - 1)$, $P_3(x) = \frac{1}{2}(5x^3 - 3x)$.

(a) As elements of $C[-1,1]$ are these four functions mutually orthogonal?

(b) Are they orthonormal?

(c) Do they span the subspace described in Exercise 8?

10. The Legendre polynomials $P_0, P_1, P_2, \cdots$ are obtained from the functions $1, x, x^2, x^3, \cdots$ by a process similar to the Gram-Schmidt process, using the inner product

$$(f|g) = \int_{-1}^{1} f(x)g(x)\,dx,$$

except that they are "normalized" by the requirement $P_n(1) = 1$ instead of $\|P_n\| = 1$. The first four Legendre polynomials are given in Exercise 9. Find the next two, i.e., P_4 and P_5.

11. Find the polynomial function of degree two, $ax^2 + bx + c$, providing the best approximation to e^x in the least-squares sense over the interval $[-1,1]$.

12. Which of the following statements are true and which are false?

(a) proj $(f:\phi)$ is a scalar.

(b) proj $(f:\phi)$ is a scalar multiple of f.

(c) proj $(f:\phi_1,\phi_2)$ is a linear combination of ϕ_1 and ϕ_2.

(d) Every orthonormal sequence is linearly independent.

(e) It is impossible for an element f to be orthogonal to every term of an orthonormal approximating basis.

13. Prove that $\sin x, \sin 2x, \sin 3x, \cdots$ is *not* an approximating basis for $C[0,2\pi]$.

***14.** Explain why you would expect $\lim_{n\to\infty} \int_0^\pi x \sin nx\,dx$ to equal zero, and verify by direct calculation.

15. Assuming $\sin x, \sin 2x, \sin 3x, \cdots$ to be an approximating basis for $C[0,\pi]$, write out Parseval's equality for the special case $f(x) = x$. (Begin by normalizing the sequence.)

16. If $g = \text{proj}\,(f:\phi_1, \phi_2, \cdots, \phi_n)$, what is proj $(g:\phi_1, \phi_2, \cdots, \phi_n)$?

17. Write out (14) in detail, taking $u_k(x) = \sin kx$, if V is the space $C[0,\pi]$.

***18.** Is the Corollary to Theorem 11 valid if the word "orthonormal" is deleted?

***19.** Prove that, if α and β are elements of R^∞, the series (25) defining $(\alpha|\beta)$ must be convergent.

20. Is the sequence $(1, \frac{1}{2}, \frac{1}{3}, \frac{1}{4}, \cdots)$ an element of Hilbert coordinate space?

2.4 • DIFFERENTIAL OPERATORS

The purpose of this section is to show, in one of the simplest possible contexts, how orthogonal sequences of functions can arise in practice. At the same time, we will introduce the notion of a *boundary-value problem* for functions of a single variable. (In later chapters, this notion will be generalized to include functions of more than one variable.)

We begin by considering three examples. Although elementary, they underlie the entire theory of Fourier series.

EXAMPLE 1: Determine for what values of λ the differential equation

$$\frac{d^2y}{dx^2} + \lambda y = 0 \tag{1}$$

has nontrivial solutions which satisfy the boundary conditions

$$y(0) = 0, \quad y(\pi) = 0 \tag{2}$$

and find these solutions.

We proceed as follows. First we note that the "general solution" of (1) is

$$y(x) = C_1 \sin \sqrt{\lambda} x + C_2 \cos \sqrt{\lambda} x. \tag{3}$$

To satisfy $y(0) = 0$ we must take C_2 to be zero. Since $\sin \theta = 0$ only when θ is an integer multiple of π, the condition $y(\pi) = 0$ requires that $\sqrt{\lambda} x = n\pi$ when $x = \pi$. Therefore nontrivial solutions (i.e., solutions other than the function identically equal to zero) exist only for $\lambda = n^2$ $(n = 1, 2, 3, \cdots)$.

Therefore every such solution is a scalar multiple of one of the functions in the sequence

$$\sin x, \sin 2x, \sin 3x, \cdots \tag{4}$$

and the corresponding values of λ are

$$1, \quad 4, \quad 9, \quad \cdots. \tag{5}$$

It will be noted that, when λ is negative, the functions $\sin \sqrt{\lambda}x$ and $\cos \sqrt{\lambda}x$ are complex-valued, and are most conveniently written in terms of hyperbolic functions. The reader can verify for himself that these solutions cannot satisfy the boundary conditions (2) unless both C_1 and C_2 are taken to be zero.

EXAMPLE 2: Determine for what values of λ the differential equation (1) has nontrivial solutions which satisfy the boundary conditions

$$y'(0) = 0, \quad y'(\pi) = 0 \tag{6}$$

and find these solutions.

Answer: Each solution must be a scalar multiple of one of the functions in the sequence

$$1, \cos x, \cos 2x, \cos 3x, \cdots \tag{7}$$

and the corresponding values of λ are

$$0, 1, \quad 4, \quad 9, \quad \cdots. \tag{8}$$

The verification is easy and is left to the reader.

EXAMPLE 3: Determine for what values of λ the differential equation (1) has solutions satisfying the boundary conditions

$$y(0) = y(2\pi), \quad y'(0) = y'(2\pi) \tag{9}$$

and find these solutions.

Answer: The permissible values of λ are the same as those in Example 2, and any solution corresponding to $\lambda = n^2$ must be a linear combination of the functions $\cos nx$ and $\sin nx$.

We immediately recognize that each of these boundary-value problems has "generated" an orthogonal sequence of functions. The sequence (4) is orthogonal relative to the interval $(0,\pi)$, and so also is the sequence (7). The sequence

$$1, \cos x, \sin x, \cos 2x, \sin 2x, \cos 3x, \cdots \tag{10}$$

is orthogonal relative to the larger interval $(0,2\pi)$. In the ensuing analysis, we will see that this is no accident, and in a later chapter we will find that a similar phenomenon occurs for boundary-value problems in two and three dimensions when the second derivative in (1) is replaced by the Laplacian operator.

Each of these examples is a special case of the more general "Sturm-Liouville problem" which we will now proceed to discuss. We will not discuss the theory in full generality; the proof of the central theorem is beyond the scope of this book. On the other hand, to omit a discussion of the Sturm-Liouville theory would deprive the reader of the pleasure of an important idea that unifies a number of diverse topics discussed elsewhere in this book.

An equation of the form

$$\frac{d}{dx}\left[p(x)\frac{dy}{dx}\right] + [q(x) + \lambda r(x)]y = 0 \tag{11}$$

is called a *Sturm-Liouville equation.* We assume that the functions p, q, and r are real and continuous in some interval $a \leqq x \leqq b$, that p is continuously differentiable, and that both $p(x)$ and $r(x)$ are positive for all x in this interval. (These assumptions greatly restrict the generality of the following discussion, as we will see later.)

In some applications, the parameter λ is a number that is proportional to either a frequency or an angular velocity; in other applications, the physical significance of λ is more obscure. At this point, we wish to direct attention to the fact that (11) is not really a *single* differential equation, even when we have chosen particular functions p, q, and r, because of the presence of the arbitrary parameter λ.

Although (11) seems to be a very special type of differential equation, a large class of second order differential equations can be written in this form (see Exercise 1).

The so-called *Sturm-Liouville problem* consists in studying nontrivial solutions of (11) which satisfy certain boundary conditions, to be specified presently, at the endpoints of the interval $a \leqq x \leqq b$. Such solutions are called *eigenfunctions,* and the corresponding values of λ are called *eigenvalues.*

This terminology is not standard. For instance, instead of the term "eigenfunction," the terms *characteristic function, proper function,* and *autofunction* are also used.

To explain the nature of the boundary conditions it is convenient to introduce the formal linear differential operator L, defined by

$$Ly = \frac{d}{dx}\left[p(x)\frac{dy}{dx}\right] + q(x)\cdot y. \tag{12}$$

In terms of this operator, (11) can be written in the form

$$Ly + \lambda ry = \theta. \tag{13}$$

If y is a twice continuously differentiable function, Ly will be continuous. In other words, L operates on functions y of class $C^2[a,b]$ to produce functions Ly of class $C[a,b]$.

The boundary conditions, which we have not yet specified, must be chosen in such a manner that the class of all functions satisfying them is a *linear space.* In particular, the class of all functions y of class $C^2[a,b]$ which satisfy these boundary conditions must be a *subspace* of $C^2[a,b]$. We denote this subspace by $BC^2[a,b]$; the letter B is intended to suggest the word "boundary."

Notice that this notation is ambiguous; corresponding to different kinds of boundary conditions, we have different subspaces $BC^2[a,b]$. This will cause no difficulty, provided the boundary conditions are carefully specified in each instance.

The second requirement we make is that the boundary conditions be chosen so that the operator L is *self-adjoint* in $BC^2[a,b]$. This means that, whenever f and g are any two functions of class $BC^2[a,b]$, whether or not they satisfy (13), we have

$$(Lf|g) = (f|Lg). \tag{14}$$

We are now in a position to describe various specific boundary conditions that will give rise to a Sturm-Liouville problem. We simply write out the expression $(Lf|g) - (f|Lg)$ and ascertain what boundary conditions will cause this expression to vanish. We have

$$\begin{aligned}(Lf|g) - (f|Lg) &= \int_a^b Lf(x)\overline{g(x)}\,dx - \int_a^b f(x)\overline{Lg(x)}\,dx \\ &= \int_a^b [(pf')' + qf]\bar{g}\,dx - \int_a^b [(p\bar{g}')' + q\bar{g}]f\,dx \\ &= \int_a^b [pf''\bar{g} + p'f'\bar{g} - p'\bar{g}'f - p\bar{g}''f]\,dx.\end{aligned} \tag{15}$$

We observe that the integrand is the derivative of the function $p[f'\bar{g} - \bar{g}'f]$, and therefore

$$\begin{aligned}(Lf|g) - (f|Lg) = p(b)[f'(b)\overline{g(b)} - \overline{g'(b)}f(b)] \\ -p(a)[f'(a)\overline{g(a)} - \overline{g'(a)}f(a)].\end{aligned} \tag{16}$$

EXAMPLE 4: The right side of (16) will vanish if at *each* endpoint both f and g satisfy *one* of the following boundary conditions:

$$y = 0. \tag{17}$$

$$\frac{dy}{dx} = 0. \tag{18}$$

$$y + \alpha\frac{dy}{dx} = 0 \quad (\alpha \text{ is an arbitrary constant; see Exercise 2}). \tag{19}$$

It will be noticed that in Example 1 a condition of type (17) was imposed at both endpoints; in Example 2 one like (18) was imposed at both endpoints. It is permissible to mix these conditions, however; for example, we can require that the functions vanish at the left endpoint and that their derivatives vanish at the right endpoint. (Examples 1 and 2 correspond to the case in which $p(x)$ and $r(x)$ are identically equal to unity, and $q(x)$ is identically zero.)

EXAMPLE 5: If the function p has the property $p(b) = p(a)$ then (16) vanishes if both f and g satisfy the conditions

$$y(a) = y(b) \quad \text{and also} \quad y'(a) = y'(b). \tag{20}$$

(Example 3 is a special case.)

We see from these examples that a variety of different boundary conditions can be chosen, and therefore many different subspaces of $C^2[a,b]$ can qualify to be denoted $BC^2[a,b]$.

So far, we have not required that the parameter λ be a real number. Let us dispose of this at once:

Lemma 1. *Every eigenvalue of a self-adjoint operator is real.*

Proof: Suppose that λ_0 is an eigenvalue, i.e., that there exists a function y, not identically zero, such that $Ly + \lambda_0 ry = \theta$. If L is self-adjoint, by (14) [taking $f = g = y$] we have $(Ly|y) = (y|Ly)$ and therefore $(\lambda_0 ry|y) = (y|\lambda_0 ry)$. Hence $\lambda_0(ry|y) = \bar{\lambda}_0(y|ry)$. In integral form this can be written

$$\lambda_0 \int_a^b r(x)|y(x)|^2\,dx = \bar{\lambda}_0 \int_a^b r(x)|y(x)|^2\,dx$$

(keep in mind that r is a real function). Since r is positive and y is not identically zero, the integrals are not zero, and it follows that $\lambda_0 = \bar{\lambda}_0$, i.e., λ_0 is real.

In the following Lemma we begin to see the significance of the self-adjoint property (14).

Lemma 2. *Eigenfunctions corresponding to distinct eigenvalues of a self-adjoint operator are orthogonal with respect to the weighting factor $r(x)$.*

Proof: Suppose that $Lf + \lambda_1 rf = \theta$ and also $Lg + \lambda_2 rg = \theta$, where $\lambda_1 \neq \lambda_2$. By (14) we have, since L is self-adjoint by hypothesis,

$$(-\lambda_1 rf|g) = (f|-\lambda_2 rg).$$

Writing this out in integral form and simplifying, we obtain (since λ_2 is real, by Lemma 1)

$$(\lambda_1 - \lambda_2) \int_a^b r(x)f(x)\overline{g(x)}\, dx = 0,$$

and since $\lambda_1 \neq \lambda_2$ it follows that

$$\int_a^b r(x)f(x)g(x)\, dx = 0,$$

as stated in the lemma.

The following theorem is much deeper.

Theorem (Sturm-Liouville). *If the boundary conditions defining $BC^2[a,b]$ are such that L is self-adjoint in $BC^2[a,b]$, i.e., $(Lf|g) = (f|Lg)$ for all functions f and g of class $BC^2[a,b]$, then there exists an infinite sequence of eigenfunctions*

$$\phi_1, \phi_2, \phi_3, \cdots \tag{21}$$

which are mutually orthogonal with respect to the weighting function r:

$$\int_a^b r(x)\phi_j(x)\overline{\phi_k(x)}\, dx = 0, \qquad \text{whenever } j \neq k. \tag{22}$$

Every (bounded) function f that is integrable over $[a,b]$ can be expanded in a series

$$f(x) = C_1\phi_1(x) + C_2\phi_2(x) + C_3\phi_3(x) + \cdots \tag{23}$$

which converges in the mean to f. If f is of class $BC^2[a,b]$, we can say even more: this series converges pointwise; in fact, it converges uniformly and absolutely to the function f, throughout the interval $[a,b]$.

A number of theorems proved later are special cases of this theorem. The general proof of this theorem is beyond the scope of this book. The following remarks are intended to clarify some of the hypotheses that we have imposed.

The reason for requiring r to be a positive function should be clear; otherwise the integral $\int_b^a r(x)f(x)\overline{g(x)}\, dx$ would not satisfy the axioms required of an inner product.

It is obvious that we would not be interested in the case in which $p(x)$ vanishes identically; (11) would not even be a differential equation. The reason for not allowing it to vanish at isolated points is less obvious, unless the reader is familiar with existence theorems for differential equations. This hypothesis was imposed to ensure that all solutions of (11) are *bounded* throughout the interval $a \leqq x \leqq b$. A number of the integrals occurring in earlier paragraphs might not make sense for unbounded functions.

EXAMPLE 6: In many important special cases $p(x)$ is positive for $a < x < b$ but $p(x) = 0$ when $x = a$ or when $x = b$. In this case there may be unbounded solutions to the differential equation. If, however, we restrict ourselves to functions that *are* bounded, and for which either dy/dx is finite or $p\, dy/dx$ tends to zero at the endpoint, then our calculations make sense and the right side of (16) vanishes without imposing any further boundary conditions. With this restriction to bounded solutions, both Lemma 1 and Lemma 2 are valid, but the Sturm-Liouville theorem may not be valid. (The situation is too complicated to discuss here in detail.) Both Legendre's equation and Bessel's equation, to be discussed later, fall in this category.

The following example shows that it would be unreasonable to expect the series (23) to converge to $f(x)$ for every x, if f is not of class $BC^2[a,b]$.

EXAMPLE 7: According to Example 1 and the Sturm-Liouville theorem, every function that is twice continuously differentiable and vanishes at $x = 0$ and $x = \pi$ can be expanded in a series of the form $C_1 \sin x + C_2 \sin 2x + C_3 \sin 3x + \cdots$ which will converge to $f(x)$ whenever $0 \leqq x \leqq \pi$. If f does not satisfy these conditions, this is manifestly impossible, since every term in this series vanishes at $x = 0$ and at $x = \pi$. However, as we will see later, it is quite possible for a series of this form to converge *in the mean*, in the interval $0 \leqq x \leqq \pi$, to a function that does not vanish at either endpoint.

• EXERCISES

1. **(a)** Show that any equation of the form

$$a_0(x)\frac{d^2y}{dx^2} + a_1(x)\frac{dy}{dx} + [a_2(x) + \lambda a_3(x)]y = 0$$

can be rewritten formally in the form of (11).

(b) What restrictions are needed in order for this to make sense?

2. Explain in detail how the vanishing of (19) at an endpoint ensures the vanishing of the corresponding term in (16) at that endpoint.

3. **(a)** Show by direct calculation that if k_1 and k_2 are distinct positive roots of $\tan kL = -k$, then $\int_0^L \sin k_1x \sin k_2x\, dx$ must equal zero.

(b) Find all the eigenfunctions for the boundary-value problem $d^2y/dx^2 + \lambda y = 0$, $y(0) = 0$, $y(L) + y'(L) = 0$.

(c) Hence, show that the result of part (a) follows from a lemma in this section.

4. Find an orthogonal sequence containing (to within a scalar factor) all eigenfunctions of the boundary-value problem $d^2y/dx^2 + \lambda y = 0$ for each of the following sets of boundary conditions.
   (a) $y'(0) = 0$, $y'(L) = 0$.
   (b) $y(a) = y(b)$, $y'(a) = y'(b)$.
   (c) $y(0) = 0$, $y'(\pi) = 0$.

5. With the notation used in this section, show in detail that $uLv - vLu = (puv' - pvu')'$ whenever u and v are of class $C^2[a,b]$.

6. Show that, if u and v are solutions of $Ly = \theta$, then $p(uv' - vu')$ is a constant.

*7. (a) Show that, if u and v are solutions of $Ly = \theta$, then the function $uv' - vu'$ can equal zero at one point in $[a,b]$ if and only if it is identically zero throughout the interval $[a,b]$.
   (b) Show that, if u and v are solutions of $Ly = \theta$, they span the linear space of *all* solutions of $Ly = \theta$ if and only if $uv' - vu'$ is nonzero.
   (c) Show that, if u and v are linearly independent solutions of $Ly = \theta$, the following function f is the unique solution of $Lf = h$ for which $f(a) = 0$ and $f'(a) = 0$. (We assume that h is a continuous function.)

$$f(x) = \int_a^x \frac{u(t)v(x) - u(x)v(t)}{p[uv' - vu']} h(t)\, dt$$

   where, as noted in a previous exercise, the denominator is a constant.

*8. A linear operator T transforms each element f in a linear space V into an element Tf of a linear space W. Moreover, every element in W is the transform of some element of V. Is this possible if V is a proper subspace of W?

2.5 • INTEGRAL OPERATORS

Since this section is essentially a continuation of the preceding section, we continue to use the same notation.

We begin by considering the following question: Is it possible to find an integral operator that will provide an "inverse" in some sense of a differential operator L (of the type considered in the preceding section)?

We recall that, if y is a function of class $C^2[a,b]$, the function Ly must be of class $C[a,b]$. We would like to find an integral operator that will play the role of L^{-1}. In other words, we would like to find an integral operator J with the property that $Jh = y$ if and only if $h = Ly$.

Without further restrictions, it is manifestly impossible to find such an operator. For example, if L is the differential operator d^2/dx^2 and if y is the function $3x + 4$, then Ly is identically zero, and it is impossible to construct an integral operator that, operating on the zero function, can reconstruct the function y. This is obvious; only a mind-reader could ascertain that we started with $3x + 4$ rather than (say) $2x - 8$.

On the other hand, if we restrict the domain of definition of L to a suitably chosen subspace of $C^2[a,b]$, then it is sometimes possible to find an integral operator that provides an inverse for L. This brief statement will be clarified shortly.

To fix ideas, we assume that $BC^2[a,b]$ is chosen as in the preceding section. In other words, the subspace $BC^2[a,b]$ is defined by boundary conditions so chosen that the differential operator L is self-adjoint in $BC^2[a,b]$.

We say that L is *nonsingular* in $BC^2[a,b]$ if the only function y of class $BC^2[a,b]$ for which $Ly = \theta$ is the function $y = \theta$. The reader can see that this is equivalent to the requirement that L admits no eigenfunctions in $BC^2[a,b]$ that correspond to the eigenvalue $\lambda = 0$. (Keep in mind that, by definition, the zero function is never called an eigenfunction, but some operators admit the number zero as an eigenvalue—such operators are said to be *singular*.)

The following lemma is absolutely trivial.

Lemma 1. *If L is nonsingular in $BC^2[a,b]$, then $Ly_1 = Ly_2$ for functions y_1 and y_2 of class $BC^2[a,b]$ if and only if $y_1 = y_2$.*

Proof: If $y_1 = y_2$ then obviously $Ly_1 = Ly_2$. If $Ly_1 = Ly_2$ then (since L is a linear differential operator) $L(y_1 - y_2) = \theta$ and (since L is nonsingular) $y_1 - y_2 = \theta$, i.e., $y_1 = y_2$.

According to a fundamental existence theorem of differential equations, for any prescribed function h of class $C[a,b]$, the differential equation $Ly = h$ has many solutions. According to Lemma 1, at most one solution is an element of $BC^2[a,b]$. In the exercises, we will show (by explicit construction) that such a solution does in fact exist. Indeed, we will see that this solution is given by an integral operator:

Lemma 2. *If the differential operator L is nonsingular and self-adjoint in $BC^2[a,b]$, there exists a function $G(x,t)$, continuous in the square $a \leqq x \leqq b$, $a \leqq t \leqq b$, with the property that, for functions h of class $C[a,b]$ and functions y of class $BC^2[a,b]$, $Ly = h$ if and only if*

$$y(x) = \int_a^b G(x,t)h(t)\,dt. \tag{1}$$

More explicitly, if the function $\int_a^b G(x,t)h(t)\,dt$ is denoted Jh, we have $L(Jh) = h$ for every function h of class $C[a,b]$, and we have $J(Ly) = y$ for every function y of class $BC^2[a,b]$.

The function G that is used to define an integral operator is called the *kernel* of the operator. We will see in the exercises that the kernel in this case is a real-valued function with the property

$$G(x,t) = G(t,x). \tag{2}$$

It is interesting to note that integral operators defined via such kernels have the property $(Jf|g) = (f|Jg)$ whenever f and g are of class $C[a,b]$. This is the self-adjoint property; the operator J is a *self-adjoint integral operator.*

Because of the formal similarity between the right side of (1) and the rule for multiplying a matrix with a column vector, such a kernel G is sometimes called a "continuous matrix."

Since the integral operator J is, in a sense, an inverse for the differential operator L, it would be surprising if there were not some kind of expansion theorem related to J in the same way that the Sturm-Liouville theorem is related to L. We now state such a theorem. It should be noted that this theorem is of independent interest quite apart from those kernels that are related to differential operators in the manner described by Lemma 2.

Theorem (Hilbert-Schmidt). *If $f(x) = \int_a^b G(x,t)h(t)\,dt$, where h is bounded and integrable over $[a,b]$, $G(x,t) = \overline{G(t,x)}$, and G is continuous (or at least bounded and integrable over the square $a \leqq x \leqq b$, $a \leqq t \leqq b$), then f can be expanded in a series*

$$f(x) = C_1\phi_1(x) + C_2\phi_2(x) + C_3\phi_3(x) + \cdots \tag{3}$$

of orthogonal functions each having the property

$$\int_a^b G(x,t)\phi_j(t)\,dt = \alpha_j\phi_j(x) \tag{4}$$

where each α_j *is real and nonzero. The series converges pointwise, indeed it converges uniformly and absolutely, to the function* f *throughout the interval* $[a,b]$.

Note: As we will see by examples, it is possible for some choices of the kernel G that only a finite number of mutually orthogonal functions ϕ_j satisfy (4). The theorem is still valid, but (3) is not an infinite series in this case, and the statements about convergence are then unnecessary. In particular, this applies to the Dirichlet kernel, which will be discussed in the next section.

• EXERCISES

1. Let H denote the *Heaviside function* defined by $H(x) = 1$ when x is positive and $H(x) = 0$ when x is negative (its value when $x = 0$ is immaterial here).
   (a) Is the function $H(x - t)$ continuous in the square $a \leqq x \leqq b$, $a \leqq t \leqq b$?
   (b) State necessary and sufficient conditions that must be satisfied by a continuous function $K(x,t)$ in order that the function $K(x,t)H(x - t)$ be continuous in this square.

2. Let $K(x,t) = \dfrac{u(t)v(x) - u(x)v(t)}{p[uv' - vu']}$ where the notation is that of Exercise 7, Section 2.4. Show that $H(x - t)K(x,t)$ is continuous in the square $a \leqq x \leqq b$, $a \leqq t \leqq b$.

3. Let $BC^2[a,b]$ be defined by the requirements that $y(a) = 0$ and $y(b) = 0$. Suppose that L is nonsingular on $BC^2[a,b]$. Let u and v be nontrivial real solutions of $Ly = \theta$ chosen so that $u(a) = 0$ and $v(b) = 0$. Show that u and v must be linearly independent.

4. Let $G(x,t) = H(x - t)K(x,t) + \dfrac{u(x)v(t)}{p[uv' - vu']}$ where the notation is the same as that in the preceding exercise.
   (a) Show that, if h is of class $C[a,b]$, and we define f by $f(x) = \int_a^b G(x,t)h(t)\,dt$, then f is of class $C^2[a,b]$. (Make use of Exercise 7, Section 2.4.)

(b) Show that $Lf = h$.
(c) Show that $f(a) = 0$ and $f(b) = 0$.
[The point is that, in Exercise 7, Section 2.4, we obtained a function f such that $f(a) = 0$ and $f'(a) = 0$. By modifying the integrand, we now have an integral operator that yields a function f satisfying the *boundary* conditions $f(a) = 0$, $f(b) = 0$, instead of these *initial* conditions.]

***5.** Generalize the preceding construction of $G(x,t)$ to the case where $BC^2[a,b]$ satisfies more general boundary conditions than $y(a) = 0$ and $y(b) = 0$.

6. Show that, if G is defined as in Exercise 4,

$$G(x,t) = G(t,x).$$

7. Show that if G is continuous and $G(x,t) = \overline{G(t,x)}$, the integral operator J having kernel G is self-adjoint, in the sense that $(Jf|g) = (f|Jg)$ for every f and g of class $C[a,b]$.

8. Let L denote the differential operator d^2/dx^2.
(a) Find linearly independent functions u and v satisfying $Ly = \theta$.
(b) Write down $K(x,t)$ for this special case.
(c) Hence, find a single integral equivalent to the iterated integral $f(x) = \int_0^x [\int_0^t h(s)\,ds]\,dt$ which is what one would obtain from integrating $d^2f/dx^2 = h$ twice to obtain a solution for which $f(0) = 0, f'(0) = 0$.
(d) Construct $G(x,t)$ for the special case of the conditions $f(0) = 0, f(1) = 0$.

9. **(a)** By integration, find a function y such that $d^2y/dx^2 = x$, $y(0) = 0$, $y(1) = 0$.
(b) Use the function $G(x,t)$ obtained in part (d) of Exercise 8 to obtain the same solution.

10. Is the operator d^2/dx^2 nonsingular on the space $BC^2[a,b]$ if the boundary conditions defining $BC^2[a,b]$ are:
(a) $y(a) = 0, y(b) = 0$?
(b) $y(a) = 0, y'(b) = 0$?
(c) $y'(a) = 0, y'(b) = 0$?
(d) $y(a) = y(b), y'(a) = y'(b)$?

11. Verify the assertion that $p(uv' - vu')$ is a constant in the special case that u and v are linearly independent solutions of $d^2y/dx^2 + dy/dx = 0$.

12. Is it possible to construct the inverse of the differential operator d^2/dx^2 acting in the space $BC^2[0,L]$ defined by $y'(0) = 0$, $y'(L) = 0$?

13. Taking $G(x,t) = \cos(x - t)$ in the Hilbert-Schmidt theorem, what functions f can be expanded in the "series" (3)?

14. Find all functions ϕ_j satisfying (4) for nonzero α_j in the special case $G(x,t) = xt$. Determine the corresponding α_j in each case.

2.6 • CONVOLUTION AND THE DIRICHLET KERNEL

In this section, we will restrict attention entirely to the sequence of functions

$$1, \cos x, \sin x, \cos 2x, \sin 2x, \cdots. \tag{1}$$

This is an orthogonal sequence relative to the interval $0 \leqq x \leqq 2\pi$, or indeed relative to any interval of length 2π. This is by virtue of the integrals

$$\int_0^{2\pi} \cos nx \cos mx \, dx = \begin{cases} 0 & \text{if } n \neq m \\ \pi & \text{if } n = m \neq 0 \\ 2\pi & \text{if } n = m = 0 \end{cases} \tag{2}$$

$$\int_0^{2\pi} \sin nx \cos mx \, dx = 0 \tag{3}$$

$$\int_0^{2\pi} \sin nx \sin mx \, dx = \begin{cases} \pi & \text{if } n = m \neq 0 \\ 0 & \text{otherwise} \end{cases} \tag{4}$$

valid whenever n and m are nonnegative integers.

It will be observed that the sequence is not orthonormal. It can, of course, be normalized by dividing each function by its mean-square norm:

$$\frac{1}{\sqrt{2\pi}}, \frac{1}{\sqrt{\pi}} \cos x, \frac{1}{\sqrt{\pi}} \sin x, \frac{1}{\sqrt{\pi}} \cos 2x, \frac{1}{\sqrt{\pi}} \sin 2x, \cdots. \tag{5}$$

Now suppose we are given a function f, which we assume to be bounded and integrable and of period 2π, and we seek to find that trigonometric polynomial of the form

$$h_n(x) = A_0 + A_1 \cos x + B_1 \sin x + \cdots + A_n \cos nx + B_n \sin nx \tag{6}$$

that provides the best least-squares approximation to $f(x)$ in any interval of length 2π.

This problem was "solved" in Section 2.3, and we will not review it here beyond remarking that one must use some care since these functions are not of unit norm. According to the results of that section, we must choose the coefficients in the following manner:

$$A_0 = \frac{1}{2\pi}\int_0^{2\pi} f(x)\,dx \tag{7}$$

$$A_n = \frac{1}{\pi}\int_0^{2\pi} f(x)\cos nx\,dx, \quad n = 1, 2, 3, \cdots \tag{8}$$

$$B_n = \frac{1}{\pi}\int_0^{2\pi} f(x)\sin nx\,dx, \quad n = 1, 2, 3, \cdots \tag{9}$$

Our first aim is to find an integral operator that will transform the function f into the function h_n. That is, we desire to find a function $G(x,t)$ such that

$$h_n(x) = \int_0^{2\pi} G(x,t)f(t)\,dt. \tag{10}$$

This function will, of course, depend on n, as well as on x and t. (Actually, we will not use the notation $G(x,t)$ below.)

To compute this function, we substitute the values given in (7), (8), and (9) into (6), and simplify the results as much as possible. To avoid confusing the "current" variable x in (6) with the variable of integration in the other expressions, we change the latter to t. Thus, when $k = 1, 2, \cdots, n$, we have

$$\begin{aligned} &A_k \cos kx + B_k \sin kx \\ &= \left[\frac{1}{\pi}\int_0^{2\pi} f(t)\cos kt\,dt\right]\cos kx + \left[\frac{1}{\pi}\int_0^{2\pi} f(t)\sin kt\,dt\right]\sin kx \\ &= \frac{1}{\pi}\int_0^{2\pi} [\cos kt\cos kx + \sin kt\sin kx]f(t)\,dt \\ &= \frac{1}{\pi}\int_0^{2\pi} f(t)\cos k(x-t)\,dt, \end{aligned} \tag{11}$$

and therefore

$$h_n(x) = \frac{1}{2\pi}\int_0^{2\pi} f(t)\,dt + \sum_{k=1}^{n}\frac{1}{\pi}\int_0^{2\pi} f(t)\cos k(x-t)\,dt. \tag{12}$$

This can be "simplified" by introducing the function

$$D_n(x) = \frac{1}{2\pi} + \frac{1}{\pi}(\cos x + \cos 2x + \cos 3x + \cdots + \cos nx), \tag{13}$$

called the *Dirichlet kernel.* The integral (12) becomes

$$h_n(x) = \int_0^{2\pi} D_n(x - t)f(t)\,dt. \tag{14}$$

To further simplify (13), we use the "collapsing principle." Multiplying both sides by $\sin(x/2)$ and using the formula $\sin\alpha\cos\beta = \frac{1}{2}\sin(\beta + \alpha) - \frac{1}{2}\sin(\beta - \alpha)$, we have

$$\begin{aligned} \sin\frac{x}{2}\cdot D_n(x) &= \frac{1}{2\pi}\left[\sin\frac{x}{2} + \left(\sin\frac{3x}{2} - \sin\frac{x}{2}\right)\right. \\ &\qquad \left. + \cdots + \left(\sin\frac{(2n+1)x}{2} - \sin\frac{(2n-1)x}{2}\right)\right] \\ &= \frac{1}{2\pi}\sin\frac{(2n+1)x}{2}, \end{aligned} \tag{15}$$

and therefore

$$D_n(x) = \frac{\sin(n + \frac{1}{2})x}{2\pi\sin\frac{1}{2}x}. \tag{16}$$

The limit convention must be used to give meaning to this expression at points where the denominator vanishes.

We can therefore write (14) in the form

$$h_n(x) = \int_0^{2\pi} f(t)\,\frac{\sin(n + \frac{1}{2})(x - t)}{2\pi\sin\dfrac{x - t}{2}}\,dt \tag{17}$$

or, with a slight change of variables (keep in mind that f is of period 2π)

$$h_n(x) = \int_0^{2\pi} f(x - t)\,\frac{\sin(n + \frac{1}{2})t}{2\pi\sin\frac{1}{2}t}\,dt. \tag{18}$$

Integrals similar to (14) occur so often in analysis that it is useful to think of the integral as defining a new type of product of two functions. We will not digress here to discuss the physical significance of this type of "multiplication" (see Section 3.1).

In the following discussion, we assume that all functions mentioned are bounded, integrable, and have period 2π.

By the *convolution product* (the terms "resultant" and "Faltung" are used by some authors) of two functions f and g we mean the function $f * g$ defined by

$$(f * g)(x) = \int_0^{2\pi} f(x - t)g(t)\,dt. \tag{19}$$

It is easy to see that this can also be written in the equivalent form

$$(f * g)(x) = \int_0^{2\pi} f(t)g(x - t)\, dt. \tag{20}$$

[Compare (17) and (18).]

It is easy to verify that convolution multiplication is linear and homogeneous in each factor. By this we mean that, for functions f, g, and h, and scalars α and β, we have

$$(\alpha f + \beta g) * h = \alpha(f * h) + \beta(g * h) \tag{21}$$

$$f * (\alpha g + \beta h) = \alpha(f * g) + \beta(f * h). \tag{22}$$

This operation is also commutative [this is the essence of the equivalence of (19) and (20) above]:

$$f * g = g * f \tag{23}$$

and associative

$$(f * g) * h = f * (g * h) \tag{24}$$

so there is no ambiguity in writing $f * g * h$. (The proof is left as a tedious exercise.)

From an algebraic viewpoint, certain convolution products are extremely interesting. To show this, we introduce the following notations:

$$\phi_0(x) = \frac{1}{2\pi} \tag{25}$$

$$\phi_n(x) = \frac{1}{\pi} \cos nx \quad (n = 1, 2, 3, \cdots) \tag{26}$$

$$\psi_n(x) = \frac{1}{\pi} \sin nx \quad (n = 1, 2, 3, \cdots) \tag{27}$$

These functions, which differ by only scalar factors from those listed in (1) and (5), play a fundamental role.

Let S_0 denote the one-dimensional linear space of all constant functions, and for each positive integer n let S_n denote the two-dimensional space consisting of all functions of the form $A_n \cos nx + B_n \sin nx$. Let H_n denote the $(2n + 1)$-dimensional space consisting of all functions of the form (6). It will be noted that $S_0, S_1, \cdots, S_n$ are distinct subspaces of H_n.

The functions $\phi_0, \phi_1, \psi_1, \phi_2, \psi_2, \cdots, \phi_n, \psi_n$ provide an orthogonal basis for H_n. For each positive k (but not for $k = 0$) S_k has a basis consisting of two elements, ϕ_k and ψ_k, and can therefore be "visualized" as a plane. S_0 has ϕ_0 as a basis; it can be visualized as a line.

Geometrically, we think of the linear space of all bounded, integrable, real-valued functions of period 2π as containing an infinite number of planes S_k together with a line S_0. Each of these subspaces is orthogonal to all the others, in the sense that if f is an element of one of them and g is an element of another, then f and g are orthogonal. It is, of course, impossible to draw a diagram including all these subspaces, but in Figure 2.6 we show two of them, S_0 and S_1.

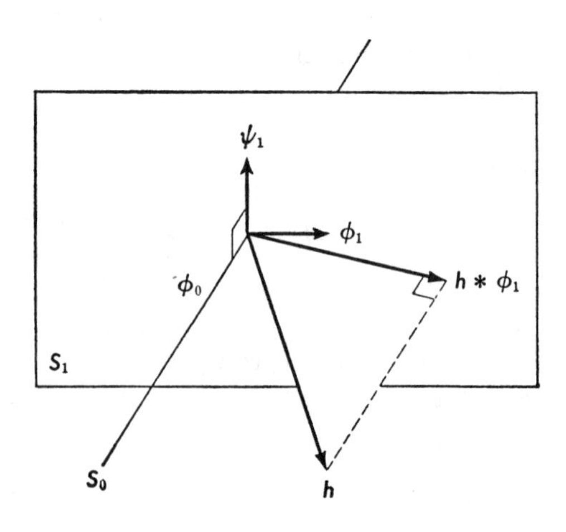

Figure 2.6

The projection of an element f into S_k (for any positive k) is, as we saw in Section 2.3, precisely the function given by (11) above. In terms of convolution multiplication and the definition of ϕ_k we see that this is precisely $f * \phi_k$.

Lemma 1. *The projection of f into S_k, for any k, is $f*\phi_k$, where $\phi_k(x) = (\cos kx)/\pi$ for $n = 1, 2, 3, \cdots$ and $\phi_0(x) = 1/2\pi$. An element f is in S_k if and only if $f * \phi_k = f$. An element f is orthogonal to S_k if and only if $f * \phi_k = \theta$, where θ (as usual) denotes the zero function.*

The proof for positive k is contained in the above remarks; the modification for $k = 0$ is trivial.

As an immediate consequence of Lemma 1, we can write down the following "multiplication laws" without further computation except in one instance:

$$\phi_n * \phi_n = \phi_n \tag{28}$$

(29) $$\phi_n * \psi_n = \psi_n$$

(30) $$\psi_n * \psi_n = -\phi_n$$

and, whenever $n \neq m$,

(31) $$\phi_n * \phi_m = \theta, \quad \phi_n * \psi_m = \theta, \quad \psi_n * \psi_m = \theta.$$

Proof: (28), (29), and the first two products in (31) follow directly from Lemma 1; thus, $\phi_n * \phi_m$ equals θ since ϕ_n is orthogonal to S_m. To obtain the last product in (31) we write $\psi_n * \psi_m = \psi_n * (\phi_m * \psi_m) = (\psi_n * \phi_m) * \psi_m = \theta * \psi_m = \theta$. (Obviously, $\theta * f = \theta$ for every f.) It therefore remains only to derive (30). This can be done by direct calculation (see Exercise 5).

We see from (28), (29), and (30) that, with respect to convolution multiplication, ϕ_n plays a role similar to that of the number one, and ψ_n similar to that of the complex unit $i = \sqrt{-1}$. Indeed, from a purely algebraic viewpoint, we can consider each of the planes S_n to be a replica of the complex plane. This greatly facilitates calculation, since we are familiar with multiplication of complex numbers.

EXAMPLE 1: (a) Find a function f such that

$$f * \sin 3x = \cos 3x + 2 \sin 3x.$$

(b) Is this solution unique?

Solution: (a) In our notation, we have $f * \pi\psi_3 = \pi\phi_3 + 2\pi\psi_3$. We solve this the same way that we would solve $(z)(\pi i) = \pi + 2\pi i$ for the complex number z. (This would give $z = 2 - i$). The solution is $2\phi_3 - \psi_3$, i.e.,

$$f(x) = \frac{2 \cos 3x}{\pi} - \frac{\sin 3x}{\pi}.$$

(b) The solution is not unique; because of (31) we could obtain many other solutions by adding extra terms involving $\sin kx$ and $\cos kx$ for $k \neq 3$.

EXAMPLE 2: What is the geometrical significance of $f * \psi_k$?

Answer: Since $f * \psi_k = f * \phi_k * \psi_k$, we can think of $f * \psi_k$ as the result of first projecting f into S_k and then rotating the resulting vector through 90°. Indeed, by Lemma 1, $f * \phi_k$ is the projection of f into S_k, and the effect of multiplying an element in S_k by ψ_k is to rotate it by 90° (the same as the effect of multiplying a complex number by i).

In any algebraic system, an element equal to its own square is called an *idempotent*. In the complex number field, the only idempotents $z^2 = z$ are the numbers $z = 0$ and $z = 1$. Similarly, in each of the subspaces S_k the only idempotents are θ and ϕ_k. Because of Lemma 1, an idempotent ϕ_k is said to generate the subspace S_k.

It is easy to see that any sum of distinct idempotents ϕ_k is again an idempotent (but not one of the ϕ_k's; it will be a new idempotent). For example, if we let D_n denote the sum of the first $n + 1$ idempotents,

$$D_n = \phi_0 + \phi_1 + \cdots + \phi_n, \tag{32}$$

then

$$\begin{aligned} D_n * D_n &= (\phi_0 + \phi_1 + \cdots + \phi_n)(\phi_0 + \phi_1 + \cdots + \phi_n) \\ &= \phi_0 + \phi_1 + \cdots + \phi_n, \end{aligned}$$

the cross products vanishing because of (31). Therefore we have

$$D_n * D_n = D_n. \tag{33}$$

Recalling that D_n is our friend the Dirichlet kernel, we see that the Dirichlet kernel is an idempotent. Indeed, it is the generating idempotent of the subspace H_n, and it was not really necessary to derive (33); it follows at once from the calculations leading to (14). These calculations and remarks can be summarized in

Lemma 2. *The projection of f into H_n, for any n, is $f * D_n$, where D_n is the Dirichlet kernel (16). An element f is in H_n if and only if $f * D_n = f$, and is orthogonal to H_n if and only if $f * D_n = \theta$.*

Despite the notation $f * D_n$ for (14), the reader must not lose sight of the fact that (14) defines an integral operator. Nor must he think this is the only integral operator that will transform a function f into a function in the subspace H_n. The significance of this particular operator is that it always yields that element of H_n that provides the best least-squares approximation to f.

We will see later that this is not the only kind of approximation that one might be led to consider; the reader may wish to scan Section 3.7 at this point. We will now direct our attention to another integral operator that yields a function $V_n(x)$ having a different relationship to $f(x)$ than the function $h_n(x)$ obtained via (14). The advantage of $V_n(x)$ is that, in certain cases, we can make an assertion about $V_n(x)$ being "close" to $f(x)$ for each value of x, whereas all we can say about $h_n(x)$ is that it is close to $f(x)$ in the sense of a mean-square average. Indeed, we have proved no theorems at all that state anything about $|f(x) - h_n(x)|$ for any individual value of x, but we will be able to make some such statement about $|f(x) - V_n(x)|$. We must first introduce the following definition.

A function f is said to satisfy a *Lipschitz condition* if

$$|f(x_2) - f(x_1)| \leqq K|x_2 - x_1|$$

for every pair of numbers x_2 and x_1 and some constant K.

Obviously, any function satisfying such a condition is continuous; because of the inequality, we can make $|f(x_2) - f(x_1)|$ as small as we like by taking x_2 sufficiently close to x_1.

Examples of functions which do (and some which do not) satisfy a Lipschitz condition are given in the Exercises.

We introduce the function E_n defined by

$$E_n(x) = \frac{1}{\alpha_n} \cos^{2n} \frac{x}{2} \tag{34}$$

where $\alpha_n = \int_0^{2\pi} \cos^{2n} \frac{x}{2}\, dx$.

Given a function f, we define V_n by

$$V_n = f * E_n \tag{35}$$

and we prove the following theorem:

Theorem. *The function V_n is a trigonometric polynomial of order at most n. If f has period 2π and satisfies a Lipschitz condition with constant K, then*

$$|f(x) - V_n(x)| \leqq K\pi/\sqrt{n} \quad \textit{for every } x.$$

The proof is easy but somewhat technical, and will therefore be split up into several lemmas.

Lemma 3. *The function E_n is a cosine polynomial of order at most n; in other words, it is a linear combination of the $n + 1$ functions $\phi_0, \phi_1, \cdots, \phi_n$.*

Proof: We have $\cos^{2n} \frac{x}{2} = \left[\frac{1 + \cos x}{2}\right]^n$. Since $\cos px \cos qx =$ $\frac{1}{2} \cos (p + q)x + \frac{1}{2} \cos (p - q)x$, we see that the ordinary product (not convolution product) of a cosine polynomial of order at most n_1 and another of order at most n_2 is a cosine polynomial of order at most $n_1 + n_2$. In particular, $1 + \cos x$ is a cosine polynomial of order one, so by induction it follows that $(1 + \cos x)^n$ is a cosine polynomial of order at most n. This property is not destroyed by the scalar factors 2^{-n} or $1/\alpha_n$, so the Lemma is proved.

Lemma 4. *V_n is a trigonometric polynomial of order at most n; in other words, it is in the linear space H_n.*

Note: Unlike the function h_n, the function V_n is not the *projection* of f into H_n.

Proof: Since $V_n = f * E_n$ and E_n is a linear combination of $\phi_0, \phi_1, \cdots, \phi_n$ (by Lemma 2) it follows that V_n is a linear combination of the functions $f * \phi_k$ $(k = 0, 1, \cdots, n)$. By Lemma 1, each of these functions is in the linear space H_n and therefore V_n is also.

Lemma 5. *In the range $0 < x < \pi/2$, we have*

$$x \leqq \frac{\pi}{2} \sin x \tag{36}$$

and in the range $0 < x < \pi$,

$$\frac{1}{\pi \sin \frac{1}{2}x} \leqq \frac{1}{x}. \tag{37}$$

Proof: Both inequalities follow easily from the observation that the derivative of $x/\sin x$ is positive in $0 < x < \pi/2$ and therefore $x/\sin x$ is increasing in this interval.

Lemma 6. $\alpha_{n+1} < \alpha_n$ $(n = 1, 2, 3, \cdots)$.

Note: The sequence $\alpha_1, \alpha_2, \alpha_3, \cdots$ actually tends to zero, although we make no explicit use of this fact.

Proof: $\alpha_n = \displaystyle\int_0^{2\pi} \cos^{2n} \frac{x}{2}\, dx = 4 \int_0^{\pi/2} \cos^{2n} x\, dx.$

Since $\cos x$ has values between 0 and 1 in the range $0 < x < \pi/2$, it follows that $\cos^{2n+2} x < \cos^{2n} x$ in this interval, from which the statement follows immediately.

Lemma 7. $\displaystyle\int_0^{\pi/2} \sin x \cos^{2n} x\, dx \leqq \sqrt{\frac{\alpha_n \alpha_{n+1}}{16(2n + 1)}}.$

Proof: We use the Schwarz inequality (Section 2.1) which for real functions g and h and the interval $(0, \pi/2)$ is

$$\left|\int_0^{\pi/2} g(x)h(x)\,dx\right| \leqq \sqrt{\int_0^{\pi/2} [g(x)]^2\,dx \cdot \int_0^{\pi/2} [h(x)]^2\,dx}. \tag{38}$$

We let $g(x) = \sin x \cos^n x$ and $h(x) = \cos^n x$. As noted in the proof of Lemma 6, the second of these integrals is then $\alpha_n/4$. The other integral can be written in the form $\int_0^{\pi/2} (\sin x)(\sin x \cos^{2n} x\,dx)$ which on integration by parts equals $\alpha_{n+1}/4(2n+1)$. This yields the desired result.

Proof of the Theorem: By the definition of α_n and observing that the functions involved are all of period 2π, we have

$$1 = \frac{1}{\alpha_n}\int_{-\pi}^{\pi} \cos^{2n}\frac{t}{2}\,dt. \tag{39}$$

Multiplying both sides by $f(x)$ (which for any fixed x is a constant), we obtain

$$f(x) = \frac{1}{\alpha_n}\int_{-\pi}^{\pi} f(x)\cos^{2n}\frac{t}{2}\,dt \tag{40}$$

and therefore [from the definition (35)]

$$\begin{aligned} |f(x) - V_n(x)| &= \left|\frac{1}{\alpha_n}\int_{-\pi}^{\pi} [f(x) - f(x-t)]\cos^{2n}\frac{t}{2}\,dt\right| \\ &\leqq \frac{1}{\alpha_n}\int_{-\pi}^{\pi} |f(x) - f(x-t)|\cos^{2n}\frac{t}{2}\,dt. \end{aligned} \tag{41}$$

Since f satisfies a Lipschitz condition with constant K, we have $|f(x) - f(x-t)| \leqq K|t|$, and (41) becomes

$$\begin{aligned} |f(x) - V_n(x)| &\leqq \frac{K}{\alpha_n}\int_{-\pi}^{\pi} |t|\cos^{2n}\frac{t}{2}\,dt \\ &= \frac{2K}{\alpha_n}\int_0^{\pi} t\cos^{2n}\frac{t}{2}\,dt \\ &= \frac{8K}{\alpha_n}\int_0^{\pi/2} t\cos^{2n} t\,dt \\ &\leqq \frac{4K\pi}{\alpha_n}\int_0^{\pi/2} \sin t\cos^{2n} t\,dt \end{aligned} \tag{42}$$

(here we used Lemma 5) and by Lemma 7

$$\leqq \frac{4K\pi}{\alpha_n}\sqrt{\frac{\alpha_n\alpha_{n+1}}{16(2n+1)}},$$

which by Lemma 6 cannot exceed $K\pi/\sqrt{n}$, which proves the theorem.

This theorem will be used later, in the proofs of Theorem 12, Section 3.7, and Theorem 2, Section 3.8. The reason for giving it here is to emphasize that, despite the technical details involved in its proof, it is elementary in the sense that it does not depend on any deep results from the theory of Fourier series.

• EXERCISES

1. Derive (2), (3), and (4) in detail, giving the answers to the following questions:

(a) Are these formulas valid without the restriction that n and m be nonnegative?

(b) Do you see any immediate geometrical reason for the fact that the integrals $\int_0^{2\pi} \sin^2 nx\,dx$ and $\int_0^{2\pi} \cos^2 nx\,dx$ both equal π? [Hint: Draw the graphs of $\sin^2 nx$ and $\cos^2 nx$ and see if they fit together in some sense.]

2. How is it that square roots occur in (5) but not in (7), (8), or (9)?

3. Use the "collapsing principle" to find the sum of

$$\sin x + \sin 3x + \sin 5x + \cdots + \sin (2n - 1)x.$$

***4.** Derive (24).

5. **(a)** Calculate the value of $\int_0^{2\pi} \sin n(x - t)\cdot\sin nt\,dt$.

(b) Hence, derive (30).

6. Find the convolution product $f * g$ if

$$f(x) = \sin 2x + \cos 3x - 7 \sin 5x + 9 \cos 7x$$

and

$$g(x) = \cos x + \cos 2x + 2 \sin 3x + 4 \cos 7x.$$

(This is easy if you make use of the multiplication rules (28), (29), (30), and (31); keep track of the π's!)

7. Find the value of

$$\int_0^{2\pi} \cos m(x - t)\,\frac{\sin (n + \frac{1}{2})t}{2 \sin \frac{1}{2}t}\,dt$$

where m and n are arbitrary positive integers. (No integration is necessary.)

8. Given that $f * D_n = \theta$ but $f * D_{n+1} \neq \theta$, what can you say about f? (Be careful not to say too much.)

9. Write (35) in the form of an integral.

10. Explicitly evaluate the constants α_n.

11. Show that, if a function f has a continuous derivative throughout an interval $a \leqq x \leqq b$, it must satisfy a Lipschitz condition in this interval. [Hint: Use the mean-value theorem.]

12. Does the function $f(x) = |x|$ satisfy a Lipschitz condition? If so, for what constant K?

13. Does the function $f(x) = \sqrt{x}$ satisfy a Lipschitz condition in the interval $0 \leqq x \leqq 1$?

14. If a function does not have a derivative at $x = 0$, can it satisfy a Lipschitz condition in an interval containing $x = 0$?

15. A *broken-line function* is a continuous function whose graph consists of finitely many straight line segments joined end-to-end. Show that such a function satisfies a Lipschitz condition in any interval.

*16. Show that, if g is of class $C[0,2\pi]$ and ϵ is a positive number, no matter how small, there is a function f satisfying a Lipschitz condition in the interval $[0,2\pi]$ such that $|f(x) - g(x)| < \epsilon$ for every x in this interval.

17. **(a)** For given $K > 0$, does the class of all functions satisfying the Lipschitz condition with constant K form a linear space?
    (b) Does the class of all functions satisfying some Lipschitz condition form a linear space? [The difference between (a) and (b) is that in (b) we permit K to be different for different functions.]

*18. Deduce from the theorem in this section that it is possible to approximate a function satisfying the conditions of the theorem, as closely as we please, *in the mean-square sense.* Hence, deduce that the sequence 1, $\cos x$, $\sin x$, $\cos 2x$, $\sin 2x$, $\cdots$ is an approximating basis in some linear space, and state explicitly what linear space you are talking about.

*19. Show in detail that 1, $\cos x$, $\sin x$, $\cos 2x$, $\sin 2x$, $\cdots$ provides an approximating basis (in the mean-square sense; see Section 2.3) in the space of all continuous functions of period 2π. (Use the result of Exercise 16.)

CHAPTER 3

FOURIER SERIES

This chapter can be read independently of the other chapters in the book. However, the reader is advised to scan the first two chapters before reading this one. *In particular, he should read the introduction to Chapter 2.*

The central idea of this chapter is to investigate to what extent it is possible to expand a prescribed function f in an infinite series of the form

$$f(x) = A_0 + \sum_{n=1}^{\infty} (A_n \cos nx + B_n \sin nx).$$

When considered in great generality, this is an extremely difficult type of investigation, far beyond the scope of an introductory text. Therefore we shall restrict our attention mostly to functions that are *integrable* over intervals of finite length. By this, we mean that the function f is required to be *bounded* in every interval (a,b), i.e., there exists some number M (possibly depending on the interval) such that $|f(x)| < M$ for $a \leqq x \leqq b$, and also that $\int_a^b f(x)\,dx$ exists in the usual sense of elementary calculus. (See Section 1.5.)

Since the terms on the right side of this expansion are periodic, all having period 2π, it is obvious from the beginning that the series can converge to $f(x)$ for all values of x only if the function f has period 2π. If f does not have period 2π, then it is possible for the series to converge to $f(x)$ only for certain values of x, say in the interval $0 \leqq x \leqq 2\pi$, and outside this interval it will represent a periodic extension of the values of f within the interval. This will be discussed in Section 3.2.

Since the terms on the right side of the expansion are continuous, it might appear that it is necessary for the prescribed function to be continuous. This is true if we demand that the series converge *uniformly*; the meaning of "uniform convergence" will be explained in Section 3.7. If we do not require this kind of convergence, then it is possible for the series to represent a discontinuous function, which seems rather remarkable at first (and undoubtedly astonished mathematicians in an earlier era). Some rather peculiar phenomena occur in this case, as we will see in Section 3.5.

3.1 • MOTIVATION

Some readers will want to know immediately *why* we should be interested in such expansions. We cannot digress here to give a detailed explanation, but we will give one example, drawn from the field of electrical engineering. Other examples will be given in Chapters 5 and 6.

Let us suppose that we have a box, containing resistances, capacitors, and various coils connected together in some manner (the construction is irrelevant here). There are two input leads and two output leads. Suppose we attach the input leads to some device capable of producing a voltage which is given as a function of time x by a periodic function $f(x)$. We measure the output voltage, which is also a function of the time x; let this function be $g(x)$. Initially there will be certain "transient" effects which will tend to disappear; it is reasonable to suppose that eventually we will obtain an output $g(x)$ which is also periodic, having the same period as the input $f(x)$. For simplicity, we take this period to be 2π.

The relationship between g and f may be very complicated. If we vary the input function f, the steady-state output g will presumably also vary. We write this relationship in operator form, $g = L(f)$.

In the simplest instances, the operator L will be *linear*, meaning that $L(\alpha_1 f_1 + \alpha_2 f_2)$ is the same as $\alpha_1 L(f_1) + \alpha_2 L(f_2)$ for functions

f_1 and f_2 and numbers α_1, α_2. If this is the case, for reasonable input functions f_1 and f_2 and reasonably restricted values of α_1 and α_2, electrical engineers call the innards of the box a *linear* system.

In many simple cases, the output will be a sinusoidal function whenever the input is a sinusoidal function. The only difference between the two is one of amplitude and phase; the frequencies will be the same. Such systems must be distinguished from other systems containing transistors and other devices which might (say) double the frequency.

If this situation prevails, it is quite clear that trigonometric expansions will be of fundamental importance in studying the response of the system to an arbitrary input signal. We first study its response to sinusoidal inputs, attaching the box to a signal generator and measuring the output for a reasonable range of input frequencies. We then predict what the output will be for an input f that is not sinusoidal by expanding f in a series of sinusoidal functions, computing the output due to each term in the series, and summing all these outputs to obtain g.

This rather over-simplified discussion should indicate why trigonometric expansions are of fundamental importance in electrical engineering. Similar remarks could be made in connection with the response of mechanical systems to vibrations.

Let us carry this discussion a little bit further. (The following analysis is not rigorous.) Suppose we apply to the system an input voltage which is a succession of *unit impulses*, each of short duration, spaced 2π seconds apart, at $x = 0, 2\pi, 4\pi, \cdots$. (By a "unit impulse" we mean that the magnitude of the impulse times its time duration is unity.) Let us suppose that the steady-state output due to this input is given by $g(x)$. Since $g(x)$ is the steady-state response due to a train of unit impulses at 0, 2π, 4π, $\cdots$ the steady-state response due to unit impulses at t, $t + 2\pi$, $t + 4\pi$, $\cdots$ will be $g(x-t)$.

Now suppose that $h(x)$ is an arbitrary applied voltage of period 2π. Considering the input h as a superposition of periodic impulses of magnitude $h(t)\,dt$, the steady-state response due to each of these impulse trains will be $g(x - t)h(t)\,dt$, and therefore the steady-state response will be $\int_0^{2\pi} g(x - t)h(t)\,dt$. This is precisely the convolution product of the functions h and g, introduced in Section 2.6.

Many other "source and effect" integrals of applied mathematics are of convolution type, and we will find that such integrals play an important role in the theory that follows. The significance of the convolution product from the viewpoint of modern algebra is discussed in the Appendix.

3.2 • DEFINITIONS

A series of the form

$$\frac{1}{2}A_0 + \sum_{n=1}^{\infty} (A_n \cos nx + B_n \sin nx) \tag{1}$$

is called a *trigonometric series*. It is called a *Fourier series* if all the coefficients A_n, B_n can be obtained from a single integrable function f in the following manner:

$$A_n = \frac{1}{\pi}\int_0^{2\pi} f(x) \cos nx\, dx \quad (n = 0, 1, 2, \cdots) \tag{2}$$

$$B_n = \frac{1}{\pi}\int_0^{2\pi} f(x) \sin nx\, dx \quad (n = 1, 2, 3, \cdots) \tag{3}$$

Similar expressions were introduced earlier in several places. Notice that we have written the constant terms as $A_0/2$ rather than A_0, so that the expression for A_0 is given by taking $n = 0$ in (2) and need not be given as a separate formula. This is simply a matter of convenience.

If (1) is a Fourier series, we say that it *represents* the function f in the interval $(0,2\pi)$, or that it is the *Fourier expansion* of f in this interval.

The interval $(0,2\pi)$ was chosen rather arbitrarily. Later on, we will consider other intervals.

By definition, every Fourier series is a trigonometric series. However, not every trigonometric series is a Fourier series.

Theorem 1. *If a trigonometric series converges to an integrable function $f(x)$ for all except possibly a finite number of values of x in the interval $(0,2\pi)$, then it is the Fourier expansion of f in this interval.*

In other words, *if* we know that (1) converges to $f(x)$ in the interval $(0,2\pi)$, except perhaps at a finite number of points in the interval, *and* if we know that f is integrable over this interval, *then* the coefficients in the series must be related to the function f by the formulas (2) and (3).

The reader will perhaps be surprised to learn that the proof of Theorem 1 is beyond the scope of an introductory text, and that we will have no need to make use of Theorem 1 later in this book.

The beginner may be surprised to learn that the converse of Theorem 1 is not valid. The Fourier expansion of an integrable function f may fail to converge to $f(x)$ for infinitely many values of x in the interval. Indeed, the Fourier expansion of an integrable function may *diverge* for some values of x. An example is known of a *continuous* function whose Fourier series *diverges* at a finite number of points! (To the author's knowledge, it is not yet known whether there exists a continuous function, or even an integrable function, whose Fourier series diverges for all x.)

Things are not as bad as they may seem, however. Even though the Fourier series for an integrable function may be divergent, it will still represent the function to a remarkable degree, as we will see later.

We *caution* the advanced reader (the beginner should skip this paragraph): if "integrable" is interpreted to be "Lebesgue integrable," this theorem is valid as it stands, but is *not* valid if "all except possibly a finite number of" is replaced by "almost all." In other words, our reference to a finite number of exceptions is not, in this instance, a way to avoid mentioning sets of measure zero.

There is no obvious way to ascertain whether or not a given trigonometric series is a Fourier series. Tests for convergence are rather beside the point, since a Fourier series may be divergent, and a convergent trigonometric series need not be a Fourier series (the sum of the series may not be an integrable function).

In many practical problems we are given the function f, and we use it to construct a Fourier series. Therefore we shall be concerned with theorems that tell us something "nice" about the Fourier series expansion of f, provided that f itself is "nice" in some sense.

Here is a typical theorem of that kind, which we will prove later.

Theorem 2. *If f is integrable over the interval $(0, 2\pi)$, its Fourier series will converge to $f(x)$ at any point x $(0 < x < 2\pi)$ where f is differentiable.*

Caution: This theorem says *nothing* about what happens at the endpoints $x = 0$ and $x = 2\pi$, or at points where f is not differentiable. All it says is that if $0 < x_0 < 2\pi$ and if $f'(x_0)$ exists, then the series converges when $x = x_0$ and its sum is $f(x_0)$. For the proof, see Section 3.6.

EXAMPLE: Let $f(x) = x$. Using (2) and (3), we obtain

$$A_0 = \frac{1}{\pi}\int_0^{2\pi} x\,dx = 2\pi, \tag{4}$$

$$A_n = \frac{1}{\pi}\int_0^{2\pi} x\cos nx\,dx = 0 \qquad (n = 1, 2, 3, \cdots) \tag{5}$$

$$B_n = \frac{1}{\pi}\int_0^{2\pi} x\sin nx\,dx = -2/n \quad (n = 1, 2, 3, \cdots) \tag{6}$$

and therefore the Fourier series expansion of $f(x) = x$ in the interval $(0,2\pi)$ is

$$\pi - 2(\sin x + \tfrac{1}{2}\sin 2x + \tfrac{1}{3}\sin 3x + \cdots). \tag{7}$$

Since $f(x) = x$ is integrable over any interval of finite length, and is differentiable for every x, by Theorem 2 we can be certain that (7) will converge to $f(x) = x$ whenever $0 < x < 2\pi$. At $x = 0$, the sum of (7) is obviously π. It follows by the periodicity of the terms in (7) that this series converges for every x to the function whose graph is shown in Figure 3.1.

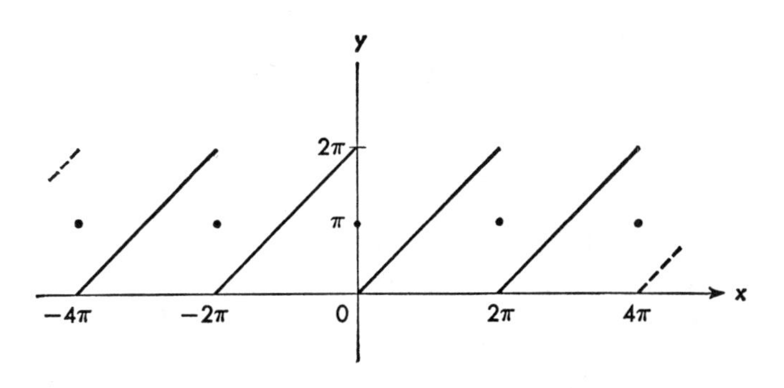

Figure 3.1

In later sections of this chapter, we will usually include in theorems like this the hypothesis that the given function f is periodic. Since most functions that arise in practice are not periodic, it might appear at first that this reduces the generality of the theorem considerably. Actually, the contrary is true. Not only will the theorems be easier to state, but they will be more general and therefore more likely to be useful! The reason for this is found in the following elementary idea, which for pedagogical reasons we dignify by calling the *extension principle. If the given function is not periodic, apply these theorems to a periodic function that is identically equal to the given function in the interior of the desired interval.*

For instance, if the interval in question is $(0,2\pi)$, we would replace the function $f(x) = x$ by the function shown in Figure 3.1. For reasons that will become clear later, it is usually convenient, when possible, to take the value of the function at the endpoints (in this example, $x = 0$ and $x = 2\pi$) to be the *average* of the limits approached by the periodic extension on either side (shown by heavy dots in Figure 3.1).

Compare the following theorem with Theorem 2.

Theorem 3. *If f is integrable and has period 2π, its Fourier series will converge to $f(x)$ at any point where f is differentiable.*

Problem: Find a Fourier series that will converge to $f(x) = x$ in the interval $-\pi < x < \pi$.

Solution: The Fourier series (7) does not converge to $f(x) = x$ when $x = 0$ or when x is negative, so it does not provide a solution to this problem. We must proceed in another manner.

Letting $f(x) = x$ in the interval $(-\pi,\pi)$ and defining f elsewhere by periodicity, we obtain the function whose graph is shown in Figure 3.2.

According to Theorem 3, the Fourier series expansion of *this* function will converge to $f(x) = x$ in the interval $(-\pi,\pi)$, and is therefore the desired series. In computing the coefficients, we observe that all the relevant functions in (2) and (3) will be of period 2π, and therefore we can integrate over any interval of length 2π. This points out another advantage in using the extension principle, for in this case it is easier to integrate over $(-\pi,\pi)$.

Calculating the coefficients, we obtain

$$(8)\quad A_n = \frac{1}{\pi}\int_{-\pi}^{\pi} x \cos nx \, dx = 0 \qquad (n = 0, 1, 2, \cdots)$$

$$(9)\quad B_n = \frac{1}{\pi}\int_{-\pi}^{\pi} x \sin nx \, dx = (-1)^{n+1}\, 2/n \quad (n = 1, 2, 3, \cdots)$$

and therefore the desired series is

(10) $$2(\sin x - \tfrac{1}{2}\sin 2x + \tfrac{1}{3}\sin 3x - \cdots).$$

Observe that, when x is a multiple of π, the sum of this series is 0. The series converges for all x to the function shown in Figure 3.2.

Now let us consider what happens at a point where the derivative does not exist; for example, at a point of discontinuity. If $f(x)$

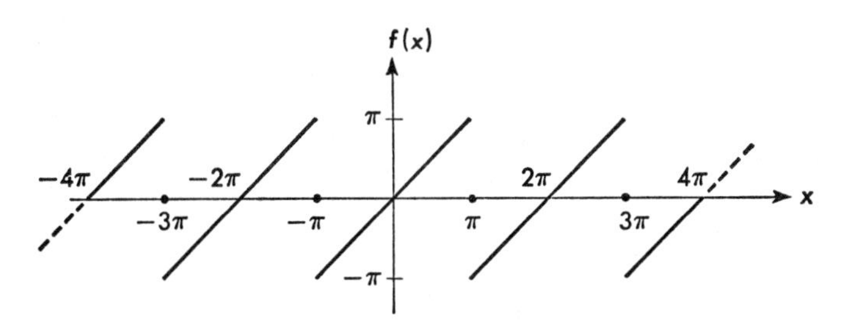

Figure 3.2

tends to a finite limit as x tends to x_0 from the left, we denote this limiting value by $f(x_0-)$. Similarly, we let $f(x_0+)$ denote the limit, if it exists, of the values $f(x)$ as x tends to x_0 from the right. If f is continuous at x_0, then $f(x_0+) = f(x_0-)$. If $f(x_0+)$ and $f(x_0-)$ both exist but are not equal, we say that f has a *jump discontinuity* at x_0.

At this point the reader is advised to review the definition of *piecewise continuity* given in Section 1.5.

If $f'(x)$ exists in an interval to the left of x_0 and if $f'(x)$ tends to a finite limit as x approaches x_0 from the left, we call the limit the left-hand derivative of f at x_0 and denote it $f'(x_0-)$. The right-hand derivative $f'(x_0+)$ is defined similarly.

A function f is said to be *piecewise smooth* if it is piecewise continuous, and, in every interval (a,b) of finite length, it has a derivative (except perhaps at a finite number of points) that is piecewise continuous. If a function is piecewise smooth, it has right-hand and left-hand derivatives at every point, and in every interval of finite length these two derivatives are equal except at a finite number of points.

The graph shown in Figure 3.3 is supposed to represent the kind of behavior a piecewise smooth function might exhibit in a typical interval. At the point x_1 the function is continuous, but its derivative is discontinuous there; $f'(x_1-)$ is negative, but $f'(x_1+)$

is positive; $f'(x_1)$ is not defined. The function has a jump discontinuity at x_2 and x_3. At x_2 we have $f(x_2+) = f(x_2)$, but looking at x_3 we find neither $f(x_3+)$ or $f(x_3-)$ is equal to $f(x_3)$. Notice that $f'(x_4+) = 0$.

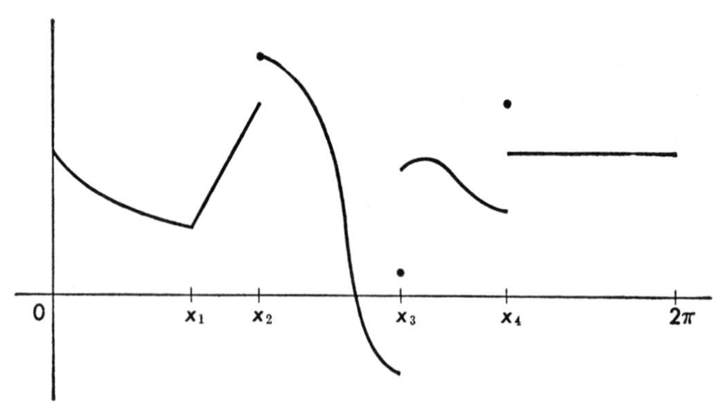

Figure 3.3

It can be shown that, if a function f is piecewise continuous, then in any interval of finite length it is bounded, and its integral $\int_a^b f(x)\,dx$ exists over any such interval. In other words, *every piecewise continuous function is integrable.*

If f is piecewise smooth, then $\int_a^b f'(x)\,dx$ can also be defined, even though $f'(x)$ may fail to exist for a finite number of values of x in the interval. It is defined simply by summing the values of the integral over the subintervals in which $f'(x)$ *does* exist and ignoring the exceptional points; no problem arises since $f'(x)$ is bounded and continuous in each of these subintervals (by virtue of the requirement that $f'(x)$ tend to a finite limit at the endpoints of each of the subintervals).

Theorem 4. *If f is an integrable function of period 2π, its Fourier series converges to $[f(x+) + f(x-)]/2$ at every point x where it has both a right-hand and a left-hand derivative.*

In particular, if f is a piecewise smooth function of period 2π, its Fourier expansion will converge at every point x. It may, however, not converge to $f(x)$ at points of discontinuity.

As an example, consider the period 2π extension of the piecewise smooth function shown in Figure 3.3. Since $f(0) = f(2\pi)$ for this function, its periodic extension will be continuous at both $x = 0$ and $x = 2\pi$. Therefore its Fourier series will converge to $f(x)$ when $x = 0$ and when $x = 2\pi$. Although $f'(x_1)$ is not defined, the series will converge to $f(x_1)$ at $x = x_1$, since both the right-hand and left-hand derivatives exist at this point and $f(x_1-)$ and $f(x_1+)$ both equal $f(x_1)$. At x_2 the series will converge, but not to $f(x_2)$. Instead, according to Theorem 4, it will converge to a value midway between $f(x_2-)$ and $f(x_2+)$. At $x = x_3$ the series will converge to $f(x_3)$, since $f(x_3) = [f(x_3+) + f(x_3-)]/2$. Again, at $x = x_4$, the series will converge, but not to the value $f(x_4)$ indicated by the heavy dot. Many more examples will occur later in this chapter. Theorem 4 will be proved in Section 3.6.

• EXERCISES

1. Find the Fourier series representing the function $f(x) = x(2\pi - x)$ in the interval $0 < x < 2\pi$.

2. For what values of x does the Fourier series obtained in Exercise 1 converge to $f(x)$?

3. Find the Fourier series representing the function $f(x) = e^x$ in the interval $0 < x < 2\pi$.

4. For what values of x does the Fourier series obtained in Exercise 3 converge to $f(x)$?

5. Sketch the period 2π extension of the function defined in the interval $0 \leqq x < 2\pi$ by

$$f(x) = \begin{cases} 1 & x \leqq \pi \\ -1 & x > \pi \end{cases}.$$

6. Find the Fourier series representing the function defined in Exercise 5, in the interval $(0,2\pi)$.

7. Does the Fourier series obtained in Exercise 6 converge to the period 2π extension you sketched in Exercise 5?

8. Find a Fourier series converging to $f(x) = |\sin x|$ for *every* x.

9. Find a Fourier series converging to $f(x) = x^2$ in the interval $0 < x < 2\pi$.

10. **(a)** Find a Fourier series converging to $f(x) = |x|$ in $-\pi < x < \pi$.
    (b) Does the series converge to $|x|$ at $x = -\pi$ and $x = \pi$?

11. Consider the function

$$f(x) = \begin{cases} x^n \sin(1/x) & (x \neq 0) \\ 0 & (x = 0) \end{cases}$$

(a) If $n = 0$, is this function bounded?
(b) If $n = 0$, is this function bounded in every interval of finite length?
(c) If $n = 0$, is this function piecewise continuous?
(d) If $n = 0$, is this function piecewise smooth?
(e) Answer parts (a) through (d), taking $n = 1$ instead.
(f) Answer parts (a) through (d), taking $n = 2$ instead.

12. If f is piecewise smooth, is it necessarily true that $\int_a^b f'(x)\,dx = f(b) - f(a)$?

13. Is every piecewise continuous function integrable over every interval of finite length?

14. Are these following classes of functions linear spaces?
(a) The class of all piecewise continuous functions.
(b) The class of all piecewise smooth functions.

15. Does every function that is continuous in the interval $0 \leqq x \leqq 2\pi$ have a continuous periodic extension? If not, state necessary and sufficient conditions that will ensure that such a function has a continuous periodic extension.

3.3 • EXAMPLES OF TRIGONOMETRIC SERIES

In this section we list a number of standard trigonometric series. Most of them are Fourier series which will be used as examples later in this chapter. By listing them here we can avoid a number of digressions later, and by using them the reader can save himself a certain amount of useless drudgery in solving problems.

The main usefulness of this list will appear later, in connection with improving the rate of convergence of a given series. It is with that usefulness in mind that the first part of the table is organized by the type of series, rather than the type of function providing the sum of the series. At this elementary stage, the reader may find the second part of the table more useful.

At this point the reader is not expected to understand the comments made in connection with some of these series. For example, the meaning of "Cesaro summable" is not explained until Section 3.9.

Series Which Terminate

These are finite sums, so no problem of convergence arises. The limit convention (page 8) is understood to give meaning to expressions wherever a denominator vanishes.

$$\sin\alpha + \sin\beta = 2[\sin\tfrac{1}{2}(\alpha+\beta)][\cos\tfrac{1}{2}(\alpha-\beta)] \tag{1}$$

$$\sin\alpha - \sin\beta = 2[\cos\tfrac{1}{2}(\alpha+\beta)][\sin\tfrac{1}{2}(\alpha-\beta)] \tag{2}$$

$$\cos\alpha + \cos\beta = 2[\cos\tfrac{1}{2}(\alpha+\beta)][\cos\tfrac{1}{2}(\alpha-\beta)] \tag{3}$$

$$\cos\alpha - \cos\beta = -2[\sin\tfrac{1}{2}(\alpha+\beta)][\sin\tfrac{1}{2}(\alpha-\beta)] \tag{4}$$

$$\sum_{m=1}^{n} \cos mx = \frac{\left[\sin n\frac{x}{2}\right]\left[\cos(n+1)\frac{x}{2}\right]}{\sin\frac{x}{2}} \tag{5}$$

$$\tfrac{1}{2} + \sum_{m=1}^{n} \cos mx = \frac{\sin(2n+1)\frac{x}{2}}{2\sin\frac{x}{2}} \tag{6}$$

$$\sum_{m=1}^{n} \sin mx = \frac{\left[\sin n\frac{x}{2}\right]\left[\sin(n+1)\frac{x}{2}\right]}{\sin\frac{x}{2}} \tag{7}$$

$$\sum_{m=1}^{n} \cos(\alpha+mx) = \frac{\left[\sin n\frac{x}{2}\right]\left[\cos\left(\alpha+\overline{n+1}\frac{x}{2}\right)\right]}{\sin\frac{x}{2}} \tag{8}$$

$$\sum_{m=1}^{n} \sin(\alpha+mx) = \frac{\left[\sin n\frac{x}{2}\right]\left[\sin\left(\alpha+\overline{n+1}\frac{x}{2}\right)\right]}{\sin\frac{x}{2}} \tag{9}$$

$$\cos(\alpha+x) + \cos(\alpha+3x) + \cdots + \cos(\alpha+\overline{2n-1}\,x) = \frac{[\sin nx][\cos(\alpha+nx)]}{\sin x} \tag{10}$$

(11) $$\sin(\alpha + x) + \sin(\alpha + 3x) + \cdots + \sin(\alpha + \overline{2n-1}\,x) = \frac{[\sin nx][\sin(\alpha + nx)]}{\sin x}$$

Divergent Trigonometric Series

(12) $$\sin x + 2\sin 2x + 3\sin 3x + \cdots$$

This series is Abel summable to zero for all x.

(13) $$\tfrac{1}{2} + \cos x + \cos 2x + \cos 3x + \cdots$$

Cesaro summable to zero in the interval $0 < x < 2\pi$.

(14) $$\sin x + \sin 2x + \sin 3x + \cdots$$

Cesaro summable to $\frac{1}{2}\cot\frac{x}{2}$ in the interval $0 < x < 2\pi$.

Trigonometric Series with Unbounded Sums

(15) $$\sum_{n=1}^{\infty} \frac{\cos nx}{n} = -\ln\left(2\sin\frac{x}{2}\right) \qquad (0 < x < 2\pi)$$

(16) $$\sum_{n=1}^{\infty} (-1)^{n+1} \frac{\cos nx}{n} = \ln\left(2\cos\frac{x}{2}\right) \qquad (-\pi < x < \pi)$$

(17) $$\sum_{n=0}^{\infty} \frac{\cos(2n+1)x}{2n+1} = -\frac{1}{2}\ln\tan\frac{x}{2} \qquad (0 < x < \pi)$$

(18) $$\sum_{n=0}^{\infty} (-1)^n \frac{\sin(2n+1)x}{2n+1} = -\frac{1}{2}\ln\tan\left(\frac{\pi}{4} - \frac{x}{2}\right) \left(-\frac{\pi}{2} < x < \frac{\pi}{2}\right)$$

Fourier Series (Standard Forms)

These standard forms are especially useful in improving the convergence of other series (see page 181).

(19) $$\sum_{n=1}^{\infty} \frac{\sin nx}{n} = \frac{\pi - x}{2} \qquad (0 < x < 2\pi)$$

For the same form with $\sin nx$ replaced by $\cos nx$, see (15).

(20) $$\sum_{n=1}^{\infty} (-1)^{n+1} \frac{\sin nx}{n} = \frac{x}{2} \qquad (-\pi < x < \pi)$$

For the same form with $\sin nx$ replaced by $\cos nx$, see (16).

$$\sum_{n=0}^{\infty} \frac{\sin (2n+1)x}{2n+1} = \frac{\pi}{4} \qquad (0 < x < \pi) \tag{21}$$

For the same form with $\sin (2n+1)x$ replaced by $\cos (2n+1)x$, see (17).

$$\sum_{n=0}^{\infty} (-1)^n \frac{\cos (2n+1)x}{2n+1} = \frac{\pi}{4} \qquad \left(-\frac{\pi}{2} < x < \frac{\pi}{2}\right) \tag{22}$$

For the same form with $\cos (2n+1)x$ replaced by $\sin (2n+1)x$, see (18).

$$\sum_{n=1}^{\infty} \frac{\cos nx}{n^2} = \frac{3x^2 - 6\pi x + 2\pi^2}{12} \qquad (0 \leqq x \leqq 2\pi) \tag{23}$$

$$\sum_{n=1}^{\infty} \frac{\sin nx}{n^2} = -\int_0^x \ln \left(2 \sin \frac{t}{2}\right) dt \qquad (0 \leqq x \leqq 2\pi) \tag{24}$$

This integral exists as an improper Riemann integral and defines a function that is bounded and continuous throughout the interval (including the endpoints) although it cannot be expressed in closed form.

$$\sum_{n=1}^{\infty} (-1)^{n+1} \frac{\cos nx}{n^2} = \frac{\pi^2 - 3x^2}{12} \qquad (-\pi \leqq x \leqq \pi) \tag{25}$$

$$\sum_{n=1}^{\infty} (-1)^{n+1} \frac{\sin nx}{n^2} = \int_0^x \ln \left(2 \cos \frac{t}{2}\right) dt \qquad (-\pi \leqq x \leqq \pi) \tag{26}$$

Except when $x = \pi$ and $x = -\pi$, this is a proper Riemann integral. The function thus defined tends to zero at both of these endpoints.

$$\sum_{n=0}^{\infty} \frac{\cos (2n+1)x}{(2n+1)^2} = \frac{\pi^2 - 2\pi x}{8} \qquad (0 \leqq x \leqq \pi) \tag{27}$$

$$\sum_{n=0}^{\infty} \frac{\sin (2n+1)x}{(2n+1)^2} = -\frac{1}{2}\int_0^x \ln \tan \frac{t}{2}\, dt \qquad (0 \leqq x \leqq \pi) \tag{28}$$

See remarks in connection with (24).

$$\sum_{n=0}^{\infty} (-1)^n \frac{\cos (2n+1)x}{(2n+1)^2} = -\frac{1}{2}\int_0^{(\pi/2)-x} \ln \tan \frac{t}{2}\, dt \qquad \left(-\frac{\pi}{2} \leqq x \leqq \frac{\pi}{2}\right) \tag{29}$$

This is an improper integral, even when $x = 0$.

$$\sum_{n=0}^{\infty} (-1)^n \frac{\sin (2n+1)x}{(2n+1)^2} = \frac{\pi x}{4} \qquad \left(-\frac{\pi}{2} \leqq x \leqq \frac{\pi}{2}\right) \tag{30}$$

(31) $$\sum_{n=1}^{\infty} \frac{\cos nx}{n^3} = \int_0^x \left[\int_0^p \ln\left(2 \sin \frac{t}{2}\right) dt\right] dp + \sum_{n=1}^{\infty} \frac{1}{n^3} \qquad (0 \leqq x \leqq 2\pi)$$

[The numerical value of $\sum_{n=1}^{\infty} n^{-3}$ to seven decimal places is 1.2020569.]

(32) $$\sum_{n=1}^{\infty} \frac{\sin nx}{n^3} = \frac{x^3 - 3\pi x^2 + 2\pi^2 x}{12} \qquad (0 \leqq x \leqq 2\pi)$$

(33) $$\sum_{n=1}^{\infty} (-1)^{n+1} \frac{\cos nx}{n^3} = \sum_{n=1}^{\infty} (-1)^{n+1} \frac{1}{n^3} - \int_0^x \left[\int_0^p \ln\left(2 \cos \frac{t}{2}\right) dt\right] dp \qquad (-\pi \leqq x \leqq \pi)$$

[$\sum_{n=1}^{\infty} (-1)^{n+1} n^{-3} = 0.9015427$, approximately.]

(34) $$\sum_{n=1}^{\infty} (-1)^{n+1} \frac{\sin nx}{n^3} = \frac{\pi^2 x - x^3}{12} \qquad (-\pi \leqq x \leqq \pi)$$

(35) $$\sum_{n=0}^{\infty} \frac{\cos (2n+1)x}{(2n+1)^3} = \frac{1}{2} \int_0^x \left[\int_0^p \ln \tan \frac{t}{2}\, dt\right] dp + \sum_{n=0}^{\infty} \frac{1}{(2n+1)^3} \qquad (0 \leqq x \leqq \pi)$$

[$\sum_{n=0}^{\infty} (2n+1)^{-3} = 1.0517998$, approximately.]

(36) $$\sum_{n=0}^{\infty} \frac{\sin (2n+1)x}{(2n+1)^3} = \frac{\pi^2 x - \pi x^2}{8} \qquad (0 \leqq x \leqq \pi)$$

(37) $$\sum_{n=0}^{\infty} (-1)^n \frac{\cos (2n+1)x}{(2n+1)^3} = \frac{\pi^3 - 4\pi x^2}{32} \qquad \left(-\frac{\pi}{2} \leqq x \leqq \frac{\pi}{2}\right)$$

(38) $$\sum_{n=0}^{\infty} (-1)^n \frac{\sin (2n+1)x}{(2n+1)^3} = \frac{1}{2} \int_0^{(\pi/2)-x} \left[\int_0^p \ln \tan \frac{t}{2}\, dt\right] dp + \sum_{n=0}^{\infty} \frac{1}{(2n+1)^3} \qquad \left(-\frac{\pi}{2} \leqq x \leqq \frac{\pi}{2}\right)$$

[See also (35).]

Standard Forms with Even Harmonics

If the given series contains terms in $\sin nx$ and/or $\cos nx$ for $n = 0, 2, 4, 6, \cdots$, and no terms with *odd* n, replace $2x$ by t to obtain a series in sin nt and/or cos nt that does not contain these gaps. Don't forget afterwards to replace t by $2x$.

For example, to sum $\frac{\sin 2x}{2} + \frac{\sin 4x}{4} + \frac{\sin 6x}{6} + \cdots$ we let $t = 2x$ to obtain

$$\frac{\sin t}{2} + \frac{\sin 2t}{4} + \frac{\sin 3t}{6} + \cdots = \frac{1}{2}\left(\frac{\pi - t}{2}\right) = \frac{\pi - t}{4}.$$

The desired sum is therefore $(\pi - 2x)/4$. Several other frequently occurring examples are given in the exercises at the end of this section.

Note that the sum obtained is valid in a smaller interval than indicated in the above list. For example, (19) is valid in the interval $(0,2\pi)$, and therefore the sum just obtained is valid only in the interval $(0,\pi)$.

Change of Interval

Replacing x by $\pi t/L$ will transform a series convergent in $0 < x < \pi$ to another series convergent in $0 < t < L$. This is only one example of how we can change the interval of convergence, but it is an important example. Be sure to keep in mind that both the *series* and its *sum* must be transformed.

For example, suppose we wish to obtain a Fourier series converging to $f(t) = t$ in the interval $0 < t < L$. This is equivalent to finding a series converging to $g(x) = Lx/\pi$ in the interval $0 < x < \pi$. We have the choice of using either (19) or (20) above. These two choices correspond to the two possibilities shown in Figures 3.1 and 3.2, Section 3.2. In either case, we obtain a series converging in an even larger interval than $(0,L)$. If this is not desirable, we can replace x by $2\pi t/L$, which will transform a series convergent in $0 < x < 2\pi$ to another convergent in $0 < t < L$. If we do this, (19) becomes

$$\sum_{n=1}^{\infty} \frac{\sin (2\pi nt/L)}{n} = \frac{\pi - 2\pi t/L}{2} \qquad (0 < t < L).$$

Solving for t, we obtain

$$(39) \quad t = \frac{L}{2} - \frac{L}{\pi}\left(\sin \frac{2\pi t}{L} + \frac{1}{2} \sin \frac{4\pi t}{L} + \frac{1}{3} \sin \frac{6\pi t}{L} + \cdots\right) \qquad (0 < t < L).$$

Observe that, if we replace L by 2π, this reduces to (7), Section 3.2.

Replacing x by $\pi t/L$ in (20), we obtain

$$(40)\quad t = \frac{2L}{\pi}\left(\sin\frac{\pi t}{L} - \frac{1}{2}\sin\frac{2\pi t}{L} + \frac{1}{3}\sin\frac{3\pi t}{L} - \cdots\right) \qquad (-L < t < L).$$

Applying the same treatment to (21) yields the following important series:

$$(41)\quad 1 = \frac{4}{\pi}\left(\sin\frac{\pi t}{L} + \frac{1}{3}\sin\frac{3\pi t}{L} + \frac{1}{5}\sin\frac{5\pi t}{L} + \cdots\right) \qquad (0 < t < L).$$

Since the terms in this series are odd functions of period $2L$, this series converges to the function shown in Figure 3.4.

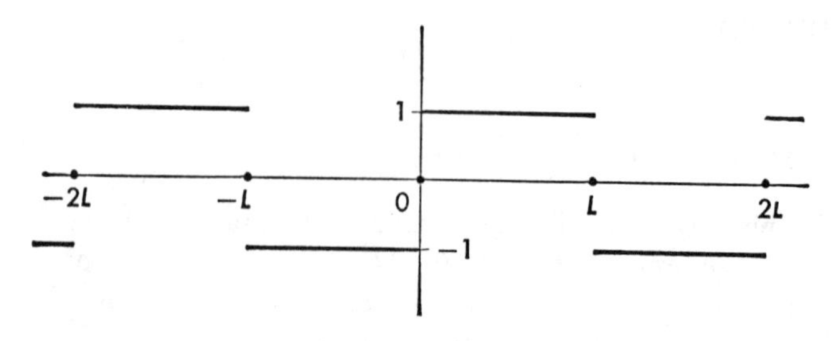

Figure 3.4

We hope that this example will convince even the most "practical-minded" reader that it is seldom enough to look at only the *sum* of the given series. The interval of convergence must not be overlooked. If the length of this interval is only half of the period of the functions in the series, one must immediately ask: what happens in the other half of each interval of periodicity? If the functions are all of the form $\sin kx$, the sum of the series must be an odd function; if they happen to be all of the form $\cos kx$, the sum is an even function. The sum of the series is then either an odd periodic extension or an even periodic extension of the sum that is indicated.

Even the preceding remarks are not enough to enable an indolent student to sketch the graphs of the sums of these series without making use of his mental faculties. If the series contains terms in $\sin nx$ and/or $\cos nx$, for only *odd* values of n, he will need to use the fact that these functions have the property $f(x + \pi) = -f(x)$

for all x. Such a function is said to possess only *odd harmonics*; this use of the word "odd" must not be confused with its use in connection with "odd functions," for an odd harmonic may not be an odd function (for instance, $\cos 3x$ is an odd harmonic that is an even function).

As a simple example, we show in Figure 3.5 the sum of the series (22). This is an *even* function containing only *odd* harmonics.

Both Figures 3.4 and 3.5 provide examples of functions of particular interest to electrical engineers. Mainly for their benefit,

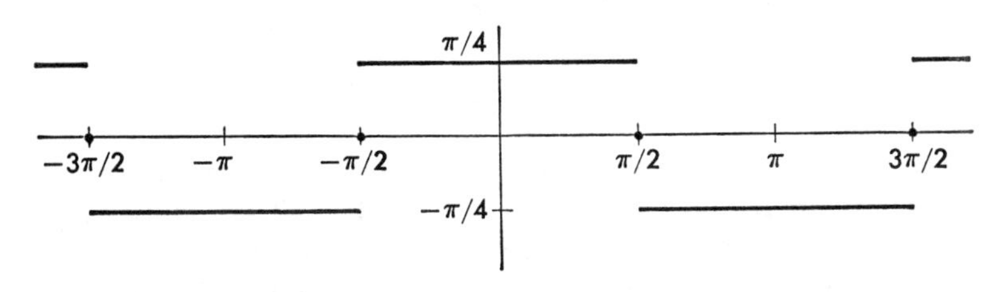

Figure 3.5

we also give a trigonometric series whose sum represents a "periodic pulse":

If
$$\begin{aligned} f(x) &= H \text{ for } c < x < c + w \\ f(x) &= 0 \text{ for } c + w < x < c + 2L \\ f(x + 2L) &= f(x), \end{aligned}$$

then

$$f(x) = \frac{Hw}{2L} + \frac{2H}{\pi} \sum_{n=1}^{\infty} \frac{1}{n} \sin \frac{n\pi w}{2L} \cos \frac{n\pi}{L}\left(x - c - \frac{w}{2}\right) \tag{42}$$

except at points of discontinuity, where the sum is $H/2$.

It is possible to combine various series to obtain series expansions for other functions. For instance, if we multiply (41) by A and (40) by B, we obtain a series representation of $A + Bt$. This representation is convergent to $A + Bt$, in general, only in the interval $(0,L)$, since (41) is valid only in this interval. We obtain (replacing t by x):

$$\begin{aligned} A + Bx = \frac{1}{\pi}\Big[&(4A + 2LB) \sin \frac{\pi x}{L} - \frac{2LB}{2} \sin \frac{2\pi x}{L} \\ &+ \frac{4A + 2LB}{3} \sin \frac{3\pi x}{L} - \frac{2LB}{4} \sin \frac{4\pi x}{L} + \cdots\Big]. \end{aligned} \tag{43}$$

Special Functions

We give here some expansions for various functions not included in the first part of the list. Several of them, however, are only trivial modifications of series already listed, and are listed here simply for convenience.

$$(44)\quad 1 = \frac{4}{\pi}\left(\sin x + \frac{\sin 3x}{3} + \frac{\sin 5x}{5} + \cdots\right) \qquad (0 < x < \pi)$$

$$(45)\quad x = \pi - 2\left(\sin x + \frac{\sin 2x}{2} + \frac{\sin 3x}{3} + \cdots\right) \qquad (0 < x < 2\pi)$$

$$(46)\quad x^2 = \frac{4\pi^2}{3} + 4\sum_{n=1}^{\infty} \frac{\cos nx}{n^2} - 4\pi \sum_{n=1}^{\infty} \frac{\sin nx}{n} \qquad (0 < x < 2\pi)$$

(47) $Ax^2 + Bx + C$ = series obtained by multiplying (46) by A, (45) by B, and adding a constant term C. The resulting series will converge to $Ax^2 + Bx + C$ in the interval $0 < x < 2\pi$. [Do *not* make use of C times (44), or the resulting series will be valid only in the smaller interval $0 < x < \pi$.]

$$(48)\quad |x| = \frac{\pi}{2} - \frac{4}{\pi}\left(\cos x + \frac{\cos 3x}{3^2} + \frac{\cos 5x}{5^2} + \cdots\right)$$
$$(-\pi \leqq x \leqq \pi)$$

$$(49)\quad e^{ax} = \frac{2 \sinh a\pi}{\pi}\left[\frac{1}{2a} + \sum_{n=1}^{\infty} (-1)^n \frac{a \cos nx - n \sin nx}{a^2 + n^2}\right]$$
$$(-\pi < x < \pi)$$

$$(50)\quad \sin ax = \frac{2 \sin a\pi}{\pi}\left[\frac{\sin x}{1^2 - a^2} - \frac{2 \sin 2x}{2^2 - a^2} + \frac{3 \sin 3x}{3^2 - a^2} - \cdots\right]$$

when $-\pi < x < \pi$, provided that a is not an integer.

$$(51)\quad \cos ax = \frac{2a \sin a\pi}{\pi}\left[\frac{1}{2a^2} + \frac{\cos x}{1^2 - a^2} - \frac{\cos 2x}{2^2 - a^2} + \frac{\cos 3x}{3^2 - a^2} - \cdots\right]$$

when $-\pi < x < \pi$, whenever a is not an integer.

$$(52)\quad \sinh ax = \frac{2 \sinh a\pi}{\pi}\left[\frac{\sin x}{1^2 + a^2} - \frac{2 \sin 2x}{2^2 + a^2} + \frac{3 \sin 3x}{3^2 + a^2} - \cdots\right]$$

when $-\pi < x < \pi$, valid for arbitrary a.

$$(53)\quad \cosh ax = \frac{2a \sinh a\pi}{\pi}\left[\frac{1}{2a^2} - \frac{\cos x}{1^2 + a^2} + \frac{\cos 2x}{2^2 + a^2} - \cdots\right]$$

when $-\pi < x < \pi$, a arbitrary.

The following four expansions are of interest to electrical engineers in connection with problems in frequency modulation. The coefficients $J_n(r)$ are Bessel functions, which will be discussed in Chapter 4.

$$\cos(r \sin x) = J_0(r) + 2[J_2(r)\cos 2x + J_4(r)\cos 4x + \cdots] \tag{54}$$

$$\sin(r \sin x) = 2[J_1(r)\sin x + J_3(r)\sin 3x + \cdots] \tag{55}$$

$$\cos(r \cos x) = J_0(r) - 2[J_2(r)\cos 2x - J_4(r)\cos 4x + J_6(r)\cos 6x - \cdots] \tag{56}$$

$$\sin(r \cos x) = 2[J_1(r)\cos x - J_3(r)\cos 3x + J_5(r)\cos 5x - \cdots] \tag{57}$$

These relations are valid for all values of x and r.

• EXERCISES

1. Find the sum of the series

$$\frac{\cos 2x}{2^2} + \frac{\cos 4x}{4^2} + \frac{\cos 6x}{6^2} + \cdots \qquad (0 < x < \pi) \tag{58}$$

2. Find the sum of the series

$$\frac{\sin 2x}{2^3} + \frac{\sin 4x}{4^3} + \frac{\sin 6x}{6^3} + \cdots \qquad (0 < x < \pi) \tag{59}$$

3. Find the sum of the series

$$\frac{\sin 2x}{2^2} + \frac{\sin 4x}{4^2} + \frac{\sin 6x}{6^2} + \cdots \qquad (0 \leqq x \leqq \pi) \tag{60}$$

4. Find a series converging to $f(x) = x(2\pi - x)$ in the interval $0 \leqq x \leqq 2\pi$, using (45) and (46).

5. **(a)** Using (45) and (46), find a series converging to $f(x) = x(\pi - x)$ in the interval $0 < x < 2\pi$.
(b) Does this series converge to $f(x)$ at $x = 0$ and $x = 2\pi$?
(c) Draw a graph of the sum of this series in the interval

$$-2\pi \leqq x \leqq 2\pi.$$

6. What formula in this section would you use to derive the following expansion?

$$x^2 = \frac{\pi^2}{3} - 4\left(\cos x - \frac{\cos 2x}{2^2} + \frac{\cos 3x}{3^2} - \cdots\right) \qquad (-\pi \leqq x \leqq \pi) \tag{61}$$

7. Use (48) and (61) to find a series of cosines converging to $f(x) = x(\pi - x)$ in the interval $0 \leqq x \leqq \pi$. Using (23), sum this series to check your answer.

8. Show how to derive the following expansion, and determine the interval in which it is valid:

$$x^2 = 2\pi\left(\sin x - \frac{\sin 2x}{2} + \frac{\sin 3x}{3} - \cdots\right) - \frac{8}{\pi}\left(\sin x + \frac{\sin 3x}{3^3} + \frac{\sin 5x}{5^3} + \cdots\right) \tag{62}$$

9. It is known that $\sum_{n=1}^{\infty} A_n \sin nx$ converges to an integrable function $f(x)$ for all x. What can you say about the coefficients A_n in each of the following instances:

(a) if $f(x) = f(x - \pi)$ for all x?

(b) if $f(x) = -f(x - \pi)$ for all x?

(c) if $f(x) = f\left(x - \frac{2\pi}{3}\right)$ for all x?

At this point, you are not expected to give a rigorous reason for your answer to part (c). (See Exercise 12.)

10. Differentiate (23) term-by-term. Is the sum of the resulting series the derivative of the sum of (23)?

11. Let S_n denote the nth partial sum of (14). Notice that S_n is given in closed form by (7).

(a) Show that S_n does not tend to a limit as $n \to \infty$.

(b) Find an expression, in closed form, for

$$(S_1 + S_2 + \cdots + S_n)/n.$$

(c) Show that the expression you found in (b) tends to a limit as $n \to \infty$, and determine this limit.

12. Let f be a continuous function of period $2\pi/3$.

(a) Do you see any simple way of proving that

$$\int_0^{2\pi} f(x) \sin x\, dx = 0?$$

(b) Define a complex number I by the formula $I = \int_0^{2\pi} f(x)e^{-ix}\, dx$. Explain how each step in the following calculation is obtained:

$$
\begin{aligned}
I &= \int_0^{2\pi} f(x)e^{-ix}dx \\
&= \int_0^{2\pi} f\left(x + \frac{2\pi}{3}\right) e^{-i(x+2\pi/3)}dx \\
&= \int_0^{2\pi} f(x)e^{-i(x+2\pi/3)}dx = e^{-2\pi i/3}I.
\end{aligned}
$$

Therefore $I = 0$.

(c) Deduce from (b) the formula given in (a).

(d) Can you generalize this result to give $\int_0^{2\pi} f(x) \sin nx\, dx$ for other values of n? [Compare Exercise 9(c).]

3.4 • SINE AND COSINE SERIES

In the preceding section, we saw by examples that several quite *different* trigonometric series can converge to the *same* function in an interval. For example, (46), (61), and (62) all converge to x^2 in the interval $0 < x < \pi$. (The first of these actually converges to x^2 in the larger interval $0 < x < 2\pi$, and the second converges to x^2 in the interval $-\pi \leqq x \leqq \pi$.) This may seem confusing at first, but actually the situation is quite easy to understand, as we shall now see. A proper grasp of the situation is very important, since in order to solve the boundary-value problems that arise in later chapters it is necessary to know which series to use.

The reader is advised to look again at (1), Section 3.2 (page 89). To distinguish this series from others obtained in a slightly different way, we call this a *complete* Fourier series. A Fourier series containing only sines will be called a Fourier *sine* series, and one containing only cosines is called a Fourier *cosine* series. We pose the questions: When does the complete Fourier series of a function turn out to be a Fourier sine series? When will it turn out to be a Fourier cosine series?

By looking at (2) and (3) on the same page, the reader can readily verify the following. We assume f is integrable and has period 2π, so that we can integrate over $(-\pi,\pi)$ rather than $(0,2\pi)$, if we like.

(1) If f is *even*, $f(x) = f(-x)$, its Fourier series contains only a constant term and cosine terms. In other words, its complete Fourier series is a Fourier cosine series.

(2) If f is *odd*, $f(x) = -f(-x)$, its Fourier series contains only sine terms. In other words, its complete Fourier series is a Fourier sine series. These remarks tell us nothing about a function like $f(x) = x^2$, because this function is *not periodic.*

Suppose now that we are given a function f, defined and integrable in the interval $(0,\pi)$. To avoid special cases, we will not use x^2 as the example, but use the function shown in Figure 3.6

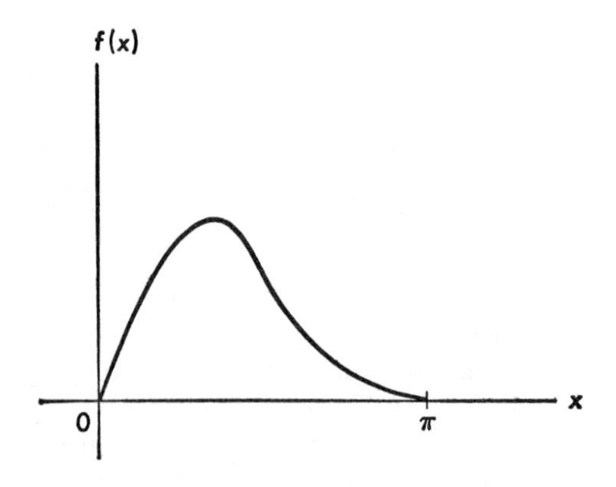

Figure 3.6

instead. There are four very useful ways in which we can obtain a periodic extension of this function that will have period 2π. These are shown in Figure 3.7(a), (b), (c), and (d), and are defined respectively by the following requirements:

(a) f is even and has period 2π.
(b) f is odd and has period 2π.
(c) f has period π.
(d) f has period 2π and alternates each half-period, i.e., $f(x + \pi) = -f(x)$ for all x.

(Strictly speaking, we should not denote any of these functions by the letter f, since they are all different from the f we started with.)

If we now use (2) and (3), Section 3.2, to form a Fourier series, we will obtain four entirely different Fourier expansions. If f is piecewise smooth, these series will all converge to the same values in $(0,\pi)$, but to entirely different values (in general) outside this interval. The reader can readily verify that in these four cases the series will have the following characteristics:

(a) the Fourier series will contain only cosine terms,
(b) the Fourier series will contain only sines,

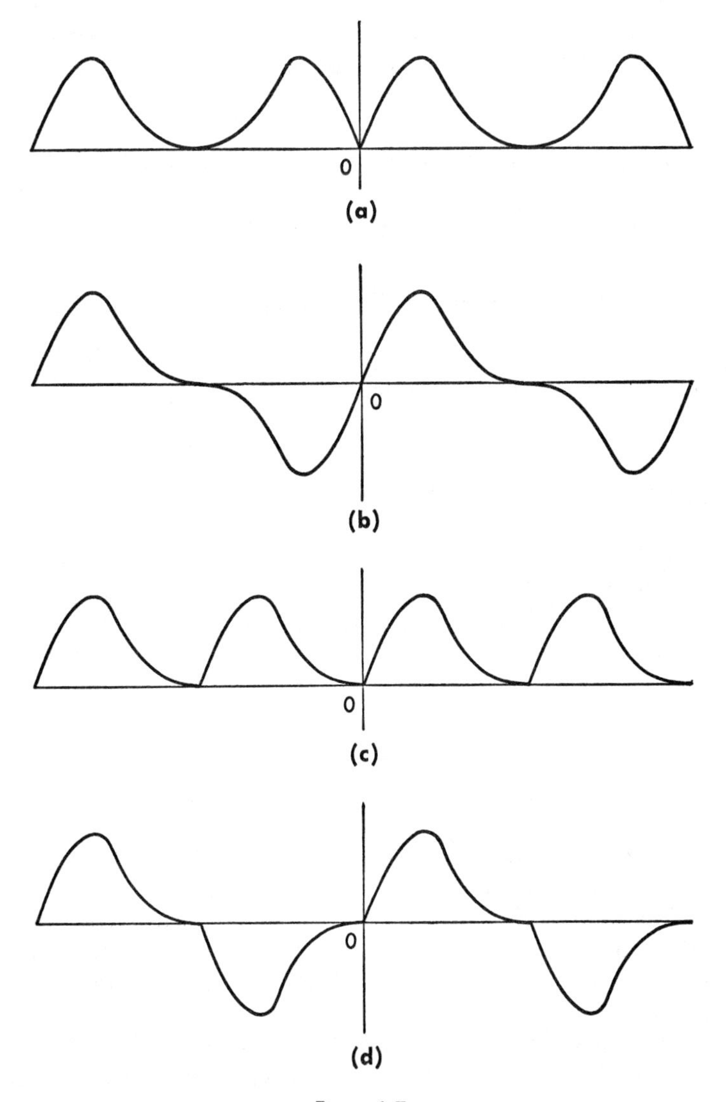

Figure 3.7

(c) the series may contain both sin nx and cos nx terms, but only for *even* values of n,

(d) the series may contain both sin nx and cos nx terms, but only for *odd* values of n.

The first two of these cases are by far the more important, so much so that the series obtained in (a) is called the *Fourier cosine series* representing f in the interval $(0,\pi)$, and that obtained in (b) is called the *Fourier sine series* representing f in the interval $(0,\pi)$.

(Notice that a constant term is considered to be a *cosine* term with $n = 0$.)

In case (a), the extended function is even, so instead of integrating the extended function over $(0,2\pi)$ or $(-\pi,\pi)$, we can take double the integral over $(0,\pi)$ using the function we had to begin with.

(3) The Fourier cosine series is of the form

$$A_0/2 + \sum_{n=1}^{\infty} A_n \cos nx$$

$$\text{where } A_n = \frac{2}{\pi}\int_0^{\pi} f(x)\cos nx\,dx \qquad (n = 0, 1, 2, \cdots).$$

In case (b), the extended function is odd, so instead of integrating the extended function over $(0,2\pi)$ or $(-\pi,\pi)$, we can take twice the integral over $(0,\pi)$, using the function we had to begin with (since the product of two odd functions is even).

(4) The Fourier sine series is of the form

$$\sum_{n=1}^{\infty} B_n \sin nx$$

$$\text{where } B_n = \frac{2}{\pi}\int_0^{\pi} f(x)\sin nx\,dx \qquad (n = 1, 2, 3, \cdots).$$

These matters are so important that we briefly summarize them here. If we are given a function f, such as that shown in Figure 3.6, defined only in $(0,\pi)$, we can form either its Fourier cosine series [which is really the complete Fourier series of the function shown in Figure 3.7(a)] or its Fourier sine series [which is really the complete Fourier series of the function shown in Figure 3.7(b)].

We take (3) to be the *definition* of the Fourier cosine series representing f in $(0,\pi)$, and (4) to be the *definition* of the corresponding Fourier sine series.

In the preceding discussion, we have treated sine series and cosine series as conceptually subordinate to the complete Fourier series. We shall now show that this is not necessary; the three different types of expansions (and others besides) may be put on an equal footing by means of the Sturm-Liouville theorem.

Since many readers find Section 2.4 rather difficult, we do not assume here any prior familiarity with the Sturm-Liouville theorem. However, we suggest that the reader look at that theorem (page 67) both before and after reading the following discussion.

Let $BC^2[0,L]$ denote the class of all functions f having continuous second derivatives in the interval $0 \leqq x \leqq L$, satisfying the boundary conditions

$$f(0) = 0, \quad f(L) = 0. \tag{5}$$

With this definition of $BC^2[0,L]$, the operator $L = d^2/dx^2$ satisfies the hypotheses of the Sturm-Liouville theorem. That theorem directs us to find the *eigenfunctions* of this operator, i.e., the nontrivial functions f of class $BC^2[0,L]$ such that $Lf = \alpha f$, for some scalar α. It is easy to see that the only nontrivial functions satisfying (5) which are constant multiples of their second derivatives are the functions $B_n \sin (n\pi x/L)$, for integer values of n. The Sturm-Liouville theorem states, among other things, that if f is of class $BC^2[0,L]$, it can be expanded in a convergent series

$$f(x) = \sum_{n=1}^{\infty} B_n \sin \frac{n\pi x}{L}. \tag{6}$$

On the other hand, if we replace the boundary conditions (5) by the conditions

$$f'(0) = 0, \quad f'(L) = 0 \tag{7}$$

then we are now speaking of quite a different class of functions when we speak of $BC^2[0,L]$. It is still true that d^2/dx^2 satisfies the conditions of the theorem, but in this space the eigenfunctions are functions of the form $A_n \cos (n\pi x/L)$ for integer values of n (including $n = 0$). In this case, the Sturm-Liouville theorem asserts the possibility of expanding any function of this class in a series similar to (6) but with cosines instead of sines.

Thus, with the boundary conditions (5), the Sturm-Liouville theorem makes a significant assertion concerning Fourier sine series, and if the boundary conditions (6) are used instead, the assertion concerns Fourier cosine series. (We usually take $L = \pi$ in both of these cases.)

With the boundary conditions

$$f(0) = f(L), \quad f'(0) = f'(L) \tag{8}$$

the eigenfunctions are of the form $A_n \cos(2n\pi x/L) + B_n \sin(2n\pi x/L)$ and the theorem yields information concerning complete Fourier series. (We usually take $L = 2\pi$ in this case.)

Since we have not given a proof of the Sturm-Liouville theorem, we will not go into further details here. No use will be made of the Sturm-Liouville theorem in this book; the convergence theorems we find necessary will be proved directly in each instance.

• EXERCISES

1. In the interval $(0,2\pi)$ let $f(x) = x^2$. Show that (46), Section 3.3, is the complete Fourier series representing f in this interval. In computing the coefficients, can the interval $(0,2\pi)$ be replaced by $(-\pi,\pi)$ in this instance?

2. In the interval $(0,\pi)$, let $f(x) = x^2$.

(a) Plot the even period 2π extension of this function.

(b) Show that (61), Section 3.3, is the complete Fourier series representing this extension.

(c) Plot the odd period 2π extension of this function.

(d) Show by actually calculating the coefficients that (62), Section 3.3, is the complete Fourier series representing the extension of part (c).

3. (a) Using Theorem 4, Section 3.2, and the assertion made in Exercise 1 of this section, show that

$$\sum_{n=1}^{\infty} \frac{1}{n^2} = \frac{\pi^2}{6}.$$

(b) Obtain the same result more simply by using (23), Section 3.3. [This is an interesting result. No similarly simple expression has ever been obtained for $\sum_{n=1}^{\infty} 1/n^3$.]

4. Why is the sum of (46), Section 3.3 not the same as the sum of (62), of the same section, when $x = 0$?

5. A function f is defined and integrable over $(0,\pi)$, and is extended to give a function having period 2π which alternates each half-period. Write down expressions for the complete Fourier series expansion of this extension, as integrals over the interval $(0,\pi)$.

6. Find the Fourier cosine series representing the function $f(x) = x$ in the interval $(0,\pi)$.

7. Find the Fourier sine series representing the function $f(x) = x$ in the interval $(0,\pi)$.

8. Draw four figures, similar to those shown in Figure 3.7,
(a) taking $f(x) = x$ in the interval $(0,\pi)$,
(b) taking $f(x) = x^2$ in the interval $(0,\pi)$,
(c) taking $f(x) = e^x$ in the interval $(0,\pi)$,
(d) taking $f(x) = \sin x$ in the interval $(0,\pi)$.

***9.** Using the Sturm-Liouville theorem, show that there is a series of the form $\sum_{n=1}^{\infty} B_n \sin k_n x$ converging to $f(x) = 1$ in the interval $(0,\pi)$, where the numbers k_n do not form an arithmetic progression but are roots of the equation $\tan k_n\pi = -k_n$. Determine the coefficients B_n in terms of the numbers k_n (it is not necessary to give numerical values for k_n). [Hint: Compare Exercise 3, Section 2.4.]

10. Is there any difference between the Fourier sine series representing $f(x) = x$ in $(0,\pi)$ and the complete Fourier series representing $f(x) = x$ in $(-\pi,\pi)$?

3.5 • THE GIBBS PHENOMENON

By discussing the Gibbs phenomenon *before* taking up a detailed analysis of convergence, we hope to convince skeptical readers that convergence *can* be interesting.

Throughout this section, we let f denote a piecewise smooth function of period 2π that has been "normalized" at points of discontinuity so that $f(x) = [f(x-) + f(x+)]/2$ for every x. According to Theorem 4, Section 3.2, the Fourier series expansion of f converges to $f(x)$ for *every* x.

At first glance this appears to be a very nice situation. It appears that, by taking enough terms in the Fourier expansion of f, we can obtain a sum of sinusoidal functions approximating f as closely as we wish. The reader may be surprised to learn that this is not at all the case.

Suppose we let S_n denote the partial sum

$$S_n(x) = A_0/2 + \sum_{k=1}^{n} (A_k \cos kx + B_k \sin kx) \tag{1}$$

where the coefficients are related to the given function f by (2) and (3), Section 3.2. With the above hypotheses, we have

$$S_n(x) \to f(x) \tag{2}$$

for every x. This means that, for each fixed x, the sequence of numbers $S_n(x)$ tends to the number $f(x)$ as $n \to \infty$.

This by no means implies that the functions S_n resemble the function f, no matter how large n may be.

Beginners tend to assume that (2) has the following (incorrect) meaning, namely, that for sufficiently large n, a roughly drawn graph of the function S_n could not be distinguished from a graph of the function f. This, however, is not so. It may happen (and always happens if f is discontinuous) that the convergence is more rapid for some values of x than for others, so that, *no matter how large n may be*, at some points $S_n(x)$ is "very nearly" the same as $f(x)$, but at other points $S_n(x)$ differs substantially from $f(x)$. The result is that the graph of S_n may appear remarkably different from that of f, no matter what value is assigned to n.

This phenomenon is not peculiar to Fourier series. It can happen even when the functions involved are not trigonometric, as the following example shows.

Example 1: Consider the sequence of functions

$$S_n(x) = \frac{nx}{1 + n^2x^2}. \tag{3}$$

As $n \to \infty$ we have $S_n(x) \to 0$ for *every* x. Yet no matter how large n may be, the graph of S_n bears no resemblance to the graph of the zero function. Indeed, there will always be some value of x for which $S_n(x) = 1/2$ (take $x = 1/n$). A typical graph of S_n is shown in Figure 3.8. There is a positive and a negative hump, each of height $1/2$, one at $x = 1/n$ and the other at $x = -1/n$. As n increases, the humps move closer to the y axis, but they never diminish in height. Despite this, for each *fixed* x we have $S_n(x) \to 0$. For example, if x is positive, the positive hump will (for sufficiently large n) be situated to the left of x, although still remaining to the right of the y axis, and the value of $S_n(x)$ will decrease thereafter as n increases.

A curious feature of this example, one that boggles the imagination at first, is: although for each x the sequence $S_n(x)$ tends to zero, it is still true that no matter how large we choose n there will be values of x for which $S_{n+1}(x)$ is farther away from zero than $S_n(x)$. (Indeed, this will be true whenever $0 < |x| < 1/2n$.)

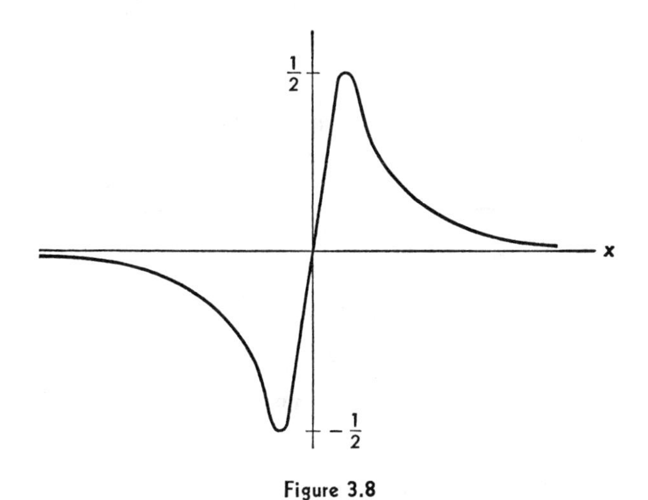

Figure 3.8

The preceding example has nothing to do with Fourier series. However, the following example shows that something quite similar can occur when we form the partial sums of a Fourier series.

EXAMPLE 2: Consider the infinite series

$$\sin x + \tfrac{1}{2} \sin 2x + \tfrac{1}{3} \sin 3x + \cdots \tag{4}$$

which by (19), Section 3.3, is the Fourier series representing

$$f(x) = \frac{\pi}{2} - \frac{1}{2}x \tag{5}$$

in the interval $(0,2\pi)$. This series converges for all x to the periodic extension of this function shown by the heavy curve in Figure 3.9.

The nth partial sum of this series is

$$S_n(x) = \sum_{k=1}^{n} \frac{1}{k} \sin kx. \tag{6}$$

A typical partial sum (for a large value of n) is shown by the light curve in Figure 3.9. This partial sum oscillates above and below the values of f. It is observed that, near each point of discontinuity, the partial sum "overshoots" the function rather noticeably in both directions.

One might think that, with increasing n, the nth partial sum would tend to overshoot to a lesser extent, but this is not the case.

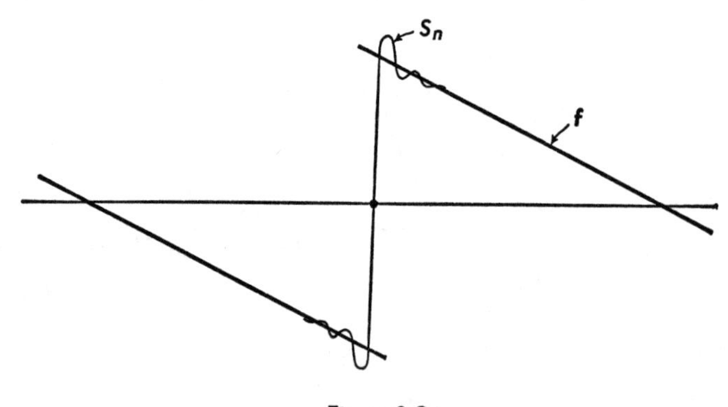

Figure 3.9

For every large n, the ripples in the graph of S_n become less evident, *except* very near the points of discontinuity, where the ripples are small in width, but have amplitudes relative to the magnitude of the discontinuity which remain practically independent of n.

As n increases, the *graph* of S_n tends *geometrically* to the object shown in Figure 3.10. By this we mean that, for sufficiently large n, the graph of S_n will be within the dotted lines shown in the figure. More precisely, given any positive ϵ, there exists an N such that, for $n > N$, every point on the graph of S_n is within ϵ in distance from some point on this geometrical figure, and every point on this figure is within ϵ in distance from a point on the graph of S_n. In this (old-fashioned) definition we allow distances to be measured in any direction, vertically, horizontally, or otherwise, whereas in all modern definitions of convergence of functions we consider only "distances" $|f(x) - S_n(x)|$ which are geometrically *vertical*.

Note that Figure 3.10 is not the graph of a function. It is impossible for the graph of a (single-valued) function to contain a vertical line segment. Electrical engineers tend to overlook this for a very understandable reason: the trace of many waves, as registered on a cathode-ray oscilloscope, sometimes look as if they contain a vertical segment at each point of relative discontinuity.

The overshoot phenomenon illustrated here is called the *Gibbs phenomenon*. In more advanced texts it is shown that for large n the magnitude of the overshoot, at points of discontinuity of a piecewise

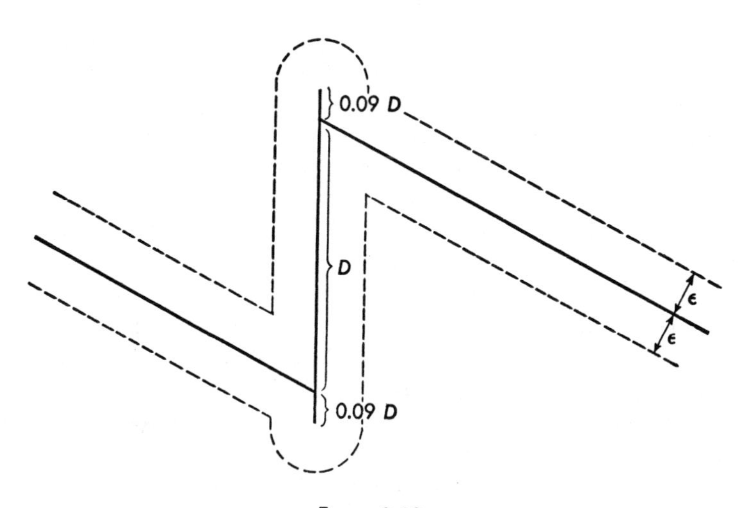

Figure 3.10

smooth function, is about nine per cent of the magnitude of the discontinuity. Instead of proving this in detail, we shall illustrate it in this example with specific numerical values. Later on, when we study the "Dirichlet kernel" in some detail, we will see clearly why the overshoot occurs.

To find the maximum value of $S_n(x)$ in this case, we differentiate (6) and set the derivative equal to zero. The derivative is

$$S_n'(x) = \sum_{k=1}^{n} \cos kx = -\frac{1}{2} + \frac{\sin (n + \frac{1}{2})x}{2 \sin \frac{1}{2}x} \tag{7}$$

where we have made use of (6), Section 3.3. With the use of a little trigonometry, we find that the smallest positive value for x for which this becomes zero is $x = \pi/(n + 1)$, and this corresponds to the first (and largest) positive maximum of $S_n(x)$. (See Exercise 1 for an easier way to obtain this value.)

If we take $n = 19$, this first maximum will come at $x = \pi/20 = 9°$. Let us study the behavior of the partial sums for this particular value of x.

$n =$	1	5	10	15	19
$S_n(x) =$	0.156	0.751	1.341	1.684	1.773
$n =$	20	25	30	32	∞
$S_n(x) =$	1.773	1.679	1.513	1.452	1.492

The value of $f(x)$ at $x = \pi/20$ is (to three decimal places) 1.492. We seem to be approaching this value at $n = 10$, but then we overshoot the mark, and reach 1.773 at $n = 19$. (The value at $n = 20$ is the same as at $n = 19$, since the 20th term in the series vanishes at $x = \pi/20$.) Notice that at $n = 32$ we have obtained a value lower than 1.492. If this table were to be extended much further, we would see still further values above and below the value 1.492, as various "peaks" and "troughs" move to the left towards the point of discontinuity $x = 0$.

At $x = \pi/20$, $f(x) = 1.492$ and $S_{19}(x) = 1.773$. The magnitude of the overshoot is $1.773 - 1.492 = 0.281$. A simple sliderule* calculation indicates this is 8.95 per cent of the magnitude of the discontinuity $D = \pi = 3.14$. If we were to take increasing values of n, measuring the maximum overshoot (at $x = 1/n$) as a percentage of the discontinuity D, the percentages would tend to a figure $8.9490 \cdots$.

• EXERCISES

1. By considering the center of gravity of the vertices of a regular polygon, or (alternatively) the vector sum of unit vectors extending from the origin, show that $\sum_{k=1}^{n} \cos kx = 0$ when $x = \pi/(n + 1)$, and that this is the smallest positive x for which this sum is zero.
2. **(a)** Suppose that f is piecewise smooth in its interval of periodicity, and has only a single discontinuity there. Show that $f = f_1 + f_2$, where f_1 is continuous everywhere and f_2 is a "sawtooth" function.

   (b) Generalize to piecewise smooth functions having more than one discontinuity in each interval of periodicity.

   [In more advanced books, the study of the Gibbs phenomenon for piecewise smooth functions is reduced in this manner to a consideration of sawtooth functions alone.]

*Here, as elsewhere, the reader may of course substitute a calculator or other device for the sliderule.

3.6 • LOCAL CONVERGENCE OF FOURIER SERIES

This section may be omitted without loss in continuity. Our purpose here is to prove Theorem 4 of Section 3.2.

Let f be an integrable function having period 2π. Let its Fourier series be

$$A_0/2 + \sum_{k=1}^{\infty} (A_k \cos kx + B_k \sin kx). \tag{1}$$

The n + first partial sum of this series,

$$S_n(x) = A_0/2 + \sum_{k=1}^{n} (A_k \cos kx + B_k \sin kx) \tag{2}$$

is a trigonometric polynomial of order at most n. If we substitute into (2) the integral formulas for the coefficients, given by (2) and (3), Section 3.2, and make a few trigonometric simplifications, we obtain

$$S_n(x) = \int_0^{2\pi} f(x - t) \frac{\sin (n + \frac{1}{2})t}{2\pi \sin \frac{1}{2}t}\, dt. \tag{3}$$

For the detailed derivation, see Section 2.6.

We need to make use of the following fact:

$$\int_0^{2\pi} \frac{\sin (n + \frac{1}{2})t}{2\pi \sin \frac{1}{2}t} = 1 \tag{4}$$

which follows trivially from (6), Section 3.3 [simply integrate both sides of (6) over this interval]. (Also see Exercise 2.)

From (4) it is trivial that

$$\int_0^{2\pi} f(x) \frac{\sin (n + \frac{1}{2})t}{2\pi \sin \frac{1}{2}t}\, dt = f(x) \tag{5}$$

since the variable of integration is t and $f(x)$ can be regarded as a constant during the integration. Letting $S_n(x)$ denote the partial sum given by (3), we obtain from (3) and (5) the important expression

$$S_n(x) - f(x) = \int_0^{2\pi} [f(x - t) - f(x)] \frac{\sin (n + \frac{1}{2})t}{2\pi \sin \frac{1}{2}t}\, dt. \tag{6}$$

For a given value of x, the Fourier series (1) converges to $f(x)$ if and only if this integral tends to zero with increasing n.

Changing variables to $u = -t$, this integral becomes

$$(7)\qquad S_n(x) - f(x) = \int_0^{2\pi} [f(x+u) - f(x)] \frac{\sin (n + \frac{1}{2})u}{2\pi \sin \frac{1}{2}u}\, du.$$

One dummy variable is as good as any other, so we can change u in (7) back to t. We then obtain, on summing (6) and (7),

$$(8)\qquad \begin{aligned} &2[S_n(x) - f(x)] \\ &= \int_0^{2\pi} [f(x-t) + f(x+t) - 2f(x)] \frac{\sin (n + \frac{1}{2})t}{2\pi \sin \frac{1}{2}t}\, dt. \end{aligned}$$

Since all the functions involved here are periodic, with period 2π, we can replace the range of integration $[0,2\pi]$ by $[-\pi,\pi]$. But the integrand in (8) is an *even* function, so we can write it as twice the integral over the range $[0,\pi]$, and this factor of 2 cancels that on the left side of (8). Therefore,

$$(9)\qquad \begin{aligned} &S_n(x) - f(x) \\ &= \int_0^{\pi} [f(x-t) + f(x+t) - 2f(x)] \frac{\sin (n + \frac{1}{2})t}{2\pi \sin \frac{1}{2}t}\, dt. \end{aligned}$$

So far we have put no restrictions on f other than what would be expected, namely that it be integrable (otherwise we could not define its Fourier series). Now if f is discontinuous at some point, we can define its value $f(x)$ rather arbitrarily at this point without in any way changing its Fourier series. Changing the value of $f(x)$ for a single point x does not change the value of its integral over the interval, nor any of the integrals defining its Fourier coefficients. Let us therefore assume, for the moment, that f is *piecewise smooth* and that, at any point x, its value is $f(x) = [f(x+) + f(x-)]/2$. This is automatically true at any point of continuity of f, and at any point of discontinuity it means we have taken the value $f(x)$ to be the average of the value it tends to on either side of the discontinuity.

With this assumption, $2f(x) = f(x+) + f(x-)$ and we substitute this into the right side of (9). Now let us split the integrand into two separate terms and write it in the following form:

$$(10)\qquad \begin{aligned} S_n(x) - f(x) &= \int_0^{\pi} [f(x-t) - f(x-)] \frac{\sin (n + \frac{1}{2})t}{2\pi \sin \frac{1}{2}t}\, dt \\ &+ \int_0^{\pi} [f(x+t) - f(x+)] \frac{\sin (n + \frac{1}{2})t}{2\pi \sin \frac{1}{2}t}\, dt. \end{aligned}$$

We desire to show that both of these integrals tend to zero under the stated assumption.

We will show that the second integral tends to zero. The proof is essentially identical for the first integral, involving nothing more than a change in notation.

Let us rewrite the integral in the form

$$\text{(11)} \qquad \frac{1}{2\pi}\int_0^\pi \frac{f(x+t)-f(x+)}{t}\cdot\frac{t}{\sin\frac{1}{2}t}\cdot\sin(n+\tfrac{1}{2})t\,dt.$$

We assume the limit convention, already discussed, to give meaning to these functions at $t = 0$. Since f is piecewise smooth, so also is $f(x+t) - f(x+)$ as a function of t; is this property destroyed by dividing it by t? The only possible difficulty might be that $\dfrac{f(x+t)-f(x+)}{t}$ would not tend to a finite limit as t tends to zero, but this difficulty does not arise here because f is piecewise smooth by hypothesis and this quotient therefore tends to the right-sided derivative $f'(x+)$ as t tends to zero. The second term gives no difficulty either; the quotient $t/[\sin(t/2)]$ tends to 2 as t tends to zero. Letting

$$\phi(t) = \frac{f(x+t)-f(x+)}{t}\cdot\frac{t}{\sin(t/2)}$$

we see that, for any fixed x (the dependence of ϕ on x is not indicated since we take x to be fixed, although quite arbitrary), the integral becomes

$$\text{(12)} \qquad \frac{1}{2\pi}\int_0^\pi \phi(t)\sin(n+\tfrac{1}{2})t\,dt$$

where ϕ is an integrable function over the interval. This may be written out in such a form that we can apply Riemann's lemma (page 55). We have

$$\text{(13)} \qquad \frac{1}{2\pi}\int_0^\pi \phi(t)\cos\frac{t}{2}\sin nt\,dt + \frac{1}{2\pi}\int_0^\pi \phi(t)\sin\frac{t}{2}\cos nt\,dt.$$

The functions $\phi(t)\cos(t/2)$ and $\phi(t)\sin(t/2)$ are integrable over $[0,\pi]$ so we can apply Riemann's lemma separately to the two integrals. The functions $\sin t$, $\sin 2t$, $\sin 3t$, $\cdots$ are orthogonal over the interval, and therefore the first integral tends to zero with increasing n, and the same argument applies to the second integral since the functions $\cos t$, $\cos 2t$, $\cos 3t$, $\cdots$ are orthogonal over the interval.

If the function f is piecewise smooth but does not satisfy the above hypothesis, we replace it by the function $g(x) = [f(x+) + f(x-)]/2$, which is identical to $f(x)$ except for a finite number of values of x, and which therefore has the same Fourier series as f.

The above argument then applies to show that $S_n(x)$ tends to $g(x)$ for every x. We have proved the following theorem.

Theorem. *The Fourier series representing a piecewise smooth function $f(x)$ in $0 \leqq x \leqq 2\pi$ converges to $[f(x+) + f(x-)]/2$ whenever $0 < x < 2\pi$ and to $[f(0+) + f(2\pi-)]/2$ when $x = 0$ or $x = 2\pi$.*

Or rather we have proved this theorem on the assumption that f is periodic with period 2π, and for nonperiodic functions the theorem follows trivially by considering the period 2π extension of f, which has the same Fourier series.

In proving that $S_n(x)$ tends to $f(x)$ we really did not make use of the piecewise smoothness of f except insofar as it guarantees that $f(x+)$ and $f(x-)$ exist, and the existence of the one-sided derivatives $f'(x+)$ and $f'(x-)$ at the particular value of x in question. Therefore the more general statement made in Theorem 4, Section 3.2, is valid. From this, Theorems 2 and 3 in that section follow as obvious corollaries. There is no point in repeating them here.

We are now in a position to understand why the Gibbs phenomenon occurs. A rough graph of the Dirichlet kernel for n approximately 20 is shown in Figure 3.11. According to (4), the shaded area is unity, provided we agree to take the area beneath the axis to be *negative*. Actually, the shaded area exceeds unity, if the area is taken (as it properly should be) always positive.

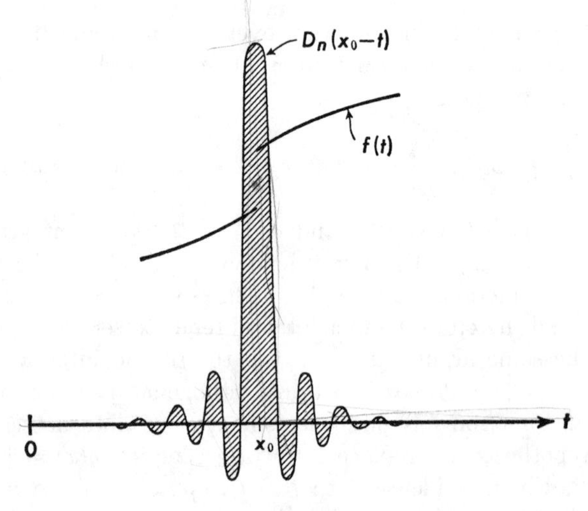

Figure 3.11

With a simple change in variables, (3) can be written in the form

$$(14)\quad S_n(x) = \int_0^{2\pi} f(t)\frac{\sin (n+\frac{1}{2})(x-t)}{2\pi \sin \frac{1}{2}(x-t)}\,dt = \int_0^{2\pi} f(t)D_n(x-t)\,dt$$

where we use D_n as a convenient abbreviation for the Dirichlet kernel, as in Section 2.6.

Let us assume that a function f has a jump discontinuity at $x = x_0$. Then the value of $S_n(x)$ at $x = x_0$ is $\int_0^{2\pi} f(t)D_n(x_0 - t)\,dt$, which is approximately the average of $f(x_0+)$ and $f(x_0-)$, since $D_n(x_0 - t)$, as a function of t, just straddles the discontinuity, with half of its area to the right of the discontinuity and half to the left. The integral yields an average value, with values on the right weighted equally with values to the left.

Now suppose we choose $x = x_1$ to be a point a little to the right of the discontinuity, as shown in Figure 3.12, so that the central

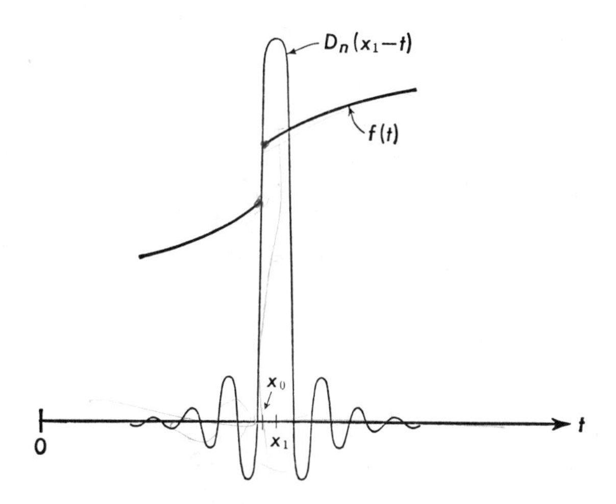

Figure 3.12

hump of $D_n(x_1 - t)$ is just to the right of the point of discontinuity. Then the total signed area under $D_n(x_1 - t)$ to the right of x_0 will exceed unity, the signed area to the left being negative (so the total area is unity). The integral $\int_0^{2\pi} f(t)D_n(x_1 - t)\,dt$ now yields an average value, with values on the right (approximately $f(x_0+)$) weighted positively, and values on the left (approximately $f(x_0-)$) weighted negatively. This yields a value (in the case shown in

Figure 3.12) that will exceed $f(x_0+)$ and accounts for the overshoot of $S_n(x_1)$ on the right side. A similar argument accounts for the excessive drop on the left side.

Another argument sometimes given asks the reader to imagine a mask constructed by drawing a graph of D_n, cutting out the shaded area. One imagines this mask sliding across the graph of $f(t)$, and imagines the value of the integral to correspond to the area thus observed. This idea leads to calling the Dirichlet kernel the *scanning function.*

Although the author finds the argument given in the preceding paragraph in no way commendable (the value of the integral will *not* be the area observed) a modified idea is useful in thinking of convolution integrals $\int_0^{2\pi} f(t)g(x - t)\,dt$ generally. Imagine the graph of $g(t)$ drawn on celluloid, which is then turned over (right to left) so that one obtains the graph of $g(-t)$. Then place the celluloid on top of a graph of $f(t)$. Sliding the celluloid to the appropriate position, one can view the graphs of $f(t)$ and $g(x - t)$ simultaneously, and perhaps estimate the value of their inner product, which is represented by the integral of their product. Varying the position of the celluloid might then enable one to estimate the integral for various values of x. In particular, one can amuse oneself by trying this in the special case $g(t) = D_n(t)$, taking $f(t)$ to be the simplest sort of piecewise smooth function with a discontinuity. One easily sees in this way how little ripples arise near the discontinuity.

• EXERCISES

1. Replace $D_n(t)$ by $F_n(t)$, defined in $-\pi < x < \pi$ to be equal to $n/2$ in the interval $-1/n < x < 1/n$, and zero elsewhere in the interval, and to be periodic with period 2π.
   (a) If f is continuous and of period 2π, what can you say about
   $$\lim_{n\to\infty} \int_0^{2\pi} f(t)F_n(x - t)\,dt?$$
   (b) If f is piecewise smooth, what is your answer to (a) when x is a point of discontinuity?
   (c) Does the Gibbs phenomenon occur here?
2. By taking $f(x) = 1$ for all x, show that (4) follows from (1) and (3). (The Fourier series for f, in this instance, consists of only a constant term.)
3. If $f(x) = |x|$, what is $f(0+)$? What is $f'(0+)$? What is $f'(0-)$?

*4. A function f is said to have a *generalized second derivative* at x if $\dfrac{f(x - t) + f(x + t) - 2f(x)}{t^2}$ tends to a finite limit as t tends to zero.

(a) Does this imply f has a derivative $f'(x)$?

(b) Does this imply f is continuous at x?

(c) What can you say about the convergence of the Fourier series representing f?

5. Show that the integrand in (8) is an even function of t.

*6. (a) Suppose that f is an integrable function with period 2π, and that $f(x)$ is identically zero in a small interval. Show that its Fourier series converges to zero for values of x interior to this interval, but not necessarily at the endpoints of the interval.

(b) Suppose that g and h are integrable functions, periodic with period 2π, that are identically equal $g(x) = h(x)$ for values of x in a small interval. Show that, if the Fourier series representing either of these functions converges at a point x within (not on the boundary) of the interval, the other must also, and they must converge to the same value. (Apply (a) to the function $f = g - h$.)

(c) Hence, prove the *localization theorem:*
If f is an integrable function, and we form the Fourier series representing f in the interval $0 < x < 2\pi$, then the convergence of its Fourier series at a specified point x in this interval depends only on the behavior of the function in the immediate vicinity of this point.

(d) Generalize the theorem of part (c) slightly to include the possibility that $x = 0$ or $x = 2\pi$.

3.7 • UNIFORM CONVERGENCE

In Section 3.5, we showed that $S_n(x)$ can converge, for every x, to $f(x)$, even though the graphs of the functions S_n do not present the same appearance as the graph of f, even for large values of n. The student can therefore feel quite justified in thinking that "convergence" of a sequence of functions, when defined in the usual way, does not correspond to one's "intuitive" notions of convergence. In this section we discuss *uniform convergence*, which is designed to match more closely what most beginning students feel "convergence" really means.

Roughly speaking, a sequence of functions S_n converges uniformly to a function f, in an interval, if for sufficiently large n it would be impossible to distinguish between the graph of S_n and the graph of f, if these graphs were plotted (for values of x in the interval) with a pencil on the same diagram. More specifically, we can find an integer N (whose value will depend on the sharpness of the pencil) such that, for every $n > N$, the graph of S_n will be so nearly identical to the graph of f that one would not recognize any difference between them. (A precise definition is given later.)

From a practical engineering viewpoint, this rough explanation is quite sufficient. For example, it is obvious that a sequence of *continuous* functions cannot converge uniformly to a function having a jump discontinuity in the interval, since the graph of such a function is always disconnected. (This will not be obvious to students who mistakenly think that a geometrical object like that shown in Figure 3.10 represents the graph of a function!)

Before giving the *precise* definition of uniform convergence, we digress to review some elementary ideas concerning convergence in general. *These ideas are of fundamental importance in analysis.* The reader will find that time spent *carefully* studying this discussion will be very well rewarded later, not only in understanding this book but in reading any advanced text in analysis.

Definition 1. A sequence c_n of real or complex numbers is said to converge to zero, written $c_n \to 0$ or $\lim\limits_{n\to\infty} c_n = 0$, if and only if for every prescribed positive real number ϵ, no matter how small, there exists an integer N such that $|c_n| < \epsilon$ whenever $n > N$.

This is sometimes expressed by saying "c_n tends to zero with increasing n."

Definition 1 is quite important, since all our later definitions of convergence depend on it.

Definition 2. A sequence a_n of real or complex numbers is said to converge, and have limit b, if and only if $|a_n - b| \to 0$. We indicate this by writing $a_n \to b$.

From a geometrical viewpoint, $|a_n - b|$ is the *distance between* a_n and b, considered as points on the real line (or in the complex plane).

In general, if p and q are points, we let $d(p,q)$ denote the distance between p and q. If they are points on the real line, or in the complex plane, $d(p,q)$ is the same as $|p - q|$. If p and q are points in space, then

$$(1) \qquad d(p,q) = [(p_x - q_x)^2 + (p_y - q_y)^2 + (p_z - q_z)^2]^{1/2}$$

is the distance between p and q, in terms of their x, y, and z coordinates (p_x,p_y,p_z) and (q_x,q_y,q_z). We generalize Definition 2 to sequences of *points* in the following manner.

Definition 3. A sequence p_n of points is said to converge, and have limit q, if and only if $d(p_n,q) \to 0$.

Henceforth we will give definitions and theorems in generalized form, so they will apply to points. Notice, however, that they apply to numbers as a special case, simply by replacing $d(p,q)$ by $|p - q|$.

Definition 4. A sequence p_n of points is said to be a *Cauchy sequence*, or to satisfy the *Cauchy criterion*, if and only if for every prescribed positive real number ϵ, no matter how small, there exists an integer N such that $d(p_n,p_m) < \epsilon$ whenever both n and m are greater than N.

EXAMPLE 1: The sequence of real numbers $1, 1\frac{1}{2}, 1\frac{2}{3}, 1\frac{3}{4}, \cdots$ satisfies the Cauchy criterion.

EXAMPLE 2: The sequence of real numbers $1, -1, 1, -1, 1, -1, \cdots$ does not satisfy the Cauchy criterion. (The conditions cannot be met, for example, if $\epsilon = \frac{4}{3}$.)

EXAMPLE 3: The sequence $1, 1 + \frac{1}{2}, 1 + \frac{1}{2} + \frac{1}{3}, 1 + \frac{1}{2} + \frac{1}{3} + \frac{1}{4}, \cdots$ does not satisfy the Cauchy criterion. *It is not enough for the difference between adjacent terms to tend to zero.* (Compare Exercise 3.)

Cauchy's Convergence Theorem. *A sequence of real or complex numbers, or a sequence of points in any finite-dimensional Euclidean space, is convergent if and only if it satisfies the criterion of Definition 4.*

In other words, every convergent sequence is a Cauchy sequence, and every Cauchy sequence is a convergent sequence. (This theorem is actually valid in any finite-dimensional linear space equipped with a norm, if we take $d(p,q)$ to be $\|p - q\|$. It is not valid in some infinite-dimensional linear spaces.)

In particular, it follows from this theorem that if we can prove a sequence of real or complex numbers is a Cauchy sequence, then we know the sequence must have a limit.

Definition 5. A sequence of functions S_n, all having the same domain of definition, is said to *converge* in this domain, or (more explicitly) to be *pointwise convergent*, if and only if there is a function $g(x)$ (with the same domain of definition) such that $S_n(x) \to g(x)$ for every x in the domain.

This is the "ordinary" definition of convergence. The function g is said to be the *limit* of the sequence S_n.

It follows from the Cauchy convergence theorem that if, for every fixed x, $S_n(x)$ is a Cauchy sequence of numbers, a function g satisfying the requirements of Definition 5 must exist. We obtain the function g by taking $g(x)$, for each x, to be the limit of the sequence $S_n(x)$.

Definition 6. A function f is said to be *bounded* in an interval if there is a number M such that $|f(x)| \leqq M$ for every x in the interval. The least value of M for which this inequality is valid is denoted $N(f)$ and is called the *uniform norm* of f in the interval.

In more technical language, $N(f)$ is the least upper bound of the values $|f(x)|$ for x in the interval. Although $N(f)$ depends, in general, on the choice of the interval, there is seldom any ambiguity in the notation $N(f)$, since, throughout any discussion, the interval in question is always kept fixed.

EXAMPLE 4: Let $f_n(x) = x^n$. In the interval $(-1,1)$ the uniform norm of f_n is $N(f_n) = 1$. In the interval $(-\frac{1}{2},\frac{1}{2})$ the uniform norm is $N(f_n) = (\frac{1}{2})^n$.

EXAMPLE 5: Let $S_n(x) = \dfrac{nx}{1 + n^2x^2}$. The uniform norm of S_n relative to the interval $(-1,1)$ is $N(f) = \frac{1}{2}$. (Compare the discussion on page 114.)

EXAMPLE 6: Let $T_n(x) = \dfrac{n^2x}{1 + n^3x^2}$. In the interval $0 \leqq x \leqq 1$ the maximum value of $T_n(x)$ occurs at $x = n^{-3/2}$ and this maximum value is $\sqrt{n/2}$. Therefore the uniform norm of T_n relative to this interval is $N(T_n) = \sqrt{n/2}$.

EXAMPLE 7: We are given a function f, defined on an interval, but all we know about the function g is that $N(f - g) < \epsilon$. We can conclude that the graph of g must be contained within the dotted lines shown in Figure 3.13.

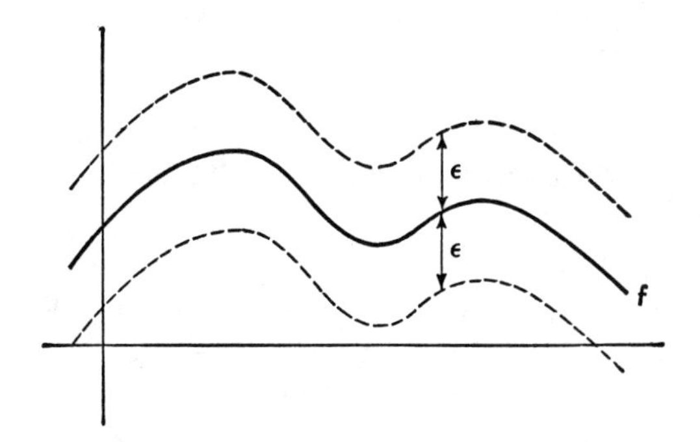

Figure 3.13

Definition 7. A sequence S_n of functions is said to *converge uniformly* to g in an interval if and only if (relative to this interval) $N(S_n - g)$ is defined, except possibly for a finite number of terms S_n (which we ignore), and $N(S_n - g) \to 0$.

Under these conditions, we say that g is the *uniform limit*, in the interval, of the sequence S_n.

EXAMPLE 8: We see from Example 4 that the sequence of functions x^n converges uniformly to the zero function in the interval $(-\frac{1}{2},\frac{1}{2})$, but not in the larger interval $(-1,1)$. In Example 5, the functions S_n do not tend uniformly to the zero function in $(-1,1)$, although they do converge pointwise to the zero function. Example 6 is interesting because it shows that $N(T_n - g)$ can tend to infinity even though $T_n(x) \to g(x)$ for all x. In all of these examples, $g(x)$ is identically zero.

EXAMPLE 9: Turn to the last theorem in Section 2.6 (page 81). We see that the sequence $V_n(x)$ converges uniformly to $f(x)$, since $N(f - V_n) \leqq K\pi/\sqrt{n}$ which tends to zero with increasing n.

EXAMPLE 10: "For every prescribed positive number ϵ, no matter how small, there exists an integer n_0 such that $|S_n(x) - g(x)| < \epsilon$ whenever $n > n_0$." If this holds for every x in an interval, then S_n converges *pointwise* to g in the interval, and *if the choice of* n_0 *depends only on* ϵ *and is independent of* x, then S_n converges *uniformly* to g in the interval.

Theorem 1. *The limit of a uniformly convergent sequence of continuous functions is a continuous function.*

Proof: Suppose that a sequence S_n of continuous functions converges uniformly to g in an interval. If x and $x + h$ are both in this interval, we have

$$\begin{aligned} &|g(x + h) - g(x)| \\ &\quad = |g(x + h) - S_n(x + h) + S_n(x + h) - S_n(x) + S_n(x) - g(x)| \\ &\quad \leqq |g(x + h) - S_n(x + h)| + |S_n(x + h) - S_n(x)| + |S_n(x) - g(x)| \\ &\quad \leqq N(g - S_n) + |S_n(x + h) - S_n(x)| + N(S_n - g) \\ &\quad = 2N(g - S_n) + |S_n(x + h) - S_n(x)|. \end{aligned}$$

The first term can be made as small as we please by taking n sufficiently large, and the second term tends to zero as $h \to 0$ since S_n is continuous. The left side $|g(x + h) - g(x)|$ is independent of n. It follows that the left side tends to zero as $h \to 0$, and therefore g is continuous.

Theorem 2. *If a sequence* S_n *of integrable functions tends uniformly to an integrable function* g, *then*

$$\int_a^b |g(x) - S_n(x)|^2\, dx \to 0. \tag{2}$$

In other words, uniform convergence implies convergence in the mean (see page 56).

Proof: Directly from the definitions of $\|f\|$ and $N(f)$, relative to an interval (a,b), we have

$$\|f\| \leqq [N(f)](b - a)^{1/2}. \tag{3}$$

Theorem 2 follows at once if we replace f by $g - S_n$, since by (3) it is clear that $N(g - S_n) \to 0$ implies $\|g - S_n\| \to 0$.

Theorem 3. *If a sequence of integrable functions S_n converges uniformly to an integrable function g, or (more generally) converges in the mean to g, then*

$$\int_a^b S_n(x)\,dx \to \int_a^b g(x)\,dx. \tag{4}$$

Proof: Because of Theorem 2, it suffices to prove the more general assertion, that $\|g - S_n\| \to 0$ implies (4). By the Schwarz inequality, we have (for any two integrable functions f and h):

$$|(f|h)| \leqq \|f\|\|h\|. \tag{5}$$

Letting $f(x) = |g(x) - S_n(x)|$ and $h(x) = 1$, this gives

$$\int_a^b |g(x) - S_n(x)|\,dx \leqq \|g - S_n\|(b - a)^{1/2} \to 0 \tag{6}$$

and therefore

$$\left|\int_a^b g(x)\,dx - \int_a^b S_n(x)\,dx\right| = \left|\int_a^b [g(x) - S_n(x)]\,dx\right|$$
$$\leqq \int_a^b |g(x) - S_n(x)|\,dx \to 0,$$

which proves the assertion.

Notice that (4) can be written

$$\lim_{n\to\infty} \int_a^b S_n(x)\,dx = \int_a^b [\lim_{n\to\infty} S_n(x)]\,dx = \int_a^b g(x)\,dx. \tag{8}$$

This is sometimes expressed by the statement that *passage to the limit under the integral sign is justified whenever a sequence of integrable functions is uniformly convergent or convergent in the mean to an integrable function.*

If, in the above proof, we had not taken $h(x)$ to be identically equal to unity, but left it arbitrary, we would have the following more general result.

Theorem 4. *If a sequence of integrable functions S_n converges uniformly to an integrable function g, or (more generally) converges in the mean to g, then for any integrable function h we have*

$$\int_a^b S_n(x)h(x)\,dx \to \int_a^b g(x)h(x)\,dx. \tag{9}$$

These ideas are usually applied to the sequence of partial sums of an infinite series. Thus, if we are given an infinite series

$$\sum_{k=1}^{\infty} u_k(x) \tag{10}$$

and we let S_n denote the nth partial sum of this series,

(11) $$S_n(x) = u_1(x) + u_2(x) + \cdots + u_n(x),$$

we say that (10) *converges* in an interval if $S_n(x) \to g(x)$ for all x in the interval. The function g is then called the *sum* of the series. If such a function g does not exist, the series is said to be *divergent.* This terminology is somewhat unfortunate, because the word "divergent" suggests to some students that $S_n(x) \to \infty$, which is not necessarily the case. For example, the series (13), Section 3.3 (page 98) diverges for all x, but we see that its partial sum [(6), page 97], does not "tend to infinity" in the interval $0 < x < 2\pi$, but simply oscillates without approaching any limit. (At $x = 0$ and $x = 2\pi$, $S_n(x) \to \infty$.)

If the sequence of partial sums converges uniformly, then we say that the series is *uniformly convergent.* If the series

(12) $$\sum_{k=1}^{\infty} |u_k(x)|$$

converges, we say that (10) is *absolutely convergent.* Uniform convergence obviously implies convergence. Absolute convergence also implies convergence (see any standard text on calculus). A series may be uniformly convergent without being absolutely convergent (see Exercise 2 for an example).

If the partial sums (11) converge *in the mean* to g in an interval, i.e., $\|S_n - g\| \to 0$, then (10) is said to *converge in the mean.* If a series is uniformly convergent then it must also be convergent in the mean. On the other hand, it is possible for a series to converge in the mean and yet be divergent for every value of x (see Exercise 1).

The following theorem is proved in all advanced calculus texts:

Theorem 5. *If a Taylor series* $\sum_{n=0}^{\infty} C_n(x - x_0)^n$ *converges for all* x *in an interval* $a \leqq x \leqq b$, *then it converges uniformly in any smaller interval* $a + \epsilon \leqq x \leqq b - \epsilon$ $(\epsilon > 0)$.

Whether or not the series is a Taylor series, the following theorem is sometimes useful in proving uniform convergence. It is called the *Weierstrass M-test.*

Theorem 6. *If* $\sum_{n=1}^{\infty} M_n$ *is a convergent series, where each term is a positive constant, and if* $\sum_{n=1}^{\infty} u_n(x)$ *is a series of functions, and if for all* n

and all x in a certain interval $|u_n(x)| \leqq M_n$, then the series of functions converges uniformly and absolutely in that interval.

The proof of this theorem is outlined in Exercise 8.

Theorem 7. *If $\sum_{k=1}^{\infty} u_k(x)$ converges uniformly in an interval, or (more generally) converges in the mean to $g(x)$, then for any integrable function h we have*

$$\int_a^b g(x)h(x)\,dx = \sum_{k=1}^{\infty} \left[\int_a^b u_k(x)h(x)\,dx\right]. \tag{13}$$

In other words, we can formally multiply both sides of

$$g(x) = \sum_{k=1}^{\infty} u_k(x) \tag{14}$$

by $h(x)$, and integrate both sides over the interval, interchanging the order of summation and integration. The resulting series of constants necessarily converges and has sum $\int_a^b g(x)h(x)\,dx$.

Proof: Let $S_n(x)$ denote the nth partial sum of the series, and apply Theorem 4.

The reason this theorem is interesting is because it is valid in many instances in which (14) is not valid in the ordinary sense; perhaps the series converges in the mean to $g(x)$ but is actually a divergent series. This shows that a series can sometimes be useful even if it is divergent. (Some readers might prefer to say that the classical definition of convergence is rather irrelevant for many purposes.)

In the rest of this section, we assume the reader has studied Chapters 1 and 2 rather carefully.

It is easy to verify the following properties of the uniform norm. Keep in mind that $N(f)$ is the least upper bound of $|f(x)|$ in the interval.

(15) $N(f)$ exists for any function that is bounded in the interval,

(16) $N(f) \geqq 0$, and $N(f) = 0$ if and only if $f(x)$ is zero for every x in the interval,

(17) $N(af) = |a|N(f)$ for any constant a,

(18) $N(f + g) \leqq N(f) + N(g)$ for any two functions bounded in the interval.

Properties (16), (17), and (18) are the general properties required of any "norm," as discussed in Section 2.1.

If we replace f by $u - v$ and g by $v - w$ in (18) we obtain

$$N(u - w) \leqq N(u - v) + N(v - w). \tag{19}$$

This is similar to the triangle inequality (page 33) and shows that the uniform norm provides another way to define the notion of "distance" in a linear space of functions. If u and v are functions defined and bounded in an interval, we can take $N(u - v)$ to be the "distance" between u and v.

Numerically, this notion of distance is quite different from the distance (defined as $\|u - v\|$) discussed earlier. It is possible for $\|S_n - g\|$ to tend to zero, even though $N(S_n - g) \to \infty$. (See Exercise 6.)

Now suppose we are given two classes of functions (A) and (B). We say that functions of class (A) can be *approximated uniformly* by functions of class (B) if and only if every function of class (A) is the uniform limit of a sequence of functions of class (B).

Theorem 8. *Functions of a given class* (A) *can be approximated uniformly by functions of class* (B) *if and only if the following condition is met: given any function f of class* (A) *and any positive ϵ there exists a function g of class* (B) *such that $N(f - g) < \epsilon$.*

Proof: If functions of class (A) can be approximated uniformly by functions of class (B), then for any function f of class (A) there is a sequence g_n of functions of class (B) such that $N(f - g_n) \to 0$. Therefore, if any positive number ϵ is prescribed, there exists some n for which $N(f - g_n) < \epsilon$. Conversely, if the condition of the theorem is met, we can take $\epsilon = 1/n$ and hence for each n there must exist a function g_n such that $N(f - g_n) < 1/n$, and therefore the sequence g_n converges uniformly to f.

Theorem 9. *If functions of class* (A) *can be uniformly approximated by functions of class* (B), *and functions of class* (B) *can be uniformly approximated by functions of class* (C), *then functions of class* (A) *can be uniformly approximated by functions of class* (C).

Proof: Let u be of class (A). According to the hypothesis, we can find a function v of class (B) such that $N(u - v) < \epsilon$, and then a function w of class (C) such that $N(v - w) < \epsilon$. By (19) it follows that $N(u - w) < 2\epsilon$. Since ϵ can be as small as we please, this shows that functions of class (A) can be approximated by functions of class (C), by virtue of Theorem 8.

The reader must not lose sight of the important fact that the uniform norm is defined *relative to a particular interval.* The approximations we are concerned with here are in general valid *only* within this interval.

Theorem 10. *Trigonometric polynomials can be approximated uniformly by polynomials in any interval of finite length.*

Proof: We recall that a trigonometric polynomial is a (finite) linear combination of functions of the form $A_n \cos nx$ and $B_n \sin nx$. These functions possess power series expansions that converge for all x. Therefore every trigonometric polynomial has a power series expansion that converges for all x. By Theorem 5, the partial sums of such a power series converge uniformly in any interval of finite length. Each of these partial sums is a polynomial. It follows that any trigonometric polynomial can be approximated uniformly by a polynomial in such an interval.

Theorem 11. *Every continuous function can be approximated uniformly by a piecewise smooth continuous function in any closed interval of finite length.*

Outline of proof: Every continuous function f, defined in the interval $a \leqq x \leqq b$, can be approximated by a *broken-line function.* (A broken-line function is simply a continuous function whose graph consists of line segments, with only a finite number of such segments in any interval of finite length.) To see this, subdivide the interval into n parts, which for convenience can be taken equal in length: $a = x_0 < x_1 < x_2 < \cdots < x_n = b$. Construct a broken-line function

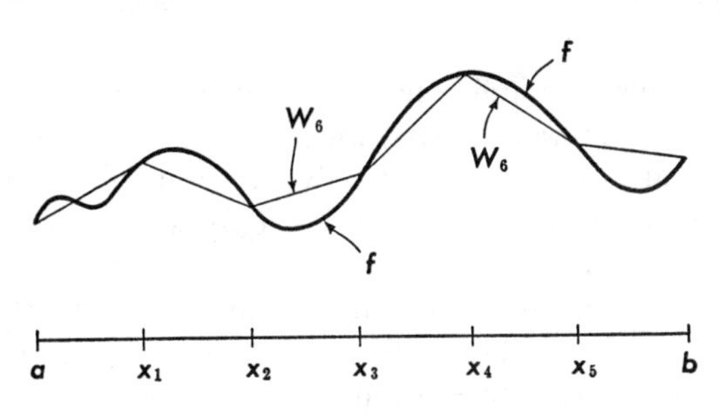

Figure 3.14

W_n by joining the successive points $(x_0,f(x_0))$, $(x_1,f(x_1))$, $\cdots$, $(x_n,f(x_n))$ with line segments; the resulting graph defines W_n in the interval. From the continuity of f, it is obvious that $W_n(x) \to f(x)$ for every x in the interval. It is somewhat less obvious, but nonetheless true, that W_n converges *uniformly* to f in the interval. Any reader familiar with the notion of "uniform continuity" can easily supply the details omitted here. For the case $n = 6$, the procedure is illustrated in Figure 3.14.

Theorem 12. *Every continuous function having period 2π can be approximated uniformly by trigonometric polynomials (in any interval).*

Proof: We simply *continue* the proof of Theorem 11, taking the interval to be $0 \leqq x \leqq 2\pi$. All of the functions W_n constructed in this manner have the property $W_n(0) = W_n(2\pi)$ in this case. Let the slopes of the n line segments be $k_1, k_2, \cdots, k_n$, and let $K = \max |k_i|$. Then obviously (the mean-value theorem is hardly needed here) we have

$$|W_n(x') - W_n(x'')| \leqq K|x' - x''|. \tag{20}$$

The period 2π extension of W_n also has this property, and the class of such functions uniformly approximate f (which also has period 2π by hypothesis) in any interval. By the theorem at the end of Section 2.6, these functions W_n can be uniformly approximated by trigonometric polynomials. The result then follows from Theorem 9. (For an alternative proof not based on Section 2.6, see the Answers and Notes section.)

Theorem 13. *Every continuous function can be approximated uniformly by polynomials in any interval of finite length.*

Note: We are assuming here that the function is continuous for all x, even though we are concerned only with an interval. The theorem is valid if f is defined only in an interval, provided the function is defined and continuous not only within the interval but also at the endpoints.

Proof: We first note that, by a change of scale, the length of the interval can be made less than 2π, and by a translation of the axis the interval can be moved to the interior of the interval $0 \leqq x \leqq 2\pi$. Then we can extend the function, as defined in this interval (a',b'), to a continuous function g so chosen that $g(0) = g(2\pi)$ [see Figure 3.15]. Apply Theorem 12 to the period 2π extension of g. This function can be uniformly approximated by trigonometric polynomials, which in turn can be uniformly approximated by poly-

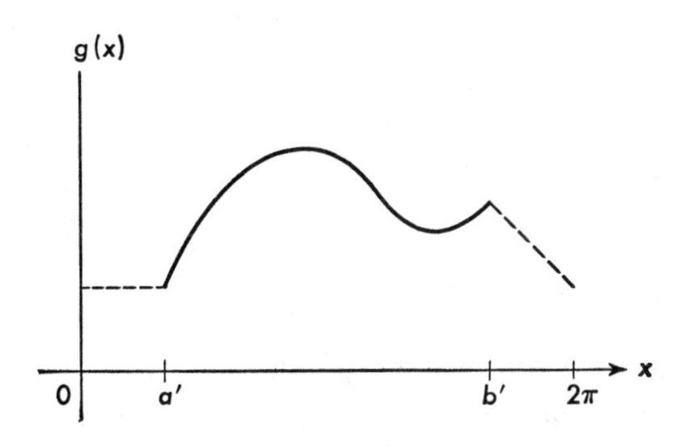

Figure 3.15

nomials (Theorem 10), and therefore (Theorem 9) g can be uniformly approximated by polynomials in the interval $(0,2\pi)$. Now we translate the axis and change the scale back the way it was to begin with; this does not make the approximation any worse. Note that, if we translate the axis and change the scale, the form of the polynomial will change, but it is still a polynomial. We leave further details to the reader.

Theorem 13 is sometimes called the *Weierstrass approximation theorem.*

For the benefit of readers having considerable mathematical maturity, we now state, without proof, a generalization of this theorem.

Theorem 14. *Let* (A) *be a class of functions having a common domain S. If* (A) *and S have the properties listed below, then every continuous function having domain S can be uniformly approximated by functions of class* (A):

(i) the domain S is a bounded set of points in n-dimensional Euclidean space, i.e. there is some fixed number d_0 such that $d(p,q) < d_0$ for every pair of points p and q in S;

(ii) S is closed, i.e. if any point p is the limit of a sequence of points p_m in S, $d(p_m,p) \to 0$, then p must also be in S;

(iii) every function of class (A) is real-valued and continuous throughout S;

(iv) (A) is a real linear space;

(v) given any point p in S, there must be some function f of class (A) such that $f(p) \neq 0$;

(vi) given any two distinct points p and q in S, there must be some function f of class (A) such that $f(p) \neq f(q)$;

(vii) products of functions of class (A) must be of class (A).

EXAMPLE 11: If (A) is the class of all (real) polynomial functions and S is the interval $a \leqq x \leqq b$, all these conditions are satisfied. Therefore Theorem 13 is a special case of this more general theorem (which is due to M. H. Stone).

EXAMPLE 12: Let S be the set of points on the circle $x^2 + y^2 = 1$. Each point in S can be designated by its polar coordinate θ. Every continuous function of period 2π can be identified with a continuous function having domain S. S satisfies the conditions (i) and (ii) above. Let (A) be the class of all trigonometric polynomials. All of the conditions of Theorem 14 are satisfied. Therefore Theorem 12 can be considered as a special case of Theorem 14.

EXAMPLE 13: Let S be the interval $0 \leqq x \leqq \pi$. Let (A) be the class of all cosine polynomials $A_0 + A_1 \cos x + \cdots + A_n \cos nx$. Since $\cos nx \cos mx = \frac{1}{2} \cos (n - m)x + \frac{1}{2} \cos (n + m)x$, it is easy to see that (vii) is satisfied. The other conditions are easily verified. Therefore any continuous function can be uniformly approximated in the interval $0 \leqq x \leqq \pi$ by cosine polynomials.

EXAMPLE 14: Let S be as in Example 13, and let (A) be the class of all sine polynomials $B_1 \sin x + B_2 \sin 2x + \cdots + B_n \sin nx$. Hypotheses (i), (ii), (iii), and (iv) are satisfied. However, (v) is not satisfied at $p = 0$ or at $p = \pi$, and (vi) is not satisfied if we take $p = 0$ and $q = \pi$. Nor is (vii) satisfied, since the product of two sine polynomials is not generally a sine polynomial. It is *not* possible to approximate an arbitrary continuous function uniformly by sine polynomials in the interval $0 \leqq x \leqq \pi$. The only functions f for which this is possible are those for which $f(0) = f(\pi) = 0$. (It is possible to approximate arbitrary continuous functions in this interval by sine polynomials in the sense of *convergence in the mean.*)

• EXERCISES

1. For x in the interval $0 \leqq x \leqq 1$, let

$$f_{nm}(x) = 1 \quad \text{whenever} \quad \frac{m-1}{n} \leqq x \leqq \frac{m}{n} \qquad (m = 1, 2, \cdots, n)$$
$$= 0 \text{ for other values of } x.$$

(a) Consider the sequence $f_{11}, f_{21}, f_{22}, f_{31}, f_{32}, f_{33}, f_{41}, f_{42}, \cdots$ obtained by ordering the functions "lexicographically." Show that this sequence of functions converges in the mean to the function identically equal to zero.

(b) Show that the sequence in part (a) does not converge uniformly to any function.

(c) Show that the sequence of functions in part (a) is divergent for every x in this interval.

2. Show that the series $\sum_{n=1}^{\infty} (-1)^n x/n$ is uniformly convergent in every interval of finite length, and find its sum.

3. Show that the series in Exercise 2 is not absolutely convergent, except at $x = 0$.

4. Let $u_n(x) = x^n(1 - x)$ in the interval $0 \leqq x \leqq 1$.

(a) Show that the series $\sum_{n=1}^{\infty} u_n(x)$ converges for all x in this interval, and find its sum.

(b) Determine the uniform norm of the nth partial sum of this series.

(c) Does this series converge uniformly?

(d) Plot a rough graph of a typical partial sum of this series.

5. (a) Is the series in Exercise 4 convergent in the mean?

(b) Is it legitimate to integrate this series term by term?

6. Let $S_n(x) = \sqrt{n}$ when $0 \leqq x \leqq 1/n^2$, and let $S_n(x) = 0$ for other values of x. Let g be the zero function.

(a) Find $N(S_n - g)$ relative to the interval $0 \leqq x \leqq 1$.

(b) Find $\|S_n - g\|$ relative to the same interval.

(c) Show that $N(S_n - g) \to \infty$ but $\|S_n - g\| \to 0$.

(d) Is this valid: $\int_0^1 S_n(x)\, dx \to \int_0^1 g(x)\, dx$?

7. If $\int_0^1 S_n(x)\,dx \to \int_0^1 g(x)\,dx$, is it necessarily true that S_n converges uniformly to g in the interval $(0,1)$?

***8.** Prove Theorem 6 by following this outline. Let

$$s_n = \sum_{k=1}^{n} M_n \quad \text{and} \quad S_n(x) = \sum_{k=1}^{n} u_k(x).$$

(a) Show that, under the hypotheses of the theorem, $s_1, s_2, s_3, \cdots$ is a Cauchy sequence.
(b) Show that $|S_n(x) - S_m(x)| \leqq |s_n - s_m|$.
(c) Deduce that, for each x, $S_n(x)$ is a Cauchy sequence, and hence converges to some value $S(x)$.
(d) Deduce that S_n converges to S uniformly in the interval by making use of the inequality

$$|S_n(x) - S(x)| \leqq |S_n(x) - S_m(x)| + |S_m(x) - S(x)|.$$

(The first term on the right side can be made small, independently of x, by using part (b). The m needed in the second term may depend on x, but it doesn't matter because m does not appear on the left side.)

***9.** By analogy with the uniform norm, one can define "distance" between points in space by the formula

$$d(p,q) = \max\,[|p_x - q_x|, |p_y - q_y|, |p_z - q_z|]. \tag{21}$$

(a) With this notion of "distance" is the triangle inequality

$$d(p,r) \leqq d(p,q) + d(q,r)$$

valid?
(b) With this notion of distance describe the set of all points whose "distance" from the origin is unity. Is it a sphere?
(c) If $p_n \to p$, i.e., $d(p_n,p) \to 0$ where distance is defined by (21), does the sequence p_n converge to p relative to the usual definition via (1) of this section?

3.8 • CONVERGENCE OF FOURIER SERIES

Very little of the preceding theory is needed to prove that the Fourier series of a piecewise smooth continuous function of period 2π converges *uniformly*. A simple proof can be based on the Bessel inequality, which we recall was quite simple to derive. If g is a real integrable function, with Fourier series

(1) $$a_0/2 + \sum_{n=1}^{\infty} (a_n \cos nx + b_n \sin nx),$$

Bessel's inequality becomes (see Exercise 1)

(2) $$a_0^2/2 + \sum_{n=1}^{\infty} (a_n^2 + b_n^2) \leqq \frac{1}{\pi} \int_0^{2\pi} [g(x)]^2 \, dx.$$

In particular, this guarantees the convergence of the infinite series $\sum (a_n^2 + b_n^2)$, which is all we need in the following argument.

Suppose that f is real-valued, has period 2π, and is continuous and piecewise smooth, so that $g(x) = f'(x)$ exists (except for a finite number of points in any interval of periodicity) and is piecewise continuous. Let the Fourier coefficients of f be denoted A_n, B_n, to distinguish them from the Fourier coefficients of g which we denote a_n, b_n. Directly from the integrals which define these coefficients we find, on integrating by parts, that

(3) $$a_n = nB_n, \qquad b_n = -nA_n$$

and therefore

(4) $$\sum_{k=1}^{n} (A_k^2 + B_k^2)^{1/2} = \sum_{k=1}^{n} \frac{1}{k} (a_k^2 + b_k^2)^{1/2}.$$

Applying the Schwarz inequality to the right side, we obtain

(5) $$\sum_{k=1}^{n} (A_k^2 + B_k^2)^{1/2} \leqq \left[\sum_{k=1}^{n} \frac{1}{k^2} \right]^{1/2} \left[\sum_{k=1}^{n} (a_k^2 + b_k^2) \right]^{1/2}.$$

Since the infinite series $\sum k^{-2}$ converges, the first factor on the right tends to a finite limit as $n \to \infty$, and so also does the second term, as already noted. It therefore follows that there exists some constant C such that

(6) $$\sum_{n=1}^{n} (A_k^2 + B_k^2)^{1/2} \leqq C$$

independently of n. The left side is positive, cannot decrease with increasing n, but is bounded above by C, so it must tend to a finite limit as $n \to \infty$. It follows that the infinite series

(7) $$\sum_{k=1}^{\infty} (A_k^2 + B_k^2)^{1/2}$$

is convergent. Notice that this is a much stronger assertion than simply saying that the infinite series $\sum (A_k^2 + B_k^2)$ converges, which we could have derived at once from Bessel's inequality (see Exercise 5).

The Fourier series expansion of f is

$$A_0/2 + \sum_{k=1}^{\infty} (A_k \cos kx + B_k \sin kx) \tag{8}$$

and we learned in Chapter 1 that

$$|A_k \cos kx + B_k \sin kx| \leqq (A_k^2 + B_k^2)^{1/2}. \tag{9}$$

Since (7) is convergent, this together with the Weierstrass M-test shows that (8) is uniformly and absolutely convergent.

We have proved the following theorem.

Theorem 1. *The Fourier series expansion of a piecewise smooth continuous function of period 2π is uniformly and absolutely convergent.*

Notice that the above derivation does not show that the sum of the function, for each x, is $f(x)$. (Perhaps the sum of the series is some other function!) This is to be expected, since the proof makes use of almost nothing beyond rather superficial properties of orthogonal expansions. [The same proof would apply if we deleted every other term from the Fourier series, and we would hardly expect the sum in that case to be $f(x)$.] However, the theorem proved in Section 3.6 completes the picture: the sum of the series must be $f(x)$ for every x.

A more direct proof, not making use of Bessel's inequality or the results of Section 3.6, can also be given. Indeed, it is possible to prove an even more general theorem:

Theorem 2. *If g is an integrable function of period 2π, satisfying a Lipschitz condition $|g(x') - g(x'')| \leqq K|x' - x''|$, the Fourier expansion of g converges uniformly to g in every interval.*

Proof: We must first analyze more closely the Dirichlet kernel

$$D_n(x) = \frac{1}{2\pi} + \frac{1}{\pi}\cos x + \cdots + \frac{1}{\pi}\cos nx = \frac{\sin\,(n + \frac{1}{2})x}{2\pi \sin \frac{1}{2} x} \tag{10}$$

It is obvious that $\int_0^{2\pi} D_n(x)\,dx = 1$, but in this proof we need to know something about $\int_0^{2\pi} |D_n(x)|\,dx$. It turns out that this integral increases without bound as n increases, but that does not concern us here. What does concern us is that this integral does not tend to infinity very rapidly. By a somewhat tedious calculation (see Exercise 7) one can show that

$$\int_0^{2\pi} |D_n(x)|\, dx \leqq C \ln n \qquad (n > 2) \tag{11}$$

for some constant C.

Now suppose that f is integrable and has period 2π. Let M be a positive number so chosen that

$$|f(x)| \leqq M \qquad \text{(for all } x\text{)}. \tag{12}$$

If T_n is the partial sum of order n of the Fourier series expansion of f,

$$\begin{aligned} |T_n(x)| &= \left| \int_0^{2\pi} f(x-t) D_n(t)\, dt \right| \leqq \int_0^{2\pi} |f(x-t)|\, |D_n(t)|\, dt \\ &\leqq M \int_0^{2\pi} |D_n(t)|\, dt \leqq MC \ln n. \end{aligned} \tag{13}$$

It follows that if $n > 2$

$$\begin{aligned} |T_n(x) - f(x)| &\leqq |T_n(x)| + |f(x)| \leqq M(1 + C \ln n) \\ &\leqq M\left(\frac{\ln n}{\ln 2} + C \ln n\right) = DM \ln n \end{aligned} \tag{14}$$

for some constant D. It is important to note that D is independent not only of n and x but of the function f as well. (The exact value of D is unimportant here.)

Now let g satisfy the hypotheses of the theorem, and let S_n denote its Fourier partial sum of order n. Let V_n be any trigonometric polynomial of order n; obviously V_n is its own Fourier series, which is finite in this case. Choose $g - V_n$ to be the function f discussed above. The Fourier series expansion of $f = g - V_n$ differs from that representing g in at most the terms of order n or less; clearly, $T_n = S_n - V_n$. Therefore $g - S_n = f + V_n - S_n = f - T_n$. By the theorem at the end of Section 2.6, we can choose V_n so that $|f(x)| \leqq K\pi/\sqrt{n}$, and therefore by (14) we have

$$|f(x) - T_n(x)| \leqq (DK\pi \ln n)/\sqrt{n} \tag{15}$$

and since $g - S_n - T_n$, it follows that

$$|g(x) - S_n(x)| \leqq (DK\pi \ln n)/\sqrt{n}. \tag{16}$$

The right side of this expression tends to zero with increasing n, independently of x. Therefore the partial sums S_n converge uniformly to g and the proof is complete.

It is possible to prove uniform convergence of a Fourier series under even weaker hypotheses than those of Theorem 2. However, if the hypotheses are weakened too much, we can no longer deduce uniform convergence. For instance, if we assume only that f is continuous and has period 2π, the Fourier series will not necessarily even be pointwise convergent. However, the series in this case will be convergent in the mean.

Theorem 3. *If f is continuous and has period 2π, the partial sums of the Fourier series expansion of f converge in the mean to the function f.*

Proof: We shall show that

$$(18)\quad \int_0^{2\pi} [f(x) - A_0/2 - \sum_{k=1}^{n} (A_n \cos nx + B_n \sin nx)]^2 \, dx \to 0$$

and from this it is easy to see that the partial sums will converge in the mean to f in *any* interval of finite length.

According to Theorem 12, Section 3.7, for any prescribed positive ϵ there exists a trigonometric polynomial $R_n(x)$ such that $|f(x) - R_n(x)| < \epsilon$ for every x. Therefore

$$(19)\qquad \int_0^{2\pi} [f(x) - R_n(x)]^2 \, dx < 2\pi\epsilon^2.$$

Since ϵ is arbitrary, this shows there is a trigonometric polynomial R_n for which the integral on the left of (19) is arbitrarily small. But we recall from Chapter 2 that the best least-squares approximation of an integrable function by a linear combination of orthogonal functions is that in which the coefficients are obtained by taking inner products; that is, if S_n is the partial sum of the Fourier series having the same order as the trigonometric polynomial R_n, replacing R_n by S_n in (19) will, if anything, *improve* the approximation. Moreover, the approximation will not be any worse for any partial sum of a higher order. It follows that

$$(20)\qquad \int_0^{2\pi} [f(x) - S_n(x)]^2 \, dx \to 0,$$

which is the same as (18) and completes the proof.

Theorem 4. *Any piecewise continuous function of period 2π can be approximated as closely as we like, in the mean-square sense, by a continuous function of period 2π.*

In lieu of a formal proof (which any interested reader can easily provide) we illustrate this by an example. In Figure 3.16 we show the graph of a function f satisfying the hypotheses of this theorem.

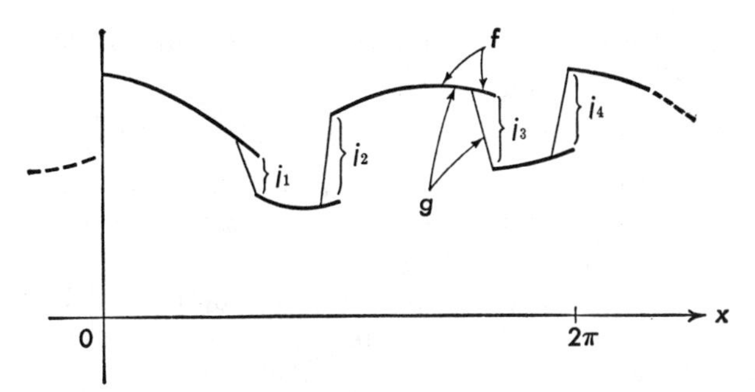

Figure 3.16

In a neighborhood of each discontinuity, we modify f to obtain a continuous function g, as shown. The graph of $f - g$ is shown in Figure 3.17. By choosing the δ's (indicated in Figure 3.17) sufficiently small, we can make $\int_0^{2\pi} [f(x) - g(x)]^2\, dx$ as small as we like, since the magnitudes of the jumps (indicated by the j's) are independent of the δ's. Notice that we have chosen g in the interval to be continuous, and also $g(0) = g(2\pi)$, so that the period 2π extension of g is also continuous.

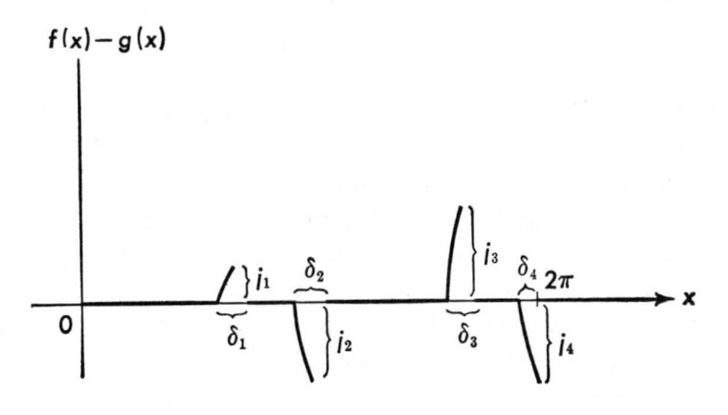

Figure 3.17

Theorem 5. *The Fourier series expansion of a piecewise continuous function of period 2π converges in the mean to the function.*

Proof: Let g satisfy the hypothesis and let f be a continuous function of period 2π. According to Theorem 3, we can find a trigonometric polynomial S_n approximating f as closely as we please; $\|f - S_n\| < \epsilon$. (We can take S_n to be a partial sum of the Fourier series representing f.) By Theorem 4, we can choose f so that $\|f - g\| < \epsilon$, and hence

$$\|g - S_n\| = \|g - f + f - S_n\| \leqq \|g - f\| + \|f - S_n\| < 2\epsilon.$$

If we take S_n^* to be a partial sum of the Fourier series representing g, having the same order as S_n, then by the same argument that led from (19) to (20) we see that $\|g - S_n^*\| \to 0$.

It is even possible to replace "piecewise continuous" in Theorems 4 and 5 by "integrable"; for the proof, see the Answers and Notes section. (If we were to extend our definition of integrability to include unbounded functions, it would be necessary to include in the hypothesis the requirement that $[f(x)]^2$ is also integrable.)

We shall now discuss the term-by-term integration of Fourier series. Let f be an integrable function of period 2π, and let its Fourier series be

$$a_0/2 + \sum_{n=1}^{\infty} (a_n \cos nx + b_n \sin nx). \tag{21}$$

We have made no assumptions beyond the integrability of f, so this series may not converge to $f(x)$. Despite this, the following theorem is valid:

Theorem 6. *Any Fourier series, whether convergent or not, can be integrated term-by-term between any limits.*

This means that we can integrate the separate terms, and the resulting series will converge to the integral of the function f. (This would follow at once from Theorem 5, and Theorem 7 of the preceding section, if we had proved Theorem 5 for arbitrary integrable functions.)

Proof: First, note that $\int_a^b f(x)\,dx = \int_0^b f(x)\,dx - \int_0^a f(x)\,dx$, so it suffices to prove the assertion for integrals over intervals of the form $(0,x)$.

We wish to prove

(22)

$$\int_0^x [f(t)]\,dt = \int_0^x \tfrac{1}{2}a_0\,dt + \sum_{n=1}^{\infty} \left[\int_0^x (a_n \cos nt + b_n \sin nt)\,dt\right],$$

so it suffices to prove

$$\int_0^x [f(t) - \tfrac{1}{2}a_0]\,dt = \sum_{n=1}^{\infty} \frac{a_n \sin nx + b_n(1 - \cos nx)}{n}. \tag{23}$$

Define $F(x)$ to be

$$F(x) = \int_0^x [f(t) - \tfrac{1}{2}a_0]\, dt. \tag{24}$$

We recall that an integrable function f is bounded. (If we were to extend the definition of integrability to include unbounded functions, this proof would need modification, but the theorem would still be valid.) Therefore there exists some number K such that $|f(t) - \frac{1}{2}a_0| \leqq K$ for all t, and

$$|F(x') - F(x'')| = \left| \int_{x''}^{x'} [f(t) - \tfrac{1}{2}a_0]\, dt \right| \leqq K|x' - x''| \tag{25}$$

and therefore F satisfies a Lipschitz condition. It follows from Theorem 2 that the Fourier series representing F converges to $F(x)$ for every x.

Let this Fourier series be

$$F(x) = \tfrac{1}{2}A_0 + \sum_{n=1}^{\infty} (A_n \cos nx + B_n \sin nx). \tag{26}$$

This converges for every x. We calculate the coefficients; thus

$$\begin{aligned} A_n &= \frac{1}{\pi} \int_0^{2\pi} F(x) \cos nx\, dx \\ &= \frac{1}{\pi} \left[F(x) \frac{\sin nx}{n} \right]_0^{2\pi} - \frac{1}{n\pi} \int_0^{2\pi} [f(x) - \tfrac{1}{2}a_0] \sin nx\, dx \end{aligned}$$

where we have integrated by parts. From (24) we see that F has period 2π, so the first term vanishes, and the second term is $-b_n/n$. Similarly, $B_n = a_n/n$. Hence

$$F(x) = \tfrac{1}{2}A_0 + \sum_{n=1}^{\infty} \frac{a_n \sin nx - b_n \cos nx}{n}. \tag{27}$$

This must be valid for every x. In particular, for $x = 0$ we obtain, since $F(0) = 0$,

$$\tfrac{1}{2}A_0 = \sum_{n=1}^{\infty} b_n/n. \tag{28}$$

Combining (27) and (28) we obtain (23), which is what we desired to prove.

Theorem 7. *If f and g are piecewise continuous functions of period 2π, with Fourier coefficients a_n, b_n and α_n, β_n, respectively, then*

$$\frac{1}{\pi} \int_0^{2\pi} f(x)g(x)\, dx = \tfrac{1}{2}a_0\alpha_0 + \sum_{n=1}^{\infty} (a_n\alpha_n + b_n\beta_n). \tag{29}$$

Proof: By Theorem 5, the Fourier series

$$\frac{1}{2}\alpha_0 + \sum_{n=1}^{\infty} (\alpha_n \cos nx + \beta_n \sin nx) \tag{30}$$

converges in the mean to g. By Theorem 7 of the preceding section, we can multiply each term of this series by $f(x)/\pi$ and integrate over the interval $(0,2\pi)$; the resulting series will converge to $\frac{1}{\pi}\int_0^2 f(x)g(x)\,dx$. The reader can verify by direct calculation that the series obtained is precisely the right side of (29).

Corollary (Parseval's Theorem). *If f is piecewise continuous, of period 2π, then*

$$\frac{1}{\pi}\int_0^{2\pi} [f(x)]^2\,dx = \frac{1}{2}a_0^2 + \sum_{n=1}^{\infty} (a_n^2 + b_n^2). \tag{31}$$

Proof: Take $f = g$ in the theorem.

Numerical examples will be considered in the Exercises.

Theorem 7 and its Corollary, in certain applications, imply that the average energy contained in a periodic wave is the sum of the average energies of the separate harmonics. This provides further justification for the procedure of reducing the analysis of periodic waves to that of their harmonic constituents.

Parseval's formula (31) is an infinite-dimensional analog of the three-dimensional geometrical theorem asserting that the square of the length of a vector is the sum of the squares of the scalar components of the vector along the coordinate axes. As expressed here, it appears rather more complicated than was the corresponding formula given in Chapter 2 (page 57), the reason for this being that the functions $\sin nx$, $\cos nx$, have not been normalized, i.e., $\|\sin nx\|$ and $\|\cos nx\|$ are not equal to unity.

The vector analogy can be carried even further. We recall that when Newton's second law "force equals mass times acceleration" is written in vector form, in Cartesian coordinates, a simple kinematic problem can be broken up into three different problems, one for each of the directions involved. In particular, what happens in (say) the x direction may be independent of what happens in the y direction. The classical illustration is that given by the monkey hanging by his tail from a tree, who sees that a coconut has been thrown in his direction and tries to avoid being hit by releasing his hold on the branch. He collides with the coconut in mid-air, since his vertical acceleration is the *same* as that of the coconut, even though the coconut has a horizontal velocity. Similarly, the response of a passive linear (electrical) network to each harmonic constituent of a periodic input is quite independent of the other harmonics present.

Theorems similar to Theorem 7 can be proved for other expansions in orthogonal functions, and the remark in the preceding paragraph applies to these expansions as well. However, Fourier expansions have the nice property that all derivatives of functions of the form $A_n \cos nx + B_n \sin nx$ are also of this form. Therefore the response of a system governed by linear differential equations with constant coefficients will always be especially simple to analyze when the input is regarded as a superposition of functions of this form. This is the fundamental reason for the importance of harmonic analysis in electrical engineering.

• EXERCISES

1. Derive (2) by applying Bessel's inequality to the orthonormal sequence $\frac{1}{\sqrt{2\pi}}, \frac{1}{\sqrt{\pi}} \cos x, \frac{1}{\sqrt{\pi}} \sin x, \frac{1}{\sqrt{\pi}} \cos 2x, \frac{1}{\sqrt{\pi}} \sin 2x, \cdots$. Keep in mind that the expansion coefficients relative to this sequence differ by a multiplicative factor from the a_n's and b_n's in (1).

2. Derive (3). Are these relations valid when $n = 0$? Where is the periodicity of f used in your derivation?

3. Show that if $\sum_{n=1}^{\infty} a_n \cos nx$ is the Fourier series representing an integrable function, not necessarily continuous, $\sum_{n=1}^{\infty} (a_n/n) \cos nx$ must be the Fourier series expansion of a continuous function.

4. Derive (5) in detail.

5. Does the convergence of the series $\sum_{k=1}^{\infty} (A_k^2 + B_k^2)$ imply the convergence of $\sum_{k=1}^{\infty} (A_k^2 + B_k^2)^{1/2}$?

6. Show that $f(x) = x^2 \sin (1/x)$ satisfies a Lipschitz condition in the interval $(-1,1)$, but is not piecewise smooth in this interval.

***7.** **(a)** Directly from (10), show that

$$\int_0^{1/n} |D_n(x)|\, dx \leqq \frac{1}{\pi} + \frac{1}{2n\pi}. \tag{32}$$

(b) Using Lemma 6, Section 2.4, show that

$$\int_{1/n}^{\pi} |D_n(x)|\, dx \leqq \tfrac{1}{2}(\ln \pi + \ln n). \tag{33}$$

(c) Combining (32) and (33), and using the fact that D_n is an even function, show that

$$\int_0^{2\pi} |D_n(x)|\, dx \leqq \ln n + \ln \pi + \frac{2}{\pi}\left(1 + \frac{1}{2n}\right). \tag{34}$$

(d) Using the fact that $(\ln n)/(\ln 2) > 1$ when $n > 2$, deduce (11) by showing that

$$\int_0^{2\pi} |D_n(x)|\, dx \leqq \ln n[1 + (\ln \pi)/(\ln 2) + 5/(2\pi \ln 2)]. \tag{35}$$

8. Prove that, if a trigonometric series converges in the mean to an integrable function f, it is the Fourier series expansion of f. (Using the theorems we have developed, this is now quite easy.)

***9.** Show that there exist convergent trigonometric series which are not Fourier series, in the following manner:

(a) Show that $\sum_{n=2}^{\infty} \dfrac{\sin nx}{\ln n}$ converges for every x.

(b) Show that $\sum_{n=2}^{\infty} \dfrac{1}{n \ln n}$ is divergent.

(c) Compare with (28) to deduce that the series in (a) is not a Fourier series.

10. Find three Fourier series listed in Section 3.3 that can be deduced from other series listed there by applying Theorem 6.

11. At first glance, (30) and (37) of Section 3.3 appear to contradict Theorem 6, since the integral of the right side of (30) differs considerably from the right side of (37). Show that one can, in fact, obtain (37) from (30) by integrating term-by-term, if one knows that

$$1 - \frac{1}{3^3} + \frac{1}{5^3} - \frac{1}{7^3} + \cdots \tag{36}$$

converges and has sum $\pi^3/32$.

12. Integrating (37), Section 3.3, twice, and using the fact that the sum of the series

$$1 - \frac{1}{3^5} + \frac{1}{5^5} + \frac{1}{7^5} - \cdots \tag{37}$$

is $5\pi^5/1536$, find (in closed form) the sum of

$$\sum_{n=0}^{\infty} (-1)^n \frac{\cos(2n+1)x}{(2n+1)^5} \tag{38}$$

in the interval $-\pi/2 \leqq x \leqq \pi/2$.

13. If one integrates a Fourier series term by term, from 0 to x, is the resulting series necessarily a Fourier series?

14. What theorem in this section ensures that the sequence 1, $\sin x$, $\cos x$, $\sin 2x$, $\cos 2x$, $\cdots$ is an approximating basis in the linear space of piecewise continuous functions of period 2π?

15. Prove that, if f and g are piecewise continuous functions of period 2π (notation as in Theorem 7),

$$\frac{1}{\pi}\int_0^{2\pi} f(x-t)g(t)\,dt = \tfrac{1}{2}a_0\alpha_0 + \sum_{n=1}^{\infty} (a_n\alpha_n - b_n\beta_n)\cos nx + \sum_{n=1}^{\infty} (a_n\beta_n + \alpha_n b_n)\sin nx. \tag{39}$$

(Compare the discussion in Section 2.6.)

16. Derive (29) from (39).

***17.** Show that (39) takes a much simpler form if the Fourier series are written in complex form, involving the complex exponentials e^{inx}.

18. **(a)** Using Parseval's theorem, show that if all Fourier coefficients of a piecewise continuous function of period 2π are zero, the function must be zero in the interval $(0,2\pi)$ except at a finite number of points.

(b) Hence, deduce that two piecewise continuous functions of period 2π have identical Fourier series if and only if the functions are equal except at a finite number of points in every interval of finite length.

19. Apply Parseval's theorem to (48), Section 3.3, to show that

$$1 + \frac{1}{3^4} + \frac{1}{5^4} + \frac{1}{7^4} + \cdots = \frac{\pi^4}{96}. \tag{40}$$

20. **(a)** Use (25), Section 3.3, to find the Fourier cosine series representing x^2 in $-\pi \leqq x \leqq \pi$.

(b) Apply Parseval's theorem to this series to obtain

$$1 + \frac{1}{2^4} + \frac{1}{3^4} + \cdots = \frac{\pi^4}{90}. \tag{41}$$

21. Verify the assertion of Theorem 7 in the special case where the two functions are (45) and (46), Section 3.3. (Caution: the constant terms are $a_0/2$ and $\alpha_0/2$, not a_0 and α_0.)

22. Suppose we know that a certain trigonometric polynomial of order n provides the best uniform approximation to a given function of period 2π, i.e., a better approximation than all other trigonometric polynomials of the same order.

(a) Does it necessarily provide the best least-squares approximation?

(b) Are its coefficients necessarily the Fourier coefficients of f?

23. Show that if the coefficients of a trigonometric series have the property that $|A_k|$ and $|B_k|$ do not exceed $1/k^2$, the series is a uniformly convergent Fourier series.

***24.** Prove by direct calculation (and making use of Exercise 23) that the Fourier series representing a broken-line function of period 2π is uniformly convergent.

3.9 • DIVERGENT SERIES

As motivation for what follows, consider the series

$$1 - 1 + 1 - 1 + \cdots. \tag{1}$$

The sequence of partial sums of (1) is

$$1, 0, 1, 0, \cdots \tag{2}$$

which does not converge. Therefore, *by definition*, (1) is a divergent series.

On the other hand, one can argue that (1) "should" have sum $\frac{1}{2}$, even though it is divergent. For if we formally set s "equal" to (1), we obtain

$$\begin{aligned} s &= 1 - 1 + 1 - 1 + \cdots \\ &= 1 - (1 - 1 + 1 - 1 + \cdots) = 1 - s, \end{aligned} \tag{3}$$

and solving $s = 1 - s$ we obtain $s = \frac{1}{2}$.

In this section, we introduce two new definitions of "sum." According to both of these definitions, (1) has $\frac{1}{2}$ as its "sum."

Given any series

$$u_1 + u_2 + u_3 + \cdots \tag{4}$$

with partial sums

$$s_n = u_1 + u_2 + \cdots + u_n, \tag{5}$$

the nth *arithmetic mean* of these partial sums is defined by

$$\sigma_n = \frac{s_1 + s_2 + \cdots + s_n}{n}, \tag{6}$$

which is simply the average of the first n partial sums of (4).

If the sequence of arithmetic means $\sigma_1, \sigma_2, \sigma_3, \cdots$ converges to σ, we say that σ is the *Cesaro sum* of the series (4).

For the series (1), the sequence of arithmetic means is

$$1, \tfrac{1}{2}, \tfrac{2}{3}, \tfrac{2}{4}, \tfrac{3}{5}, \tfrac{3}{6}, \tfrac{4}{7}, \tfrac{4}{8}, \cdots, \tag{7}$$

which tends to $\frac{1}{2}$. Therefore the Cesaro sum of (1) is $\frac{1}{2}$.

Let us consider a less trivial example. Consider the series of functions

$$\tfrac{1}{2} + \cos x + \cos 2x + \cos 3x + \cdots. \tag{8}$$

This series diverges for every value of x. According to (6), Section 3.3, the $n + 1$st partial sum is

$$\frac{1}{2} + \sum_{m=1}^{n} \cos mx = \frac{\sin (2n+1)(x/2)}{2 \sin (x/2)}. \tag{9}$$

(See Exercise 3, Section 2.2.) Therefore its nth arithmetic mean is

$$\frac{1}{n}\sum_{k=0}^{n-1} \frac{\sin (2k+1)(x/2)}{2 \sin (x/2)} = \frac{1}{2n \sin (x/2)} \sum_{k=0}^{n-1} \sin (k + \tfrac{1}{2})x, \tag{10}$$

which by (20), Section 1.4 (taking $\theta = x/2$), can be written in closed form

$$\sigma_n(x) = \frac{\sin^2 n(x/2)}{2n \sin^2 (x/2)}. \tag{11}$$

If x is in the interval $0 < x < 2\pi$, the numerator of this expression is in the range (0,1) and the denominator increases without bound as n increases. Therefore $\sigma_n(x)$ tends to zero. It follows that the Cesaro sum of (8) is zero for every x in the interval $0 < x < 2\pi$. Observe, however, that when $x = 0$ or $x = 2\pi$, the nth arithmetic mean has the value [obtained from (11) by the limit convention, or obtained directly from (8)] $\sigma_n = n^2/2n = n/2$ (see

Exercise 10). Therefore the Cesaro sum of (8) does not exist when x is an integral multiple of 2π.

Although (8) is not a Fourier series, it is of fundamental importance in the theory of Fourier series, as we will see in the proof of the following theorem.

Theorem 1. *The Cesaro sum of the Fourier expansion of a continuous function of period 2π exists and is equal to the function for every x. Moreover, the sequence of arithmetic means converges uniformly to the function.*

The proof will be part of the following discussion.

If we regard $(A_n \cos nx + B_n \sin nx)$ as a single term, the nth partial sum of the Fourier series is obtained by convoluting the function with the Dirichlet kernel D_{n-1}. Therefore, the nth arithmetic mean is obtained by convoluting the function with

$$\frac{1}{n}(D_0 + D_1 + D_2 + \cdots + D_{n-1}). \tag{12}$$

When written out explicitly, (12) is

$$\frac{1}{n}\sum_{k=0}^{n-1} \frac{\sin (k + \frac{1}{2})x}{2\pi \sin (x/2)}. \tag{13}$$

Except for the factor $1/\pi$, this is identical to (10). Therefore the nth arithmetic mean is obtained by convoluting the function with

$$F_n(x) = \frac{\sin^2 n\,(x/2)}{2\pi n \sin^2 (x/2)}, \tag{14}$$

which is called the *Fejer kernel*.

It is easy to see that

$$\int_0^{2\pi} F_n(x)\, dx = 1 \tag{15}$$

for every n. (See Exercise 10.) Since $F_n(x)$ is never negative, we also have

$$\int_0^{2\pi} |F_n(x)|\, dx = 1. \tag{16}$$

The Dirichlet kernel *also* has property (15), but does *not* have property (16).

Let M be a constant chosen so that $|f(x)| \leq M$ for every x. Then

$$(17)\quad |\sigma_n(x)| = \left| \int_0^{2\pi} f(t)F_n(x-t)\,dt \right| \leqq \int_0^{2\pi} |f(t)||F_n(x-t)|\,dt$$
$$\leqq M \int_0^{2\pi} |F_n(x-t)|\,dt = M \int_0^{2\pi} |F_n(t)|\,dt$$

where the last step is justified because F_n has period 2π. Therefore by (16) we conclude

$$(18)\qquad |\sigma_n(x)| \leqq M.$$

This is interesting because it shows that the arithmetic means do not ever exceed the maximum value of f. This is in sharp contrast with the behavior of the partial sums, which may overshoot the maximum value of the function if the function is discontinuous. (The Gibbs phenomenon does not occur with arithmetic means.)

Now suppose that g is a function satisfying the hypotheses of Theorem 1. Let h be a trigonometric polynomial so chosen that $|h(x) - g(x)| < \epsilon$ for every x; such a polynomial exists for any positive ϵ by Theorem 12, Section 3.7. Let $f(x) = h(x) - g(x)$. Since $|f(x)| < \epsilon$, it follows from (18) that every arithmetic mean of the Fourier expansion of f is in absolute value less than ϵ.

Since $g = h + f$, the Fourier expansion of g is the sum of those for h and f, and the arithmetic means of the Fourier expansion of g are clearly the sums of the corresponding arithmetic means for h and f. It follows from the preceding paragraph that every arithmetic mean of the Fourier expansion of g can differ by at most ϵ from the corresponding arithmetic mean for h.

Since h is a trigonometric polynomial, it is its own Fourier series, and the Cesaro sums of this series converge uniformly to h (Exercise 13). Therefore, for some n_0, the nth arithmetic mean of the expansion of h is within ϵ of $h(x)$ for every x, provided that $n_0 \leqq n$. The corresponding arithmetic mean for g is therefore within 2ϵ of $h(x)$ for every x, and therefore (by the way h was chosen) within 3ϵ of $g(x)$ for every x. Since ϵ is arbitrary, this shows that the arithmetic means of the Fourier expansion of g converge uniformly to g, and completes the proof of Theorem 1.

More direct proofs of Theorem 1 can be constructed, but this one is instructive because it shows clearly the connection between Theorem 12, Section 3.7, and the convergence of the arithmetic means of a Fourier series. (See Exercise 13.)

Theorem 2 (Fejer's Theorem). *The Fourier series of an integrable function f of period 2π is Cesaro summable to $[f(x+) + f(x-)]/2$ for* *every value of x for which this expression has a meaning. In particular, the series has Cesaro sum $f(x)$ at every point where f is continuous.*

Proof: Let σ_n denote the nth arithmetic mean of the Fourier series of f. Exactly as in Section 3.6 (using the Fejer kernel instead of the Dirichlet kernel) one obtains

$$\sigma_n(x) - f(x) = \int_0^\pi [f(x-t) - f(x-)] \frac{\sin^2 n(t/2)}{2\pi n \sin^2 (t/2)}\, dt + \int_0^\pi [f(x+t) - f(x+)] \frac{\sin^2 n(t/2)}{2\pi n \sin^2 (t/2)}\, dt \tag{19}$$

where we assume (as in that section) that f has been normalized so that $f(x) = [f(x+) + f(x-)]/2$ at the single point x in question (which does not change the Fourier series in any way). It suffices to prove that one of these integrals tends to zero; the proof for the other is identical. This will prove the theorem.

The second of these integrals can be written as the sum of two integrals,

$$I_1 = \int_0^\delta [f(x+t) - f(x+)] \frac{\sin^2 n(t/2)}{2\pi n \sin^2 (t/2)}\, dt \tag{20}$$

and

$$I_2 = \int_\delta^\pi [f(x+t) - f(x+)] \frac{\sin^2 n(t/2)}{2\pi n \sin^2 (t/2)}\, dt. \tag{21}$$

By hypothesis, $f(x+)$ exists, and therefore we can choose δ sufficiently small that $|f(x+t) - f(x+)| < \epsilon$ for t in the interval $(0, \delta)$. Therefore, for this choice of $\delta > 0$,

$$\begin{aligned} |I_1| &= \left| \int_0^\delta [f(x+t) - f(x+)] \frac{\sin^2 n(t/2)}{2\pi n \sin^2 (t/2)}\, dt \right| \\ &\leqq \int_0^\delta |f(x+t) - f(x+)| \left| \frac{\sin^2 n(t/2)}{2\pi n \sin^2 (t/2)} \right| dt \\ &\leqq \epsilon \int_0^\delta \left| \frac{\sin^2 n(t/2)}{2\pi n \sin^2 (t/2)} \right| dt \leqq \epsilon \int_0^{2\pi} \left| \frac{\sin^2 n(t/2)}{2\pi n \sin^2 (t/2)} \right| dt \end{aligned} \tag{22}$$

where the last inequality is justified because the integrand is never negative. Therefore by (16), $|I_1| < \epsilon$.

On the other hand, we also have

$$|I_2| = \int_\delta^\pi [f(x+t) - f(x+)] \left| \frac{\sin^2 n(t/2)}{2\pi n \sin^2 (t/2)} \right| dt. \tag{23}$$

The numerator of the Fejer kernel never exceeds unity in absolute value, and the denominator is never less than $2\pi n \sin^2 (\delta/2)$ in the interval (δ,π), and hence

$$|I_2| \leqq \frac{1}{2\pi n \sin^2 (\delta/2)} \int_\delta^\pi |f(x+t) - f(x+)|\, dt \tag{24}$$

and therefore, for sufficiently large n_0, $|I_2| < \epsilon$ whenever $n > n_0$. It follows that $|\sigma_n - f(x)| < 2\epsilon$ whenever $n > n_0$, and since ϵ is arbitrary the proof is complete.

Theorem 3. *Cesaro sums have the following properties: If $\sum a_n$ has Cesaro sum s and $\sum b_n$ has Cesaro sum t, then*

$$\sum (a_n + b_n) \text{ has Cesaro sum } s + t, \tag{25}$$

$$\sum ka_n \text{ has Cesaro sum } ks, \tag{26}$$

$$\sum_{n=2}^{\infty} a_n \text{ has Cesaro sum } s - a_1. \tag{27}$$

Moreover, every series $\sum c_n$ which is convergent and has sum c is also Cesaro summable to the same sum c.

Proof: We leave the proofs of (25), (26), and (27) to the exercises. The last assertion is somewhat harder to prove, so we give it in detail here.

Let s_n denote the nth partial sum of $\sum c_n$ and let σ_n denote its nth arithmetic mean. Assuming that $s_n \to c$, we will prove that $\sigma_n \to c$ also. Let ϵ be an arbitrary positive number. For some n_0 we will have $|s_n - c| < \epsilon$ whenever $n > n_0$. Therefore, for any $n > n_0$, we have

$$\begin{aligned} &\sigma_n - c \\ &= \frac{s_1 + s_2 + \cdots + s_n}{n} - c \\ &= \frac{(s_1 - c) + (s_2 - c) + \cdots + (s_n - c)}{n} \\ &= \frac{(s_1 - c) + \cdots + (s_{n_0} - c)}{n} + \frac{(s_{n_0+1} - c) + \cdots + (s_n - c)}{n}. \end{aligned} \tag{28}$$

The second of these fractions is less than ϵ in absolute value, since each term in its numerator has this property and there are fewer than n such terms. Keeping n_0 fixed, we now choose $N > n_0$ sufficiently large that the first fraction will be less than ϵ in absolute value whenever $n > N$. It then follows that $|\sigma_n - c| < 2\epsilon$ for $n > N$, and since ϵ is arbitrary, it follows that $\sigma_n \to c$, which is what we desired to prove.

Theorem 4. *The Fourier series of a piecewise continuous function f of period 2π may or may not converge, but if it does converge, it must converge to $[f(x+) + f(x-)]/2$. In particular, if the Fourier series of a continuous function f of period 2π converges at a point, it must converge to $f(x)$ at that point.*

Proof: If f is piecewise continuous, then $f(x+)$ and $f(x-)$ are defined for every x. By Theorem 2, its Fourier series must be Cesaro summable to $[f(x+) + f(x-)]/2$ for every x. By Theorem 3, if it is convergent for some x, its sum must be the same as its Cesaro sum, and therefore if it is convergent it must converge to its Cesaro sum.

Theorem 4 may surprise the reader, for he probably assumed its validity earlier. Actually, however, no theorem proved in earlier sections shows that the Fourier series of a continuous function of period 2π might not converge to some value other than $f(x)$ for some values of x. (Similar theorems, proved earlier, demanded that f be piecewise smooth or satisfy a Lipschitz condition.)

The method of summation we have been discussing is called *Cesaro's method* or the *method of the first arithmetic mean.* The reader may suspect that it can be carried further, and indeed it can. If the arithmetic means do not converge, one might try taking the averages of the first 2, 3, $\cdots$, n arithmetic means, and seeing if this sequence converges! We will not discuss these more general types of Cesaro summability, but will instead turn to another method, known as *Abel's method* or the *method of convergence factors.*

Let us suppose we are given a series

$$u_0 + u_1 + u_2 + \cdots \tag{29}$$

whose terms may be numbers or functions. We form a new series

$$u_0 + u_1r + u_2r^2 + u_3r^3 + \cdots. \tag{30}$$

If it should happen that (30) converges when r is in the interval $0 \leqq r < 1$, and tends to a finite limit when $r \to 1$, then we call this limit the *Abel sum* of the series (29).

Theorem 5. *Abel sums also have the properties listed in Theorem 3.*

In other words, Theorem 3 is valid if "Cesaro sum" is replaced throughout by "Abel sum." We omit the proof.

As a simple example, let us sum (1) by the method of convergence factors. Multiplying the $n + 1$st term by r^n, we obtain the series

$$1 - r + r^2 - r^3 + \cdots \tag{31}$$

which converges in the interval $(-1,1)$ to $1/(1 + r)$ [indeed, (31) is the series obtained by formally dividing 1 by $1 + r$]. Although the series does not converge at $r = 1$, the limiting value of $1/(1 + r)$ as $r \to 1$ is $\frac{1}{2}$. Therefore the Abel sum of (31) is $\frac{1}{2}$.

As a less trivial example, let us find the Abel sum of (8). As in the preceding example, we form the series containing the convergence factors r^n, which in this case gives

$$\tfrac{1}{2} + \sum_{n=1}^{\infty} r^n \cos nx. \tag{32}$$

To write this in closed form, we observe (compare Exercises 2 and 3, Section 2.2) that (32) is the real part of the complex series

$$\tfrac{1}{2} + z + z^2 + z^3 + \cdots \qquad (z = re^{ix}) \tag{33}$$

which converges for $|z| < 1$ and has sum

$$\frac{1}{2} + \frac{z}{1 - z} = \frac{1 + z}{2(1 - z)}. \tag{34}$$

By a simple algebraic calculation, the real part of (34) is

$$\frac{1 - r^2}{2(1 - 2r \cos x + r^2)} \tag{35}$$

and therefore, in the interval $0 \leqq r < 1$, (32) converges,

$$\frac{1}{2} + \sum_{n=1}^{\infty} r^n \cos nx = \frac{1 - r^2}{2(1 - 2r \cos x + r^2)} \qquad (0 \leqq r < 1). \tag{36}$$

As $r \to 1$, this tends to zero, provided that x is in the interior of the interval $(0,2\pi)$. Therefore (32) is Abel summable to zero in the interior of this interval. At the endpoints $x = 0$ and $x = 2\pi$, the series does not have an Abel sum.

Now let us discuss the Abel summability of Fourier series. This is interesting to engineers and physicists because it has a direct physical interpretation, which is the main reason for discussing it here.

Let f be an integrable function of period 2π, with Fourier series

$$A_0/2 + \sum_{n=1}^{\infty} (A_n \cos nx + B_n \sin nx). \tag{37}$$

In applying Abel's method, we form the series

$$f_r(x) = A_0/2 + \sum_{n=1}^{\infty} r^n(A_n \cos nx + B_n \sin nx) \qquad (0 \leqq r < 1). \tag{38}$$

For each *fixed* positive $r < 1$ this series is uniformly convergent and therefore defines a continuous function $f_r(x)$, by the Weierstrass M-test (Exercise 21). If $f_r(x)$ tends to a limit as $r \to 1$, the series is Abel summable.

We recall that the term $(A_n \cos nx + B_n \sin nx)$ in (37) is obtained by convoluting f with ϕ_n (Section 2.6). Therefore, the term $r^n(A_n \cos nx + B_n \sin nx)$ is obtained by convoluting f with $r^n\phi_n$ (where r is considered a constant in evaluating the convolution product) and therefore the $n + 1$st partial sum of (38) is obtained by convoluting f with $\phi_0 + r\phi_1 + \cdots + r^n\phi_n$. Writing this out explicitly, the $n + 1$st partial sum of (38) is the convolution product of f with

$$(39) \qquad \frac{1}{2\pi} + \sum_{k=1}^{n} r^n \frac{\cos nx}{\pi} \qquad (0 \leqq r < 1).$$

By the Weierstrass M-test (using the fact that $\sum r^n$ converges for each r in this interval), (39) uniformly converges as $n \to \infty$, and indeed it converges to the function

$$(40) \qquad \delta_r(x) = \frac{1 - r^2}{2\pi(1 - 2r\cos x + r^2)} \qquad (0 \leqq r < 1)$$

as we can see from (36). We call this function the *Poisson kernel.*

Because of the uniform convergence, we can interchange the order of summation and integration (Theorem 7, Section 3.7) in the integral defining the convolution product, and therefore (38) can be obtained by convoluting f with the function $\delta_r(x)$,

$$(41) \qquad f_r = f * \delta_r \qquad (0 \leqq r < 1).$$

Writing this out quite explicitly,

$$(42) \quad f_r(x) = \int_0^{2\pi} f(t) \frac{1 - r^2}{2\pi[1 - 2r\cos(x - t) + r^2]}\, dt \qquad (0 \leqq r < 1).$$

This is known as the *Poisson integral formula.*

It is interesting to notice that the Poisson kernel shares many of the properties of the Fejer kernel. For each fixed r in the range $0 \leqq r < 1$, it is positive for every x, is periodic (in x) with period 2π, and

$$(43) \qquad \int_0^{2\pi} |\delta_r(x)|\, dx = \int_0^{2\pi} \delta_r(x)\, dx = 1$$

as we see readily if we take $f(t)$ identically equal to unity in (42) and observe that, in this case, A_n and B_n in (38) are zero except when $n = 0$ ($A_0/2 = 1$).

Theorem 6. *The Fourier series of an integrable function of period 2π is Abel summable to $[f(x+) + f(x-)]/2$ for every value of x for which this expression has a meaning.*

The proof is left as an exercise, which is extremely easy since it is essentially identical to the proof of Theorem 2, replacing Fejer's kernel by Poisson's kernel. (See Exercise 24.)

Theorem 7. *The Fourier series of a continuous function f of period 2π is uniformly summable by Abel's method.*

What this means is that the Abel sum of the series is $f(x)$ for every x, and that $f_r(x)$ tends to $f(x)$ *uniformly* as $r \to 1$. That is, given any $\epsilon > 0$, there is a number r_0 such that

$$(44) \qquad |f(x) - f_r(x)| < \epsilon \text{ whenever } r_0 < r < 1.$$

Proof: As noted in Section 2.6, convolution is a commutative operation, so we can write (41) in either of the two forms

$$(45) \qquad f_r(x) = \int_0^{2\pi} f(t)\,\delta_r(x - t)\,dt \quad \text{or} \quad \int_0^{2\pi} f(x - t)\,\delta_r(t)\,dt.$$

In (42) we chose the first of these forms; here we will use the second. Noting that the functions are periodic, of period 2π (for each *fixed* r), we can integrate over $(-\pi,\pi)$ instead of $(0,2\pi)$, and by virtue of (43) we have

$$(46) \qquad |f_r(x) - f(x)| = \left|\int_{-\pi}^{\pi} [f(x - t) - f(x)]\,\delta_r(t)\,dt\right|$$
$$= |I_1 + I_2| \leqq |I_1| + |I_2|$$

where I_1 is the integral over $(-\pi,0)$ and I_2 is the integral over $(0,\pi)$.

Since f is continuous and periodic, we can choose a sufficiently small $h > 0$ so that $|f(x - t) - f(x)| < \epsilon$ for every x, provided that $|t| < h$. Then

$$(47) \quad |I_2| = \left|\int_0^{\pi} [f(x - t) - f(x)]\,\delta_r(t)\,dt\right|$$
$$\leqq \int_0^h |f(x - t) - f(x)||\delta_r(t)|\,dt + \int_h^{\pi} |f(x - t) - f(x)||\delta_r(t)|\,dt$$
$$\leqq \epsilon \int_0^h |\delta_r(t)|\,dt + \frac{1 - r^2}{8\pi r \sin^2(h/2)} \int_h^{\pi} |f(x - t) - f(x)|\,dt$$

where we have made use of a simple inequality given in Exercise 22. Because of (43), the first term is less than ϵ and for some r_0 close enough to 1 the second term will be less than ϵ independently of x (since f is bounded) whenever $r_0 < r < 1$. Therefore $|I_2| < 2\epsilon$. Similarly, $|I_1| < 2\epsilon$, and it follows that $|f(x,r) - f(x)| < 4\epsilon$ for $r_0 < r < 1$. Since $\epsilon > 0$ is arbitrary, we have derived (44) and hence the theorem is proved.

Readers interested in applications may be surprised by the following remarks. It will be noted (Theorem 5) that if a series is convergent, and has sum s, then it is also Abel summable and its Abel sum is also s. If we could imagine a world in which these sums were different, it would actually be more distressing to physicists to have the Abel sum of a Fourier series (satisfying the hypotheses of Theorem 7) to fail to equal $f(x)$ than it would be to have its ordinary sum fail to be $f(x)$. For example, if the former situation occurred, this would imply that the steady-state temperature within a circular plate does not tend to the temperature on its boundary, as we approach the boundary (see Section 5.5). If the latter failure occurred, however, only an inconvenience in numerical calculations would result, in those instances in which it is easier to calculate the ordinary sum than the Abel sum. It is of some physical significance that Theorem 7 is valid, but it is no great handicap to physicists that a continuous function may have a divergent Fourier series.

Indeed, it is safe to say that no physical significance can be attached to the fact that some Fourier series converge and some do not, which suggests that there should be an altogether different approach to Fourier series than the classical one. This idea will be explored in the next section.

• EXERCISES

1. Find the first ten arithmetic means of

$$1 + 0 - 1 + 0 + 1 + 0 - 1 + 0 + 1 + \cdots.$$

2. Find the Cesaro sum of the series in Exercise 1.

3. Using the technique illustrated in (3), find the "sum" of the series in Exercise 1.

4. Repeat Exercises 1, 2, and 3 for the series

$$1 + 0 + 0 - 1 + 0 + 0 + 1 + 0 + 0 - 1 + \cdots.$$

Do the zeros have a more profound effect here?

5. Repeat Exercises 1, 2, and 3 for the series

$$1 + 0 - 1 + 0 + 0 + 1 + 0 - 1 + 0 + 0 + 1 + \cdots$$

(repeating every five terms). Do the zeros have an effect here?

6. Under what conditions does a series of the form

$$\mathrm{a} + \mathrm{c} + \mathrm{a} - \mathrm{c} + \mathrm{a} + \mathrm{c} + \mathrm{a} - \mathrm{c} + \cdots$$

have a Cesaro sum?

7. The nth arithmetic mean of a sequence $S_1, S_2, S_3, \cdots$ is $\sigma_n = \frac{1}{n} \sum_{k=1}^{n} S_k$. If $\sigma_n \to \sigma$, call σ the Cesaro limit of the sequence, and write (C)lim $S_n = \sigma$.

(a) Is it necessarily true that

$$(\mathrm{C})\lim (A_n + B_n) = (\mathrm{C})\lim A_n + (\mathrm{C})\lim B_n?$$

(b) Is it necessarily true that

$$(\mathrm{C})\lim (A_n B_n) = [(\mathrm{C})\lim A_n][(\mathrm{C})\lim B_n]?$$

8. Let $S_{n+1} = (14 - S_n)/(1 + S_n)$.

(a) Letting $S_1 = 3$, find S_2, S_3, and S_4.

(b) What is the Cesaro limit of S_n in part (a)?

(c) Does this limit satisfy the equation $S = (14 - S)/(1 + S)$?

9. (a) Show that the nth arithmetic mean of $\frac{1}{2} + 1 + 1 + 1 + \cdots$ is $n/2$.

(b) Obtain the same result by finding the limit of (11) as $x \to 0$.

10. (a) Show that every arithmetic mean of the series

$$1 + 0 + 0 + 0 + 0 + \cdots$$

is equal to 1.

(b) Show that if $f(x) = 1$ identically, then the convolution product $f * F_n$ is the integral $\int_0^{2\pi} F_n(x)\,dx$.

(c) Hence, deduce (15) by considering the trivial Fourier expansion of the function identically equal to unity.

11. (a) Show that the arithmetic means of a series of the form

$$a_1 + a_2 + \cdots + a_n + 0 + 0 + 0 + \cdots$$

(all terms beyond a_n are zero) converge to

$$s = a_1 + a_2 + \cdots + a_n.$$

(b) Is any arithmetic mean exactly equal to s?

12. Show that the arithmetic means of the Fourier series expansion of a trigonometric polynomial $p(x)$ converge to $p(x)$ for every x. (Compare Exercise 11.)

13. Imitate the proof of Theorem 1, replacing the Fejer kernel by the Dirichlet kernel, in an attempt to prove that the Fourier series expansion of a continuous function converges uniformly to the function. What goes wrong?

14. Prove (25), (26), and (27).

15. Find the values of

$$\int_0^{2\pi} \sin t \frac{\sin^2 \dfrac{n(x-t)}{2}}{2n\pi \sin^2\left(\dfrac{x-t}{2}\right)}\, dt \qquad (n = 1, 2, 3, \cdots).$$

16. Find the values of

$$\int_0^{2\pi} \cos 2(x-t) \frac{\sin^2 (nt/2)}{2n\pi \sin^2 (t/2)}\, dt \qquad (n = 1, 2, 3, \cdots)$$

without actually performing any integrations.

17. Is the function defined by (11) an idempotent? (In other words, when convoluted with itself, is the resulting function the same?)

18. Show that the sequence σ_n of arithmetic means of the partial sums of $1 - 2 + 3 - 4 + \cdots$ is not convergent, since $\sigma_{n+1} = (n+2)/2n$ for even n and 0 for odd n, but that a repetition of the averaging process gives the limit $\frac{1}{4}$. (This is called summation by the second arithmetic means.)

19. Show that any summability method satisfying (25), (26), and (27) will give $\frac{1}{4}$ as the sum of $1 - 2 + 3 - 4 + 5 - \cdots$, if it gives any sum at all. [Hint: You will need to use all three of these properties. Note that $(1 - 1 + 1 - 1 + \cdots) + (1 - 2 + 3 - 4 + \cdots) = (2 - 3 + 4 - 5 + \cdots)$.]

20. Is (36) valid for any negative values of r?

21. Show that (38) is uniformly convergent for any fixed r in the interval $0 \leqq r < 1$. [Hint:

$$|r^n(A_n \cos nx + B_n \sin nx)| \leqq 2Mr^n.$$

How do you know such an M exists independently of n?]

22. Show that, in the interval $0 < h \leqq x \leqq \pi$,

$$\delta_r(x) < \frac{1 - r^2}{8\pi r \sin^2(h/2)} \qquad (0 \leqq r < 1).$$

23. Show that $\delta_r(x)$ tends to zero uniformly as $r \to 1$ in any closed interval not containing an integer multiple of 2π.

***24.** Prove Theorem 6, imitating the proof of Fejer's theorem. (The only modification needed is given in Exercise 22.)

***25.** By imitating the proof of Theorem 7, give a direct proof of Theorem 1.

26. Show that the Abel sum of $1 - 2 + 3 - 4 + \cdots$ is $\frac{1}{4}$. (This example shows that a series may be Abel summable but not Cesaro summable.)

27. Find the Abel sum of $\sin x + 2 \sin 2x + 3 \sin 3x + \cdots$.

***28.** Prove that any series that is Cesaro summable is Abel summable to the same "sum."

3.10 • GENERALIZED FUNCTIONS

This section may be omitted without loss in continuity.

With $\delta_r(x)$ denoting the Poisson kernel, we have seen that

$$f(x) = \lim_{r \to 1} \int_{-\pi}^{\pi} f(t)\, \delta_r(x - t)\, dt \tag{1}$$

and if $F_n(x)$ denotes the Fejer kernel we have

$$f(x) = \lim_{n \to \infty} \int_{-\pi}^{\pi} f(t) F_n(x - t)\, dt \tag{2}$$

whenever f is continuous of period 2π.

These are examples of *singular integrals*. We will not give the general definition, but wish merely to point out that "singular integrals" are integrals in which it would not make sense to interchange the order of the operations of integrating and passing to the limit.

If it *were* possible to pass to the limit under the integral sign in (1), this would imply the existence of a function δ with the property

$$f(x) = \int_{-\pi}^{\pi} f(t) \delta(x - t)\, dt \tag{3}$$

and this function would be an "identity" element relative to the convolution product: we would have $\delta * f = f * \delta = f$ for functions f of period 2π. No such function exists.

It happens that, for some years, engineers and physicists have found it convenient to introduce fictitious functions having "ideal" properties that no actual functions can possibly possess. These "functions" are variously called "singularity functions" or "generalized functions."

As a simple example, let $\delta(x) = 1/2h$ when $|x| < h$ and let $\delta(x) = 0$ elsewhere (Figure 3.18). The area under the graph is unity. As $h \to 0$, the magnitude of $\delta(x)$ increases at $x = 0$ and the width of the interval in which this function is nonzero shrinks. It is easy to see that if f is a continuous function,

$$\int_{-\pi}^{\pi} f(x)\delta(x)\,dx \tag{4}$$

will tend to $f(0)$ as $h \to 0$. Engineers like to imagine that there is a "function" δ which is infinite at $x = 0$ and zero elsewhere, but which has the property that (4) is equal to $f(0)$ for any function f. This function represents, for instance, a "unit impulse" at time $x = 0$ (if x denotes time and $\delta(x)$ denotes force), or a "unit mass" at $x = 0$ (if x denotes position and $\delta(x)$ denotes mass density in units of mass per unit length). When integrated with another function, as in (4),

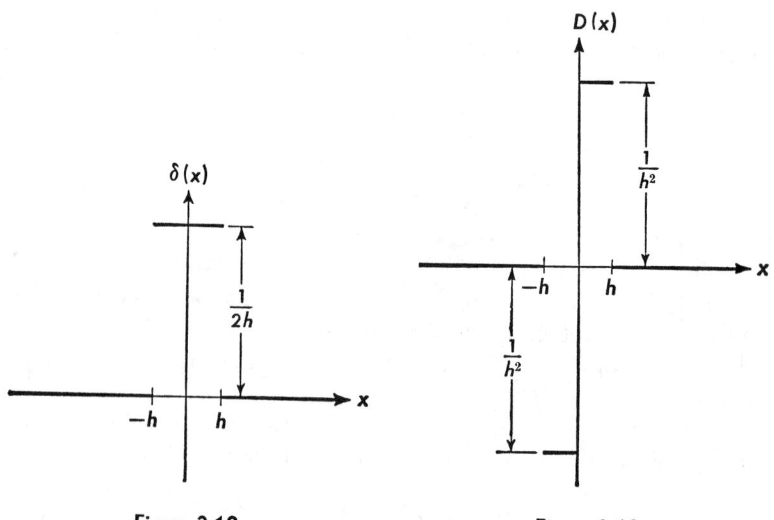

Figure 3.18

Figure 3.19

it "punches out" the value of that function at $x = 0$. Physicists also use this "function," and call it the Dirac delta function.

More generally, such a function would have property (3), since $\delta(x - t)$ would be zero except at $x = t$ and would punch out the value of f at that point.

Another example of such a fictitious function is provided by Figure 3.19. Here, $D(x) = -1/h^2$ in the interval $-h \leqq x < 0$ and $D(x) = 1/h^2$ when $0 < x \leqq h$, and is zero elsewhere. We now let h tend to zero and imagine that the limit makes sense. We would then have, for a function f,

$$\int_a^b f(x)D(x)\,dx = f'(0) \tag{5}$$

whenever (a,b) contains the origin $x = 0$. This function "punches out" the derivative of f at the origin. (We suggest that the reader give a heuristic reason for why this should be so.)

If $D(x)$ is imagined to represent charge density, in units of charge per unit length, this would represent a spread of positive charge to the right of the origin and a spread of negative charge to the left. As h tends to zero, the magnitude of both charge densities increases, but the effective distance between them decreases, so that the dipole moment remains constant. In the limit, we imagine there is no net charge, but that we have a "point charge dipole" at $x = 0$. This fictitious function is sometimes called the *doublet function.*

It is interesting to calculate formally the Fourier series expansions of the period 2π extensions of these functions. For the Dirac delta function δ, we would have

$$A_n = \frac{1}{\pi}\int_{-\pi}^{\pi} \delta(x)\cos nx\,dx = \frac{1}{\pi}\cos n0 = \frac{1}{\pi}$$

and

$$B_n = \frac{1}{\pi}\int_{-\pi}^{\pi} \delta(x)\sin nx\,dx = \frac{1}{\pi}\sin n0 = 0,$$

so the "Fourier expansion" is

$$\delta(x) = \frac{1}{2\pi} + \frac{1}{\pi}\cos x + \frac{1}{\pi}\cos 2x + \frac{1}{\pi}\cos 3x + \cdots \tag{6}$$

and a similar calculation for the doublet function yields

$$D(x) = \frac{1}{\pi}\sin x + \frac{2}{\pi}\sin 2x + \frac{3}{\pi}\sin 3x + \cdots. \tag{7}$$

We immediately recognize that the partial sums of (6) are the Dirichlet kernels, first introduced in Section 2.6. The terms in (6) are the functions ϕ_n introduced in 2.6, and we recall that $f * \phi_n$ is the projection of f into the two-dimensional subspace spanned by $\cos nx$ and $\sin nx$ $(n = 1, 2, 3, \cdots)$. If (6) made sense, we would regard it as expressing the identity projection as a sum of the projections ϕ_n.

There are several ways in which generalized functions can be removed from the sphere of fiction and given a sound mathematical basis. Insofar as Fourier series are concerned, the simplest of these is that which we will now introduce.

We *define* a generalized function of period 2π to be a trigonometric series

$$f(x) = A_0/2 + \sum_{n=1}^{\infty} (A_n \cos nx + B_n \sin nx) \tag{8}$$

whose coefficients have the property

$$A_n/n^k \to 0 \quad \text{and} \quad B_n/n^k \to 0 \quad \text{as} \quad n \to \infty \tag{9}$$

for some positive integer k. The equality in (8) means only that the left side is an abbreviation for the right side, and is not intended to imply that there is a numerical value $f(x)$ for any x. (The series need not converge.)

According to this definition, $\delta(x)$ as defined in (6) is a generalized function, since (9) is satisfied for $k = 1$. Similarly, $D(x)$ is a generalized function, since (9) is satisfied by the coefficients in (7) when $k = 2$. The value of k is not unique; if the coefficients satisfy (9) for some k, they will also satisfy (9) for any larger value of k.

Operations with generalized functions are defined formally, without any regard for matters of convergence. Linear combinations of generalized functions are obtained purely formally, and the convolution product of two generalized functions is obtained by convoluting them term-by-term. Since we know from Section 2.4 that $(1/\pi)(\cos nx) * (A_n \cos nx + B_n \sin nx) = A_n \cos nx + B_n \sin nx$ $(n = 1, 2, 3, \cdots)$ and $(1/2\pi) * A_0/2 = A_0/2$, it follows that

$$\delta * f = f * \delta = f$$

for any generalized function f. Therefore, (3) is justified with this definition of generalized function.

The derivative of any generalized function is obtained through term-by-term differentiation. Thus if f is a generalized function (8), its derivative is

$$f'(x) = \sum_{n=1}^{\infty} (-nA_n \sin nx + nB_n \cos nx). \tag{10}$$

If the coefficients in (8) satisfy (9), the coefficients in (10) will satisfy (9) for a larger value of k, so we have proved:

Theorem 1. *Every generalized function of period 2π can be differentiated, and its derivative is also a generalized function of period 2π.*

We now prove

Theorem 2. *Associated with every generalized function of period 2π there is an infinite series*

$$f_r(x) = A_0/2 + \sum_{n=1}^{\infty} r^n(A_n \cos nx + B_n \sin nx) \tag{11}$$

which is uniformly convergent for each fixed r in the range $0 \leqq r < 1$ and defines an ordinary function f_r of period 2π which is continuous and has continuous derivatives of all orders.

Proof: By virtue of (9), the series

$$\sum_{n=1}^{\infty} A_n r^n \quad \text{and} \quad \sum_{n=1}^{\infty} B_n r^n \tag{12}$$

both converge when $0 \leqq r < 1$, since the series $\sum n^k r^n$ converges, so by the Weierstrass M-test, (11) is uniformly convergent. It follows that f_r is a continuous function of period 2π. Any uniformly convergent series of continuously differentiable functions can be differentiated term-by-term, providing the resulting series converges uniformly (see Exercise 1). In this case, the resulting series is $(f')_r$ and since f' is a generalized function (by Theorem 1) the above argument applies to $(f')_r$ as well as to f_r, so it follows that term-by-term differentiation of (11) is legitimate and $(f_r)' = (f')_r$ is a continuous function. Repeating this argument n times, we see that $f_r(x)$ has a continuous nth derivative, and that $(f_r)^{(n)} = (f^{(n)})_r$.

This argument also shows that there is no ambiguity in writing $f_r^{(n)}$, i.e., the function associated with $f^{(n)}$ is the nth derivative of the function associated with f.

The significance of Theorem 2 derives in part from the following *fundamental principle of graphics: the salient features of any generalized function of period 2π are represented by the graph of the associated function f_r if r is chosen sufficiently close to* 1.

For example, if δ is the Dirac delta function (6), the associated function δ_r is the Poisson kernel, which has the typical appearance of an impulse function (something like Figure 3.18, except that δ_r is continuous). Similarly, the graph of D_r looks like a smoothed-out version of Figure 3.19 (indeed, its graph is similar to that of Figure 3.8, page 115, except that its peaks are not bounded by $\frac{1}{2}$).

If (8) happens to be the Fourier series of a continuous function f of period 2π, we know that $f_r(x)$ converges uniformly to $f(x)$ (Theorem 7 of the preceding section) as $r \to 1$. Therefore, for values of r sufficiently close to unity, one could not distinguish between the graph of f and that of f_r, which lends further support to this fundamental principle of graphics. If (8) is the Fourier series of a piecewise continuous function f which is not continuous, f_r will be continuous, but will not be greatly distorted otherwise, since no Gibbs phenomenon occurs with Abel summability.

Definition. A generalized function f of period 2π is said to be *pseudo-continuous* if the (ordinary) function $f_r(x)$ converges to a continuous function as $r \to 1$. A generalized function f of period 2π is said to be *continuous* if the (ordinary) function f_r converges uniformly to a continuous function as $r \to 1$.

The Dirac delta function is neither continuous nor pseudo-continuous, since $\delta_r(x)$ does not converge when $x = 0$. The doublet function $D(x)$ is pseudo-continuous, since (7) is Abel summable to zero for every x (see Exercise 27, Section 3.9), but is not continuous, as will be seen from the following theorem.

Theorem 3. *A generalized function of period 2π is continuous if and only if it is the Fourier series of a continuous function of period 2π.*

Proof: The "if" part follows from Theorem 7 of the preceding section. The "only if" part is almost trivial: if f_r converges to the Fourier coefficients of f, then since (11) is uniformly convergent, it is the Fourier series of f_r, and its coefficients converge to the coefficients in (8) as $r \to 1$, so it follows that (8) must be the Fourier expansion of a continuous function f.

We *identify* any continuous generalized function f with the corresponding continuous function. More generally, we will identify any generalized function with an ordinary function if it is the Fourier series of that function. There is a slight danger in this: if f is not differentiable, it will not have a derivative in the ordinary sense, but it will always have a derivative in the sense of generalized functions, obtained formally by differentiating its Fourier series term-by-term.

EXAMPLE: In the interval $0 < x < 2\pi$ we have [see (19), Section 3.3]

$$\sum_{n=1}^{\infty} \frac{\sin nx}{n} = \frac{\pi - x}{2}. \tag{13}$$

Formally differentiating the left side of (13) and comparing with (6), we obtain $\pi\delta(x) - \frac{1}{2}$. The $-\frac{1}{2}$ term was to be expected, since that is the derivative of the right side of (13). The $\pi\delta(x)$ term arises because the period 2π extension of the function on the right side of (13) has a jump discontinuity of magnitude π whenever x is an integer multiple of 2π. [In general, a piecewise continuous function having a jump discontinuity of magnitude j at $x = t$ will have a derivative containing the term $j\delta(x - t)$.] (Compare Exercise 6.)

Theorem 4. *Every generalized function of period 2π is the sum of a constant and the nth derivative of some continuous generalized function of period 2π.*

Proof: Let p be any integral multiple of 4 that exceeds $k + 2$, where k is the same k as in (9). Then by the Weierstrass M-test, on comparing with $\sum n^{-2}$, the series

$$h(x) = \sum_{n=1}^{\infty} \left(\frac{A_n}{n^p} \cos nx + \frac{B_n}{n^p} \sin nx \right) \tag{14}$$

is uniformly convergent, and therefore is a continuous generalized function. Since the fourth derivatives of $\cos nx$ and $\sin nx$ are $n^4 \cos nx$ and $n^4 \sin nx$, respectively, and p is an integer multiple of 4, the pth derivative of (14) is equal to (8), except for the constant term, which proves the theorem.

If g is an integrable function and f is a generalized function of period 2π, we define $\int_a^b f(x)g(x)\,dx$ as follows:

$$\int_a^b f(x)g(x)\,dx = \lim_{r \to 1} \int_a^b f_r(x)g(x)\,dx \tag{15}$$

provided this limit exists.

In the preceding section, we saw that δ_r converges uniformly to zero in any interval $a \leqq x \leqq b$ not containing an integer multiple of 2π, and $\int_{-\pi}^{\pi} \delta_r(x)g(x)\,dx$ tends to $g(0)$ as $r \to 1$, and therefore

(16) $$\int_{-h}^{h} \delta(x)g(x)\,dx = g(0)$$

whenever g is a continuous function and h is positive.

Similarly, it can be shown that

(17) $$\int_{-h}^{h} D(x)g(x)\,dx = g'(0)$$

whenever g is continuously differentiable and $h > 0$.

The rest of this section will be devoted to more technical matters. Let C^∞ denote the class of all functions of period 2π that are continuous and have continuous derivatives of all orders (in the usual sense). Any function of class C^∞ will be called a *testing function.* Thus, $\sin nx$ and $\cos nx$ are testing functions, and for any generalized function f, the associated function f_r is a testing function for each fixed r in the range $0 \leqq r < 1$ (by Theorem 2).

If u and v are testing functions, we obtain (on integrating by parts n times)

(18) $$\int_{-\pi}^{\pi} u^{(n)}(x)v(x)\,dx = (-1)^n \int_{-\pi}^{\pi} u(x)v^{(n)}(x)\,dx$$

(Exercise 8). This formula is useful in what follows.

Theorem 5. *If f is any generalized function of period 2π and g is a testing function, the integral defined by*

(19) $$\int_{-\pi}^{\pi} f(x)g(x)\,dx = \lim_{r\to 1} \int_{-\pi}^{\pi} f_r(x)g(x)\,dx$$

necessarily exists. Moreover, it is the sum of the numerical series obtained by multiplying every term of (8) by g and integrating term-by-term.

Proof: Let h be the function mentioned in the proof of Theorem 4. The series obtained in the formal manner described here is identical to that obtained by multiplying the Fourier expansion of h by $(-1)^p g^{(p)}(x)$ and integrating term-by-term (as we see easily on integrating by parts), plus a constant term that causes no difficulty (we ignore this constant term in the sequel). The Fourier expansion of h converges uniformly and $g^{(p)}$ is continuous, so term-by-term integration is justified and the sum of the series is

$$(-1)^p \int_{-\pi}^{\pi} h(x)g^{(p)}(x)\,dx.$$

This is the limit as $r \to 1$ of

(20) $$(-1)^p \int_{-\pi}^{\pi} h_r(x)g^{(p)}(x)\,dx$$

since h_r tends uniformly to h. Since $h = f^{(p)}$ we have, by (18),

$$\int_{-\pi}^{\pi} f_r(x)g(x)\,dx = (-1)^p \int_{-\pi}^{\pi} h_r(x)g^{(p)}(x)\,dx. \tag{21}$$

Since the right side of (21) tends to a limit as $r \to 1$, the left side does also, and since this limit is the sum of the series described in the theorem the proof is complete.

Theorem 6. *If f is a piecewise continuous function of period 2π, and g is a testing function, the integral $\int_{-\pi}^{\pi} f(x)g(x)\,dx$ defined as $\lim_{r\to 1} \int_{-\pi}^{\pi} f_r(x)g(x)\,dx$ is equal to the usual integral defined for such functions.*

Proof: If (8) is the Fourier series of a piecewise continuous f, it converges in the mean to f, by Theorem 5 of Section 3.8. If every term is multiplied by $g(x)$ the resulting series converges in the mean to $f(x)g(x)$ and therefore can be integrated term-by-term to yield a numerical series with sum $\int_{-\pi}^{\pi} f(x)g(x)\,dx$. According to Theorem 5, this sum is also equal to $\lim_{r\to 1} \int_{-\pi}^{\pi} f_r(x)g(x)\,dx$.

Theorem 7. *Two generalized functions f and h are equal, i.e., have identical coefficients, if and only if*

$$\int_{-\pi}^{\pi} f(x)g(x)\,dx = \int_{-\pi}^{\pi} h(x)g(x)\,dx. \tag{22}$$

for every testing function g.

Proof: By Theorem 5, these integrals can be obtained by formally multiplying the generalized function in question by g and integrating term-by-term. Taking g to be one of the functions $\sin nx$ $(n = 1, 2, 3, \cdots)$ or $\cos nx$ $(n = 0, 1, 2, \cdots)$, we see at once that (22) is valid for every testing function g only if the corresponding coefficients in the series are equal. (For example, if we multiply by $\sin nx$ and integrate term-by-term over $(-\pi,\pi)$, all terms vanish except that containing B_n, and (22) implies that both generalized functions have the same coefficients B_n). Conversely, if f and h have identical coefficients, then (22) follows from the way the integrals are defined in (15).

For any generalized function f, and any testing function g, we let $L_f(g) = \int_{-\pi}^{\pi} f(x)g(x)\,dx$. The operator L_f thus associates with each testing function g a number $L_f(g)$. If g and h are testing functions,

$$L_f(\alpha g + \beta h) = \alpha L_f(g) + \beta L_f(h) \tag{23}$$

and therefore we say that L_f is a *linear functional* on C^∞.

More generally, any operator L that associates a number with each g in C^∞ and has the linearity property (23), is called a linear functional on C^∞.

A sequence of functions g_m of class C^∞ is said to converge *strongly* to a function g of class C^∞ if g_m converges *uniformly* to g and if all derivatives $g_m^{(n)}$ converge uniformly to the corresponding derivative $g^{(n)}$. (For examples, see Exercises 12 and 13.)

If a linear functional L has the property that $L(g_m) \to L(g)$ whenever g_m converges strongly to g, we say that L is a *continuous linear functional* on C^∞.

Theorem 8. *If L is a continuous linear functional on C^∞, there exists a positive integer p (depending only on L) such that $L(g_m) \to L(g)$ whenever g_m converges uniformly to g and $g_m^{(n)}$ converges uniformly to $g^{(n)}$ for $n = 1, 2, \cdots, p$.*

In other words, it is not really necessary for *all* derivatives of g_m to converge uniformly; if only the first p derivatives converge uniformly, we can conclude that $L(g_m) \to L(g)$. (However, p may be different for different continuous linear functionals.)

Proof: It suffices to prove the theorem for the special case that g is identically zero, since otherwise we could replace g_m by $g_m - g$. We wish to prove that there exists a positive integer p such that $L(g_m) \to 0$ whenever $g_m, g_m', \cdots, g_m^{(p)}$ all converge uniformly to zero. We prove this by contradiction. If a sequence g_m has this property (for a fixed p) but $L(g_m)$ does not tend to zero, then there must be some positive ϵ such that $|L(g_m)| > \epsilon$ for infinitely many values of m. Since $g_m^{(n)}$ converges uniformly to zero for $n = 0, 1, \cdots, p$, so also does $\frac{1}{\epsilon} g_m^{(n)}$, so there must exist some m for which $\left|\frac{1}{\epsilon} g_m^{(n)}(x)\right| < 1/2^p$ for all x when $n = 0, 1, \cdots, p$ but for which $\left|L\left(\frac{1}{\epsilon} g_m\right)\right| > 1$. Denote this function by f_p, so we have

$$|L(f_p)| > 1 \tag{24}$$

$$|f_p^{(n)}(x)| < 1/2^p \qquad (n = 0, 1, \cdots, p) \tag{25}$$

for all x. From (25) it follows that there exists a sequence of functions $f_1, f_2, f_3, \cdots$ converging strongly to the zero function, but from

(24) it follows that $L(f_p)$ does not tend to zero, which contradicts the definition of L. Therefore, there must exist some p having the property stated, which proves the theorem.

Theorem 9. *If f is a generalized function, and L_f is defined by*

$$L_f(g) = \int_{-\pi}^{\pi} f(x)g(x)\,dx \tag{26}$$

for every g of class C^∞, then L_f is a continuous linear functional on C^∞. Moreover, every continuous linear functional on C^∞ is related to a generalized function in this way.

Proof: Using the notation established in (8) and making use of (21), we have

$$L_f(g) = \int_{-\pi}^{\pi} (A_0/2)g(x)\,dx + (-1)^p \int_{-\pi}^{\pi} h(x)g^{(p)}(x)\,dx \tag{27}$$

where h is a continuous function. Therefore, if g_m converges uniformly to g and $g_m^{(p)}$ converges uniformly to $g^{(p)}$, it is clear that $L_f(g_m)$ converges to $L_f(g)$. This shows that L_f is a continuous linear functional.

Conversely, if L is a continuous linear functional on C^∞, if p is the corresponding positive integer mentioned in Theorem 8, and if we define coefficients A_n and B_n by

$$A_n = L\left(\frac{\cos nx}{\pi}\right) \qquad (n = 0, 1, 2, \cdots) \tag{28}$$

$$B_n = L\left(\frac{\sin nx}{\pi}\right) \qquad (n = 1, 2, 3, \cdots) \tag{29}$$

then by Theorem 8, taking $g_m(x) = (\cos nx)/\pi n^{p+1}$, we see that $A_n/n^{p+1} \to 0$, and similarly (replacing $\cos nx$ by $\sin nx$) we obtain $B_n/n^{p+1} \to 0$. Using these coefficients in (8), we obtain a generalized function f. It follows from the way f was constructed, and from Theorem 5, that $L(g) = L_f(g)$ whenever g is a function of the form $\cos nx$ or $\sin nx$, for $n = 0, 1, 2, \cdots$, and by linearity this is also true when g is a trigonometric polynomial. The partial sums of the Fourier series of any testing function g converge strongly to the function g (Exercise 15), so by taking limits it follows that $L(g) = L_f(g)$ for every testing function g, which completes the proof of the theorem.

By Theorem 7, every generalized function f is completely characterized by its associated linear functional L_f, and by Theorem 9, every continuous linear functional on C^∞ determines a generalized function f. This shows that there is an alternative way to define a generalized function of period 2π: it is a continuous linear functional on C^∞. This basically returns us to our starting point. The fictitious functions mentioned at the beginning of this section are never used by themselves; they occur only as part of the integrand of a singular integral. Thus, an essential property of the Dirac delta is that $\int_{-\pi}^{\pi} \delta(x)g(x)\,dx = g(0)$. This may just as well be written $L(g) = g(0)$, but persons who use generalized functions prefer the integral notation. They like to think that $\delta_r(x)$ actually has a limit as $r \to 1$, and the theory of generalized functions provides a sense in which this is true. The usefulness of the concept of generalized functions is a matter of controversy; they are undoubtedly useful to those who like to think the Cheshire cat is still there, when in fact all that is left is the grin.

• EXERCISES

1. Using Theorem 7, Section 3.7, show that it is legitimate to differentiate, term-by-term, any convergent series of continuously differentiable functions, provided the resulting series converges uniformly.

2. If f is an integrable function, of period 2π, what is the significance of $(-n\psi_n) * f$, from the viewpoint of Section 2.6?

3. **(a)** If f is of period 2π, has a continuous derivative, and we let $M = \max |f'(x)|$, show that the Fourier coefficients satisfy $|A_n| \leqq M/n$, $|B_n| \leqq M/n$.

(b) If f has two continuous derivatives, and $M = \max |f''(x)|$, show that $|A_n| \leqq M/n^2$, $|B_n| \leqq M/n^2$.

(c) Generalize (a) and (b), and show as a consequence that a testing function has Fourier coefficients tending to zero so rapidly that $n^k A_n \to 0$, $n^k B_n \to 0$ for every positive integer k.

4. Use the results of the preceding problem to show that if (8) is multiplied term-by-term by a testing function and integrated over $(0,2\pi)$, the resulting series necessarily converges.

5. Prove that if f_r converges in the mean to an integrable function f, (8) is the Fourier series of f.

6. **(a)** What is $\int_0^x \delta(x)\,dx$?

(b) Show that $\displaystyle\int_0^x \left[\pi\delta(x) - \frac{1}{2}\right] dx = \frac{\pi}{2} - \frac{x}{2} \quad (x > 0).$

7. What is the value of (16) if g is a piecewise continuous function?

8. Verify (18) by integrating by parts.

9. Show that, if f and g are generalized functions of period 2π, $f' * g = f * g'$.

10. **(a)** Using the formulas in Section 2.4, show that

$$(-n\psi_n) * (a_n\phi_n + b_n\psi_n) = \frac{d}{dx}(a_n\phi_n + b_n\psi_n).$$

(b) Use this to show that $\delta' * f = f'$.
(c) Obtain this same result using Exercise 9.

***11.** Prove the following theorem: If an integrable function of period 2π has a derivative $f^{(m)}(x)$ of order m at a point x, then the series obtained by differentiating the Fourier series of $f(x)$ m times, term-by-term, is Abel summable to $f^{(m)}(x)$ at this point.

12. Show that the sequence $g_m(x) = \dfrac{1}{m}\sin x$ converges strongly to the zero function.

13. Show that $g_m(x) = \dfrac{1}{m^3}\sin mx$ converges uniformly to the zero function, as $m \to \infty$, but does not converge strongly to the zero function.

14. If L is a continuous linear functional on C^∞ and g_m is the sequence defined in Exercise 13, can you make any statement concerning $\lim_{m\to\infty} L(g_m)$?

15. Show that, if g is a testing function, the sequence of partial sums of the Fourier series expansion of g converges strongly to g.

16. Discuss the Abel summability of the series

$$\cos x + 2^2 \cos 2x + 3^2 \cos 3x + 4^2 \cos 4x + \cdots.$$

17. If g is a testing function and f is the generalized function defined in Exercise 16, what is the value of $\int_{-\pi}^{\pi} f(x)g(x)\,dx$?

18. The series $\sin x + 2^3 \sin 2x + 3^3 \sin 3x + 4^3 \sin 4x + \cdots$ is multiplied term-by-term by an infinitely differentiable function of period 2π, and the resulting series is integrated term-by-term over the interval $(-\pi,\pi)$. What is the significance of the sum of the numerical series thus obtained?

19. In terms of the doublet function D, show that the generalized function

$$\sin x - 2\sin 2x + 3\sin 3x - 4\sin 4x + \cdots \tag{30}$$

is $-\pi D(x-\pi)$.

20. Consider the Fourier series

$$\sum_{n=2}^{\infty} (-1)^n \frac{n \sin nx}{n^2-1}. \tag{31}$$

(a) Considering (31) as a generalized function f, show that it satisfies the differential equation

$$f''(x) + f(x) = -\sin x - \pi D(x-\pi).$$

(b) Hence, deduce that the sum of (31) in any interval not containing an odd multiple of π must be of the form

$$f(x) = C_1 \cos x + C_2 \sin x + \tfrac{1}{2}x \cos x.$$

21. Differentiate (31) term-by-term and show that the resulting series is Abel summable to $\frac{3}{4}$ at $x = 0$.

22. Deduce from Exercises 20 and 21 that the sum of (31) in the interval $-\pi < x < \pi$ is

$$f(x) = \frac{\sin x}{4} + \frac{x \cos x}{2}.$$

23. Consider the series

$$\sum_{n=1}^{\infty} \frac{n^3 \sin nx}{n^4+1}. \tag{32}$$

(a) Show that the generalized function defined by (32) satisfies the differential equation

$$f^{(4)}(x) = \pi\delta^{(3)}(x) - f(x).$$

(b) Hence, deduce that the sum of (32) is an infinitely differentiable function in any interval not containing an integral multiple of 2π.

24. Consider the series

$$\sum_{n=1}^{\infty} \frac{\cos nx}{n^2+1}. \tag{33}$$

(a) Show that the generalized function defined by (33) satisfies the differential equation

$$f''(x) - f(x) = \tfrac{1}{2} - \pi\delta(x).$$

(b) Hence, deduce that the sum of (33) in any interval not containing an integer multiple of 2π is of the form

$$f(x) = C_1e^x + C_2e^{-x} - \tfrac{1}{2}.$$

25. Consider the series

$$\sum_{n=2}^{\infty} (-1)^n \frac{n^3}{n^4 - 1} \sin nx. \tag{34}$$

(a) If $f(x)$ is defined by (34), show that

$$f'(x) = \delta(x - \pi) - \frac{1}{2} + \cos x + \sum_{n=2}^{\infty} \frac{(-1)^n \cos nx}{n^4 - 1}.$$

(b) By considering $\int_0^x f'(x)\,dx$, show that in the interval $-\pi < x < \pi$, the sum of (34) is the same as the sum of

$$-\frac{x}{2} + \sin x + \sum_{n=2}^{\infty} (-1)^n \frac{1}{n^5 - n} \sin nx. \tag{35}$$

(c) Deduce the same result without making use of generalized functions, using Equation (20), Section 3.3, and the relation

$$\frac{n^3}{n^4 - 1} - \frac{1}{n} = \frac{1}{n^5 - n}.$$

[The point of this exercise is that the infinite series in (35) converges more rapidly than (34) and is therefore more convenient for numerical work.]

26. Consider the series

$$\frac{1}{2} + \sum_{n=1}^{\infty} (-1)^n \frac{\cos nx - n \sin nx}{n^2 + 1}. \tag{36}$$

(a) Letting $f(x)$ be (36), show by differentiating term-by-term that $f'(x) = f(x) - \pi\delta(x - \pi)$.

(b) Hence deduce that $f(x)$ is a constant times e^x in any interval not containing an odd multiple of π.

(c) Using (49), Section 3.3, verify that this function does have a jump discontinuity of magnitude $-\pi$ at $x = \pi$, as indicated by the derivative in (a).

27. **(a)** Write down the trigonometric expansion of $\delta(x) - \delta(x + \pi)$.

(b) Determine $\int_0^x [\delta(x) - \delta(x + \pi)]\,dx$.

(c) Hence, derive (21), Section 3.3.

28. Devise a method, using Fourier series, for showing that

$$\frac{\pi}{2 \sinh \pi} = \frac{1}{2} + \sum_{n=1}^{\infty} (-1)^n \frac{1}{n^2 + 1}.$$

(The usual method, in complex variable theory, is to use the calculus of residues.)

3.11 • PRACTICAL REMARKS

These remarks are of an elementary nature, with particular emphasis on numerical procedures and the kinds of mistakes one might make through misunderstanding some of the theorems.

From a practical viewpoint, some of the theorems about Fourier series, as commonly quoted, are either false or misleading. One commonly quoted theorem, called the *localization theorem* (see Exercise 6, Section 3.6) states that *the behavior of a Fourier series at a point depends only on the behavior of the function in a neighborhood of the point.* The idea of the proof is as follows: if f and g are integrable functions of period 2π, and if $f(x) = g(x)$ identically in an interval $a < x < b$, then the Fourier series expansion of $f - g$ converges to zero for each point x in this interval. It follows that if f and g are functions taking the same values in $a < x < b$, no matter how small this interval may be, their Fourier series will either both diverge or they will both converge to the same values in this interval.

The reader will immediately see that the proof does not imply the general remark it is supposed to prove. In numerical work, the rate at which a series converges is an important aspect of its behavior, and the rapidity with which a Fourier series converges at a point does *not* depend on the behavior of the function in a small neighborhood.

In general, the smoother a function is, the more rapidly its Fourier coefficients tend to zero. It is important to realize that, if the function is defined only in an interval $0 \leqq x \leqq 2\pi$, it is the period 2π extension that must be smooth. For example, the function $f(x) = x(2\pi - x)$ is very smooth in the interval $0 \leqq x \leqq 2\pi$; in fact, it is infinitely differentiable there! Nevertheless, its Fourier series representation in this interval converges rather slowly. The reason is that the period 2π extension of this function, although continuous everywhere, does not have a continuous derivative at $x = 0$ and $x = 2\pi$ (or at any integer multiple of 2π).

The slow rate of convergence of most Fourier series arising in applications can be a source of great disappointment to those using them in practice. For example, in the interval $0 \leqq x \leqq 2\pi$, we have

$$(1) \qquad x(2\pi - x) = \tfrac{2}{3}\pi^2 - 4(\cos x + \tfrac{1}{4}\cos 2x + \tfrac{1}{9}\cos 3x + \cdots).$$

At $x = \pi/20$, the value of $x(2\pi - x)$ is approximately 0.962. If we begin computing partial sums of (1), we find to our delight that the sixth partial sum of (1) is 0.966, less than one per cent from this value. Unfortunately, the seventh partial sum is 0.901, and a great many partial sums must be computed before we are again within one per cent of 0.962.

In (1) it is obviously preferable to use the left side, and not the right side, to calculate any numerical values. If, as a result of theoretical work, we obtain the solution to a problem in the form of a Fourier series, from which we need to obtain numerical results, we must consider the following possibilities: (a) If the series matches one given in Section 3.3, we can write down its sum in a simple form, which is usually preferable for numerical purposes to using the series itself; (b) if the series can, by some means, be rewritten as a linear combination of series given in Section 3.3, we can proceed in a similar manner to obtain analytic expressions for the sum of the series; (c) it may be possible to write the series as the sum of series listed in Section 3.3, and another series that converges more rapidly and hence is more suitable for numerical work; and (d) one must keep an open mind to the possibility that the problem should be solved by an altogether different procedure not involving the use of Fourier series. We will discuss the last three of these possibilities very briefly.

As a simple example of (b), we consider the series

$$(2) \qquad \sum_{n=1}^{\infty} \frac{\cos nx}{n+1}$$

which is not identical to any on the list, although it is similar to $\sum_{n=1}^{\infty} (\cos nx)/n$ and can therefore be expected to have a similar behavior. This series converges very slowly; if we wish a numerical value to three decimal places, we should need to sum hundreds of terms, and even then would have a numerical result for only a single value of x; to plot the graph of the sum in a naive manner would require a forbidding amount of calculation. But if we take $m = n + 1$ in (2) we obtain

$$
\begin{aligned}
&\sum_{m=2}^{\infty} \frac{\cos (m-1)x}{m} \\
&= \left[\sum_{m=2}^{\infty} \frac{\cos mx}{m}\right] \cos x + \left[\sum_{m=2}^{\infty} \frac{\sin mx}{m}\right] \sin x \\
(3) \quad &= \left[-\ln\left(2 \sin \frac{x}{2}\right) - \cos x\right] \cos x + \left[\frac{\pi - x}{2} - \sin x\right] \sin x \\
&= \frac{\pi - x}{2} \sin x - \left[\ln\left(2 \sin \frac{x}{2}\right)\right] \cos x - 1 \qquad (0 < x < 2\pi)
\end{aligned}
$$

by making use of the list in Section 3.3 and a simple trigonometric identity.

We give two examples to illustrate (c). Particularly dramatic results are sometimes obtained when the coefficients are of the form $p(n)/q(n)$, where p and q are polynomials, and where for large values of n the ratio $p(n)/q(n)$ is nearly the same as the general coefficient in a series whose sum we already know. The procedure then is to subtract the known series from the one we are trying to sum. For example,

$$
(4) \qquad \sum_{n=1}^{\infty} \frac{n^4}{n^5 + 1} \sin nx
$$

is a very slowly converging series. For large values of n, $n^4/(n^5 + 1)$ is very nearly $1/n$, and on subtracting we obtain

$$
(5) \qquad \frac{n^4}{n^5 + 1} - \frac{1}{n} = \frac{-1}{n(n^5 + 1)}
$$

and therefore we replace (4) by

$$
(6) \qquad \sum_{n=1}^{\infty} \frac{1}{n} \sin nx - \sum_{n=1}^{\infty} \frac{1}{n(n^5 + 1)} \sin nx.
$$

The first of these series, although slowly convergent, is on our list, and therefore presents no difficulties. The second series converges very rapidly; for a numerical value to three decimal places, two terms are sufficient.

As another example, consider the series

$$
(7) \qquad \sum_{n=1}^{\infty} \frac{\cos nx}{n + a} \quad (a > 0).
$$

If a is a positive integer, we would proceed as in (2), which is the special case $a = 1$. If not, we can proceed by using the relation

$$
(8) \qquad \frac{1}{n + a} = \frac{1}{n}\left[1 - \frac{a}{n} + \frac{a^2}{n^2} - \frac{a^3}{n^2(n + a)}\right]
$$

and therefore (7) can be replaced by

$$(10)\quad \sum_{n=1}^{\infty} \frac{\cos nx}{n} - a \sum_{n=1}^{\infty} \frac{\cos nx}{n^2} + a^2 \sum_{n=1}^{\infty} \frac{\cos nx}{n^3} - a^3 \sum_{n=1}^{\infty} \frac{\cos nx}{n^3(n + a)}.$$

The first three of these series are listed in Section 3.3, and the last series is more rapidly convergent than (7).

Concerning (d) we can say little here, since the situation will depend on the particular type of problem in question. Later on we will solve certain boundary-value problems, finding answers in the form of infinite series. From a numerical viewpoint, these answers are not really "solutions" at all. For example, we will find the temperature distribution in a rectangular plate, in the form of an infinite series. In the time it would take to substitute into this series and obtain the temperature at a single point, it may be possible to obtain (by altogether different numerical methods, not discussed in this book) the temperature at an entire grid of points within the plate. Therefore if only numerical results are desired, the Fourier method may not be the method of choice. The so-called "relaxation" methods, and other iterative procedures, may be preferable.

Now let us turn to the opposite problem, in which we are given values of a function, perhaps obtained experimentally, and we wish to determine a Fourier series. Here again we must consider several possibilities.

Graphical procedures for finding Fourier coefficients directly from a carefully drawn graph have been devised. If the procedure is truly graphical, i.e., requires no supplementary calculations, it is of necessity quite complicated and probably should not be attempted by a person unskilled in graphical methods. Details will be found in several standard books on graphics, and also in *Practical Analysis*, by Dr. Fr. A. Willers (Dover, 1948).*

Both mechanical and electronic devices are available. The mechanical ones operate on a principle similar to that of a planimeter, and one reads the Fourier coefficient from a scale. Electronic synthesizers are available; here, one manipulates dials in an attempt to produce a graph (shown on an oscilloscope) which matches the given function. Many large laboratories have harmonic analyzers designed especially for a specific type of problem. To give further details is beyond the scope of this book.

Numerical procedures can be used to calculate Fourier coefficients, just as they can be used to perform any other integration. Thus one can use the rectangular rule, the trapezoidal rule, or even Simpson's rule (if the accuracy of the data warrant the extra time

*Currently out of print.

spent). These are familiar methods taught in calculus and will not be reviewed here. The rectangular rule is most commonly used, mainly because it has special theoretical significance in the Fourier theory. We will now outline this method.

If the given function has period 2π, we divide the interval $(0,2\pi)$ into m equal parts by the points

$$0, \frac{2\pi}{m}, 2 \cdot \frac{2\pi}{m}, \cdots, (m-1)\frac{2\pi}{m}, 2\pi.$$

Let the values of the function at these points be denoted $y_0, y_1, \cdots, y_m$. Since the function is periodic, $y_0 = y_m$.

According to the rectangular rule, the integrals for determining the coefficients will be replaced by sums, and the coefficients will be given approximately by

$$A_n = \frac{2}{m} \sum_{k=0}^{m-1} y_k \cos n\frac{2k\pi}{m} \qquad (n = 0, 1, 2, \cdots), \tag{11}$$

$$B_n = \frac{2}{m} \sum_{k=0}^{m-1} y_k \sin n\frac{2k\pi}{m} \qquad (n = 1, 2, 3, \cdots). \tag{12}$$

To make the work systematic and simplify the actual calculations, special schemes have been devised. These are described in detail by Whittaker and Robinson in their book *Calculus of Observations* (Blackie and Son, 1944), which contains special sheets for the cases $m = 12$ and $m = 24$. We demonstrate the general idea here, for the case $m = 6$, but we warn the reader that this is not a sufficiently large value of m for most practical work.

Suppose we are given six numbers y_0, y_1, y_2, y_3, y_4, and y_5, representing the values of a function at the points $x = 0°, 60°, 120°, 180°, 240°$, and $300°$. (The value y_6 at $360°$ is not needed, since $y_6 = y_0$.) We form a table as follows:

	y_0	y_1	y_2	v_0	v_1	w_0	w_1
	y_3	y_4	y_5		v_2		w_2
Sum	v_0	v_1	v_2	p_0	p_1	r_0	r_1
Difference	w_0	w_1	w_2		q_1		s_1

(Zero is understood to appear in positions where there is no entry.)

The coefficients are then found by using the formulas

$$6a_0 = 3A_0 = p_0 + p_1, \quad 3A_1 = r_0 + s_1/2 \quad 3A_2 = p_0 - p_1/2$$

$$6A_3 = r_0 - s_1 \quad 3B_1 = \frac{\sqrt{3}}{2} r_1, \quad 3B_2 = \frac{\sqrt{3}}{2} q_1$$

This is called the *six point method*. (See Exercise 2.)

If 0.866 is considered a sufficiently close approximation to $\sqrt{3}/2$, it is convenient to observe that $0.866 = 1 - \frac{1}{10} - \frac{1}{30}$, approximately, and therefore one can multiply by 0.866 mentally. (With a few tricks like this, the reader can easily establish a reputation as a calculating prodigy.)

This method yields a trigonometric polynomial of order three,

$$A_0/2 + A_1 \cos x + B_1 \sin x + A_2 \cos 2x + B_2 \sin 2x + A_3 \cos 3x, \tag{13}$$

which roughly approximates the first few terms of the Fourier expansion of the function.

There is hardly any point in using formulas (11) and (12) to calculate the Fourier coefficients A_n and B_n for values of n greater than $m/2$. In practice, m is always taken to be an even integer, and (12) always gives $B_n = 0$ when $n = m/2$, which accounts for the absence of the term $B_3 \sin 3x$ from (13) above.

The theoretical interest in this method derives from the fact that it yields a trigonometric polynomial whose values, in principle, are *exactly* equal to the values $y_0, y_1, \cdots, y_{m-1}$ at the m points in question. Thus, if m is even, it yields a trigonometric polynomial of order $m/2$ which exactly fits the prescribed data. (See Exercise 7, and compare this procedure with the "sine-polynomial game" of Section 1.4.)

When the scheme for $m = 12$ or $m = 24$ is used, the coefficients obtained will often be quite close to the actual Fourier coefficients, at least for the smaller values of n. In many applications, only the first few Fourier coefficients are desired. If this is the case, the scheme given by Whittaker and Robinson need not be carried out in full. Here is how we proceed if $m = 12$ and we desire only a trigonometric polynomial of degree 3. We form the tables

	y_0	y_1	y_2	y_3	y_4	y_5	y_6
		y_{11}	y_{10}	y_9	y_8	y_7	
Sum	u_0	u_1	u_2	u_3	u_4	u_5	u_6
Difference		v_1	v_2	v_3	v_4	v_5	

	u_0	u_1	u_2	u_3	v_1	v_2	v_3
	u_6	u_5	u_4		v_5	v_4	
Sum	s_0	s_1	s_2	s_3	q_1	q_2	q_3
Difference	t_0	t_1	t_2		r_1	r_2	

The coefficients are now found from

$$
\begin{aligned}
12a_0 = 6A_0 &= s_0 + s_1 + s_2 + s_3 \\
6A_1 &= t_0 + \frac{\sqrt{3}}{2}\, t_1 + \tfrac{1}{2} t_2 \\
6A_2 &= s_0 - s_3 + \tfrac{1}{2}(s_1 - s_2) \\
6A_3 &= t_0 - t_2 \\
6B_1 &= \tfrac{1}{2} q_1 + \frac{\sqrt{3}}{2}\, q_2 + q_3 \\
6B_2 &= \frac{\sqrt{3}}{2}\,(r_1 + r_2) \\
6B_3 &= q_1 - q_3.
\end{aligned}
$$

The resulting polynomial

$$
A_0/2 + \sum_{n=1}^{3} (A_n \cos nx + B_n \sin nx) \tag{14}
$$

will not, in general, match the function at the 12 points 0°, 30°, 60°, $\cdots$ corresponding to the values $y_0, y_1, y_2, \cdots$, but will more nearly approximate a partial sum of the Fourier series than (13). (The complete scheme given by Whittaker and Robinson for $m = 12$ yields a trigonometric polynomial of order 12 that, in principle, takes on exactly the values $y_0, y_1, y_2, \cdots$.)

Whittaker and Robinson also give a "12-point fast approximation" which is especially easy to use in practice because it avoids any multiplications. We will not discuss it here.

In the following table, we give the first seven Fourier coefficients of the Fourier series expansion of the function $f(x) = x(2\pi - x)$ in

Fourier Coefficients	12 Points	12 Points "Fast"	6 Points
$A_0/2 = 6.580$	6.534	6.534	6.397
$A_1 = -4.000$	-4.093	-3.373	-4.387
$A_2 = -1.000$	-1.096	-1.234	-1.462
$A_3 = -0.444$	-0.548	-0.562	-0.548
$A_4 = -0.250$	-0.365		
$A_5 = -0.160$	-0.294		
$A_6 = -0.111$	-0.137		

Note: $B_n = 0$ for all n in this case.

the interval $(0,2\pi)$. For comparison, we give in the second column the Fourier coefficients obtained by the 12-point method; the 12-point procedure just described gives only the first four values listed here. The third column gives the coefficients as calculated by Whittaker and Robinson's 12-point "fast approximation" and the last column was obtained using the six-point method we have described in detail.

As interpolation devices, these methods are very poor. For example, at 9°, the value of the function is approximately 0.962. The values obtained from these trigonometric polynomials are, from left to right, 0.901, 0.377, 1.527, and 0.186.

In Chapter 2, it was emphasized that the $n + 1$st partial sum of the Fourier series representing a function in an interval provides the best *least-squares* approximation to the function, in that interval, of any trigonometric polynomial of order n. In practical harmonic analysis it is important to recognize that this may not be the best *uniform* approximation. Therefore, if we use a harmonic synthetizer (or any other method) to construct, by trial-and-error, a trigonometric polynomial of order n whose graph "looks the most" like that of the given function, one is in principle *not* seeking Fourier coefficients. The theory of such approximations is more closely related to the theory of Tchebycheff polynomials, not discussed in this book.

Because of this confusion in ideas, the reader is advised to treat with great caution any claims he reads in experimental papers comparing "calculated" with "measured" values of the harmonics in (say) the output of a signal source. The reader should not even believe claims that a particular harmonic is "dominant" over another, unless the author explains what he means and describes the way he obtained the result; different methods have been known to give different conclusions.

Sometimes, experimental data do not give m numbers $y_0, y_1, \cdots, y_{m-1}$ corresponding to the values $f(x)$ of a function, but rather m values $z_0, z_1, \cdots, z_{m-1}$ which are *average values* over intervals of length $2\pi/m$ centered at the points $2k\pi/m$ ($k = 0, 1, \cdots, m - 1$). Since the average value of a function over an interval is usually not equal to the value of the function at the center of the interval, it is best to provide a correction to these values, if the Fourier coefficients of the function itself are to be calculated by any of the above methods. The usual rule-of-thumb is to add to each mean z_k one-twelfth of the excess of z_k over the average of z_{k-1} and z_{k+1}. That is, we take y_k to be

$$(15)\quad y_k = z_k + \frac{1}{12}\left(z_k - \frac{z_{k-1} + z_{k+1}}{2}\right) \qquad (k = 0, 1, \cdots, m - 1).$$

(See Exercise 8 for the idea behind this rule.) Note that we take $z_{-1} = z_{m-1}$ and $z_m = z_0$ in using (15), since the function is assumed to be periodic.

If the data are given at points in the interval which are not evenly spaced, the above procedures fail. One can, of course, plot a smooth graph of the given data, and read from it the values at m points that *are* evenly spaced. Another procedure, of greater interest in theory than in practice, is given in Exercise 9.

Several other observations of practical importance are mentioned in connection with Exercises 11 and 12.

• EXERCISES

1. These questions refer to the six-point method.

(a) Show that $p_1 = y_1 + y_2 + y_4 + y_5$.

(b) Show that $q_1 = y_1 - y_2 + y_4 - y_5$.

(c) Explain how $3A_3 = r_0 - s_1$ was derived.

(d) Compute (0.866)(66) by mental arithmetic, using the suggestion made in this section.

2. Derive in detail the formulas for the six-point method, using (11).

3. Use any of the methods given in this section to find an approximation, by trigonometric polynomials, to $f(x) = 1 - (x/2\pi)$, in the interval $0 < x < 2\pi$, and compare it with the Fourier series representing this function.

4. By the six-point method, find a trigonometric polynomial of order three, having values given in this table:

$x =$	0°	60°	120°	180°	240°	300°
$y =$	1.000	0.366	0.366	1.000	−1.366	−1.366

5. Show that the following six vectors constitute an orthogonal basis in six-dimensional Cartesian space.

(a) $(1,1,1,1,1,1)$.

(b) $(1,-\frac{1}{2},-\frac{1}{2},1,-\frac{1}{2},-\frac{1}{2})$.

(c) $(0,\sqrt{3}/2,\sqrt{3}/2,0,-\sqrt{3}/2,-\sqrt{3}/2)$.

(d) $(1,\frac{1}{2},-\frac{1}{2},-1,-\frac{1}{2},\frac{1}{2})$.

(e) $(1,-1,1,-1,1,-1)$.

(f) $(0,\sqrt{3}/2,-\sqrt{3}/2,0,\sqrt{3}/2,-\sqrt{3}/2)$.

6. Find the norms of each of the vectors in Exercise 5.

7. Show that the six-point method gives a trigonometric polynomial of order 3 that exactly matches the prescribed values $y_0, y_1, \cdots, y_5$. (See Exercises 5 and 6.)

8. Show that (15) provides an exact relation between the value of a function at the center of an interval and its mean values over that interval and two adjacent intervals of the same length, provided that $f(x)$ is of the form $ax^2 + bx + c$.

9. Given $2n + 1$ points entirely within an interval of length less than 2π, $x_0, x_1, \cdots, x_{2n}$, not necessarily evenly spaced, and $2n + 1$ numbers $y_0, y_1, \cdots, y_{2n}$, show that the following procedure yields a trigonometric polynomial of order n exactly equal to y_j at each of the points x_j. For each j, let $g_j(x)$ be the the product of $2n$ factors

$$g_j(x) = \sin \tfrac{1}{2}(x - x_0) \sin \tfrac{1}{2}(x - x_1) \cdots \sin \tfrac{1}{2}(x - x_{2n})$$

where the factor $\sin \frac{1}{2}(x - x_j)$ is deleted. Then form the quotients $\phi_j(x) = [g_j(x)]/[g_j(x_j)]$. The desired trigonometric polynomial is given by $T(x) = \sum_{j=0}^{2n} y_j\phi_j(x)$.

10. For some classes of series, the error made in truncating the series is no greater than the magnitude of the first omitted term. Is this true for Fourier series?

11. **(a)** Is a uniformly convergent series necessarily rapidly convergent?
(b) Do the Cesaro sums provide a rapid way to calculate the values of a Fourier series?

12. Suppose a Fourier series is known to converge uniformly to $f(x)$ in an interval, and we wish to know how many terms in the series are needed to find $f(x)$ to within a prescribed tolerance ϵ over the interval. We substitute a given value of x, and calculate the partial sums until we know we are within ϵ of the correct value. Can we assume that this partial sum is within ϵ of the correct value over the entire interval?

13. Find the sum of the series $\sum_{n=1}^{\infty} \frac{\sin nx}{n+1}$.

14. Find the sum of the series $\sum_{n=1}^{\infty} \frac{n(n+1)}{n^3} \cos nx$.

15. Improve the convergence of the series $\sum_{n=1}^{\infty} \frac{\sin nx}{n + \sqrt{2}}$.

***16.** Explore various ways of rewriting the series $\sum_{n=1}^{\infty} \frac{n}{(n+1)^2} \cos nx$. If possible, find an analytic expression for its sum. Which expression would you personally use if you needed to construct a table of values for the sum of this series?

CHAPTER 4

LEGENDRE POLYNOMIALS AND BESSEL FUNCTIONS

This chapter can be studied independently of the other chapters, although occasional reference will be made to material in earlier chapters.

The treatment is necessarily incomplete. A book three times the size of this one could be written on Bessel functions alone; indeed, such a book exists. Convergence theorems will, for the most part, be stated without proof.

This chapter is mainly theoretical. Applications will be found in Chapters 5 and 6.

4.1 • PARTIAL DIFFERENTIAL EQUATIONS

The theory of partial differential equations is beyond the scope of this book. However, some general remarks will be made here, to provide unity to certain topics arising later. *Neither this section nor the next section is intended to be mathematically rigorous.*

We recall that an operator L is said to be *linear* if $L(\alpha f + \beta g) =$

$\alpha L(f) + \beta L(g)$ for scalars α and β and functions f and g. Of particular importance in physics are linear operators that involve partial differentiation. Here are some examples:

(1)
$$\frac{\partial^2}{\partial x \partial y}$$

(2)
$$\frac{\partial^2}{\partial x^2} + \frac{\partial^2}{\partial y^2}$$

Both of these are examples of *linear partial differential operators.*

If L is such an operator, an equation of the form $Lf = h$, where h is a prescribed function, is called a *linear partial differential equation.* If the highest order derivative in Lf is of the nth order, the equation is said to have *order* n.

Thus, $\frac{\partial^2 f}{\partial x \partial y} = g$ is a second order linear partial differential equation, and so also is $\frac{\partial^2 f}{\partial x^2} + \frac{\partial^2 f}{\partial y^2} = 0$. We will be entirely concerned with second order linear partial differential equations.

Observe that the equation $\frac{\partial^2 f}{\partial x^2} = \frac{1}{c^2}\frac{\partial^2 f}{\partial t^2}$ is a second order linear partial differential equation; it can be written $Lf = 0$ by taking L to be the operator

(3)
$$\frac{\partial^2}{\partial x^2} - \frac{1}{c^2}\frac{\partial^2}{\partial t^2}.$$

We recall that a second order linear *ordinary* differential equation such as $\frac{d^2 y}{dx^2} + y = e^x$ may have a "general solution" with two arbitrary constants (in this case, $y = \frac{1}{2}e^x + C_1 \cos x + C_2 \sin x$). With partial differential equations the situation is quite different. In the cases we consider, we will have a "general solution" with two arbitrary *functions* instead.

For example, let us solve the partial differential equation

(4)
$$\frac{\partial^2 f}{\partial x \partial y} = 0.$$

This may be written in the form

(5)
$$\frac{\partial}{\partial x}\left(\frac{\partial f}{\partial y}\right) = 0$$

from which we deduce

(6)
$$\frac{\partial f}{\partial y} = p(y).$$

Observe that any function of y is considered as a constant when taking the partial derivative with respect to x. Thus, from $\partial u/\partial x = 0$, we do not deduce $u = C$ where C is a constant, but $u = p(y)$ where p is some function of y. This explains how we obtained (6) from (5).

Similarly, from $\partial f/\partial y = p(y)$, we do not deduce $f = \int p(y)\,dy + C$, but rather

$$f(x,y) = \int p(y)\,dy + q(x) \tag{7}$$

which we rewrite in the form

$$f(x,y) = r(y) + q(x). \tag{8}$$

Any constant terms can be lumped with either of the functions $r(y)$ or $q(x)$, so we don't need to indicate a constant term separately.

The process of passing from (4) to (6) and then finally (8) is sometimes called *partial integration.* Thus, we say that (6) is obtained from (4) by partially integrating with respect to x, and (8) is obtained from (6) by integrating partially with respect to y. (This should not be confused with *integration by parts.*)

Clearly, for arbitrary functions $r(y)$ and $q(x)$, the function (8) does satisfy (4), as we see by substituting into (4). Therefore, we have shown that (8) is the *general solution* of (4). Again we note that this discussion is not rigorous. Unless these functions are suitably differentiable, it would not make sense to call (8) a solution of (4).

This example demonstrates, in rudimentary form, the idea that a second order linear partial differential equation has a general solution containing two arbitrary *functions* rather than two arbitrary *constants.* However, it is not always possible to solve a partial differential equation by partial integration, so the reader should not be misled into thinking that partial differential equations are always solved as easily as this one.

There is a second important distinction between ordinary differential equations and partial differential equations. *The general solution of a partial differential equations is seldom of much practical usefulness.* This is in great contrast to the situation with ordinary differential equations, where the general solution is usually quite useful. The reason for this is that it is much more difficult to determine the precise nature of arbitrary *functions,* in adapting a general solution to a particular case, than it is to evaluate a finite number of arbitrary *constants* in matching a general solution to a specific problem.

Therefore, we shall only rarely seek general solutions in later sections. We will usually proceed directly to find particular solutions, by a procedure known as "separation of variables." This procedure is not universally applicable, although it works quite well in the special cases we will consider. Practical procedures, that work when the method of separation of variables does not, usually involve numerical methods, or theoretical procedures which either are more advanced than we can discuss in this book or are nonexistent at the present time.

Readers who are weak in partial differentiation are urged to work through the following exercises. Of special importance are expressions such as $z = f(x - vt)$. If these appear confusing at first, keep in mind that f is a function of a single variable that (in effect) becomes a function of x and t by replacing the single variable by $x - vt$. (Here v is a constant.) Thus, we can write $z = f(u)$, $u = x - vt$, if the notation $f(x - vt)$ is not clear. If the reader does not see why $\partial f/\partial t = -vf'(x - vt)$ he should work the exercises in the order in which they appear, checking his answers as he goes along.

• EXERCISES

1. Determine which of the following is a linear partial differential equation, and, if linear, the order of the equation.

(a) $6\dfrac{\partial f}{\partial x} + \dfrac{\partial^2 f}{\partial y^2} - \dfrac{\partial f}{\partial x} + x^2 f = 0$

(b) $\left(\dfrac{\partial f}{\partial x}\right)^2 + \left(\dfrac{\partial f}{\partial y}\right)^2 + \left(\dfrac{\partial f}{\partial z}\right)^2 = 4$

(c) $f\dfrac{\partial f}{\partial x} = \dfrac{\partial^2 f}{\partial t^2}$

(d) $\dfrac{\partial^3 f}{\partial x^3} + (\sin x)^2 = \dfrac{\partial^2 f}{\partial t^2}$

(e) $\dfrac{\partial^2 f}{\partial x^2} + 3xyz\dfrac{\partial f}{\partial y} - z^3\dfrac{\partial^2 f}{\partial z^2} = 0$

2. If $\partial^2 f/\partial x \partial y = 3x^2 y + y^2$, what can you say about f?

3. Find the "general solution" of $\partial^2 f/\partial x \partial y = e^{xy}$.

4. Find $\partial z/\partial x$ and $\partial z/\partial t$:

(a) $z = \sin (x + vt) + e^{(x-vt)}$.

(b) $z = (x + vt)^5 + (x - vt)^7$.

(c) $z = f(x + vt) + g(x - vt)$, in terms of f' and g'.

5. Show that each of the functions in Exercise 4 is a solution of the partial differential equation

$$\frac{\partial^2 z}{\partial x^2} = \frac{1}{v^2}\frac{\partial^2 z}{\partial t^2}.$$

6. Find functions f and g such that $f(x + y) + g(x - y) = \sin x \cos y$.

7. **(a)** What is the relationship between the graphs of $y = \sin x$ and $y = \sin (x - a)$? (Let a be a fixed positive constant.)

(b) What is the relationship between the graphs of $y = f(x)$ and $y = f(x - a)$?

(c) Show that $f(x - vt)$ represents a wave traveling to the right along the x axis, with velocity v, if t represents time. (Assume $v > 0$.)

8. **(a)** Plot $y = \sin x \cos t$ for several values of the time t, and hence convince yourself that this represents a "standing wave."

(b) Show that $y = \sin x \cos t$ can also be obtained as the sum of two traveling waves (one moving to the right and the other to the left along the x axis).

9. **(a)** Letting $u = x + vt$ and $p = x - vt$, write $\partial^2 z/\partial x^2$ and $\partial^2 z/\partial t^2$ in terms of partial derivatives of z with respect to u and/or p. (Assume v is constant.)

(b) Hence, show that with this change of variables the partial differential equation of Exercise 5 takes a simpler form. Solve this equation.

10. By taking $\alpha = x + iy$ and $\beta = x - iy$ (where $i^2 = -1$) change variables in the partial differential equation $\dfrac{\partial^2 z}{\partial x^2} + \dfrac{\partial^2 z}{\partial y^2} = 0$, and solve the resulting equation by partial integration. Write this general solution in terms of x and y.

11. **(a)** Is it possible to find a function $y = f(x - t)$ that is identically zero (for all x) when $t = 0$ but not identically zero for some $t > 0$?

(b) Find a function of the form $y = f(x - t) + g(x + t)$ that is identically zero when $t = 0$ but is not identically zero for any positive value of t.

*12. Find a function satisfying the equation $f(x+t) = f(x)f(t)$.

13. If x and y are functions of time t, find d^2y/dx^2 in terms of time-derivatives of x and y.

14. If $z = f(x,y)$, $x = r\cos\theta$, $y = r\sin\theta$, and z satisfies the differential equation $\partial^2 z/\partial x^2 + \partial^2 z/\partial y^2 = 0$, find a differential equation (involving derivatives with respect to r and θ) satisfied by the function $u(r,\theta) = f(r\cos\theta, r\sin\theta)$.

15. If $\partial x/\partial r = \cos\theta$, is it necessarily true that $\partial r/\partial x = \sec\theta$?

16. If $u = ax + by$ and $v = cx + dy$, find $\partial x/\partial u$ and $\partial x/\partial v$, assuming a, b, c, and d are constants. Is any other assumption necessary?

4.2 • THE INTUITIVE MEANING OF THE LAPLACIAN OPERATOR

The *Laplacian operator*

$$\nabla^2 = \frac{\partial^2}{\partial x^2} + \frac{\partial^2}{\partial y^2} + \frac{\partial^2}{\partial z^2} \tag{1}$$

is by far the most important partial differential operator arising in mathematical physics. This operator is discussed in every book on vector analysis, and will be treated in some detail in Section 5.4. In this section we discuss its intuitive meaning from a nonrigorous viewpoint. We will find later that Legendre polynomials and Bessel functions arise when we attempt to solve boundary-value problems associated with this operator.

If f is a function of three variables x, y, and z, then $\nabla^2 f(x,y,z)$ denotes the value of

$$\nabla^2 f = \frac{\partial^2 f}{\partial x^2} + \frac{\partial^2 f}{\partial y^2} + \frac{\partial^2 f}{\partial z^2} \tag{2}$$

at the point (x,y,z). Thus, if we think of $f(x,y,z)$ as giving us the temperature at an arbitrary point (x,y,z) in space, $\nabla^2 f(x,y,z)$ is a number that tells us something about the behavior of the temperature in the vicinity of (x,y,z). Engineers and physicists who use the Laplacian frequently think of this number as providing a measure of the difference between the average temperature in the immediate neighborhood of the point and the precise value of the temperature at the point.

The word "average" here refers to an average over a region of space, not to a time average. Variations with time t do not enter into the calculation of $\nabla^2 f$.

Thus, if $\nabla^2 f$ is positive at a point, this means that the temperature in a vicinity of the point is, on the average, greater than the temperature at the point itself. In particular, if the temperature takes its minimum value at a certain point in space, it is reasonable to expect that the value of $\nabla^2 f$ will *not* be negative at that point. In this respect, the Laplacian can be viewed as a sort of three-dimensional generalization of the ordinary operator d^2/dx^2, which is used in elementary calculus to test extreme points to see if they represent maxima or minima.

If $\nabla^2 f$ is identically zero, this suggests that the average value of f throughout any sphere (or any cube) will be exactly equal to the value of f at the center of the sphere (or cube). This will be *proved* later.

Suppose that (x,y,z) is a fixed point in space, and that $\bar{f}$ denotes the mean value of f throughout the interior of a sphere (or cube) with center at (x,y,z). If this sphere (or cube) is sufficiently small, we will have (approximately)

$$\bar{f} - f(x,y,z) = K\nabla^2 f(x,y,z) \tag{3}$$

where K is a constant depending only on the dimensions of the sphere or cube. If we take a sphere, we have $K = \frac{1}{10}R^2$, where R is the radius of the sphere. If a cube, $K = a^2/24$, where a is the length of the side of the cube. [The justification for (3) is deferred to Section 5.4.]

Relation (3) is actually correct only in one very special instance: when $\nabla^2 f$ is a constant, independent of x, y, and z. It is approximate otherwise, but the approximation is fairly good in some sense if the sphere (or cube) is sufficiently small. Numerical analysts use an expression similar to (3), but the author knows of no practical application of (3) itself; however, it is very helpful in giving an intuitive meaning to expressions containing the Laplacian.

For example, let us look at the partial differential equation satisfied by the temperature f in a homogeneous isotropic material. If t denotes time, and $k = K/s\rho$ where K is the conductivity of the material, s is its specific heat, and ρ its density, the equation is

$$\nabla^2 f = \frac{1}{k}\frac{\partial f}{\partial t} \qquad (k > 0). \tag{4}$$

It is not our purpose here to derive (4), but simply to interpret it heuristically by using (3).

Suppose that $\nabla^2 f$ is positive at a point. Then by (3) its average value in a neighborhood of that point must be greater than its value at the point itself. Because of this temperature difference, heat must be flowing towards the point in question, raising the temperature at that point. Since the temperature is rising, $\partial f/\partial t$ is positive. This suggests that $\partial f/\partial t$ may be proportional to $\nabla^2 f$. By a more careful dimensional analysis, one can derive (4) by rough reasoning of this sort. In any event, we see that (4) is consistent with the intuitive meaning of the Laplacian operator provided by (3).

Equation (4) is called the *heat equation.* It is also called the equation of diffusion since it is satisfied by the concentration f of a substance which penetrates a porous media by diffusion (the physical constants have a different interpretation in this case).

If the temperature is steady-state (i.e., time-independent) then (4) reduces to *Laplace's equation*

$$\nabla^2 f = 0, \tag{5}$$

which will be discussed in some detail later. Laplace's equation also applies to the potential f of an electrostatic field in a region where no charges are present. If charges are present, the potential of an electrostatic field is governed by

$$\nabla^2 f = -\rho/\epsilon_0 \tag{6}$$

where ρ is the charge density (units of charge per unit volume) and the positive constant ϵ_0 depends on the choice of units.

In the presence of a continuously distributed positive charge, according to (6) the Laplacian will be negative, and therefore, by (3), the value of the potential at a point will be greater than its average value in a neighborhood of the point. This result is familiar to students of physics.

Any equation of the form

$$\nabla^2 f = g \tag{7}$$

where g is a function of position in space (but not of time) is called *Poisson's equation.* Equation (6) is a special case of (7). Equation (7) applies to the steady-state temperature distribution due to a distributed heat source (where the source is not only independent of time but of temperature as well), and to the velocity potential of an incompressible, irrotational ideal fluid with continuously distributed sources or sinks.

We do not assume that the reader has any knowledge or interest in these matters; he will find, on proceeding, that only a very elementary knowledge of physics is needed to understand the applications given in this book.

Try to visualize the following situation, which is relevant to (7). Imagine a chemical process taking place whereby heat is being absorbed (converted to chemical energy) throughout the region. If the rate at which heat is absorbed (per unit volume) is independent of the temperature f, then (7) applies; g is proportional to the time rate of absorption of heat per unit volume. We are assuming that f is a function of x, y, and z alone, and is independent of time (i.e., "steady-state").

If g is positive at a point, then heat must be flowing into a small region surrounding the point (at such a rate that it is totally absorbed by the chemical reaction so that no change in temperature occurs). For heat to flow into the small region, the temperature in this region (i.e., at the point) must be lower than the average surrounding temperature (heat flows from higher to lower temperature) and therefore, by (3), $\nabla^2 f$ must be positive. From this we see that (7) is consistent with the intuitive picture, provided we take g to be proportional to the rate of *absorption* of heat, and not the rate at which heat is *generated* by the reaction. In particular, the reaction is endothermic if g is positive, and is exothermic if g is negative.

An equation of the form

$$\nabla^2 f = pf \qquad (p > 0) \tag{8}$$

may also be "pictured" by thinking of an endothermic reaction, but here the right side contains the temperature f. We must, therefore, assume that the reaction rate (the rate at which heat is absorbed) is proportional to the temperature. Presumably, the function p will depend on the density of the reacting material involved. (For a formal derivation, see Section 5.1.)

Readers with an interest in elasticity or acoustics may find the following example quite interesting. Consider the vibrations of an elastic substance. Let f denote some component of the displacement, at an instant of time, of particles within the substance. In this case, by (3), a positive value of $\nabla^2 f$ will mean that particles in the vicinity of a point are displaced, on the average, a greater amount than a particle at the point itself. Elastic forces will therefore produce a net positive acceleration of the particle. This suggests that $\nabla^2 f$ is proportional to $\partial^2 f/\partial t^2$. Writing $\nabla^2 f = \alpha(\partial^2 f/\partial t^2)$ and using the cgs system, with displacement f in units of centimeters, we see from (2) that the units of $\nabla^2 f$ will be cm^{-1}, and the units of $\partial^2 f/\partial t^2$ will be cm/sec^2, so the units of α will be sec^2/cm^2. Letting $\alpha = 1/v$, we obtain

$$\nabla^2 f = \frac{1}{v^2}\frac{\partial^2 f}{\partial t^2} \tag{9}$$

where the units of v are cm/sec. This partial differential equation is called the *wave equation* and is satisfied by waves having velocity v (independent of wavelength). It arises in theoretical elasticity, acoustics, and electromagnetic wave theory. This derivation is not rigorous; (9) will be derived more rigorously later in a special case. It is nonetheless interesting because it shows the kind of "plausible" reasoning that is possible, and often helpful, in working with the Laplacian operator.

• EXERCISES

1. Suppose that $\nabla^2 f = 0$ identically. Give an heuristic argument to support the formula $\iint \partial f/\partial n \, dS = 0$. The integral is a surface integral over a sphere with center at the origin, and $\partial f/\partial n$ is the derivative of f in the direction of the outer normal to the surface. (In other words, $\partial f/\partial n = \partial f/\partial r$ where $r^2 = x^2 + y^2 + z^2$.)
2. Would you expect the relation given in Exercise 1 to be valid for other surfaces, or is there something special about the choice of a sphere centered at the origin? (Continue to assume that $\nabla^2 f$ is identically zero.)
3. Do you expect the relation given in Exercise 1 to be valid for functions f satisfying (8) instead of Laplace's equation? (Think in terms of an endothermic chemical reaction; no deep physical insight is required.)
4. Suppose f satisfies (8) and is zero at every point on the surface of a sphere. Give reasons why you would expect f to be zero within the sphere as well. We assume f is independent of time. (It is easier to do this if you think of f as temperature in degrees absolute, so that f cannot take negative values. Somewhat deeper physical reasoning is needed if temperature is in degrees Centigrade, to take account of the possibility of negative values of f.)
5. Would your argument (Exercise 4) change drastically if p in (8) is allowed to be negative?

4.3 • LEGENDRE POLYNOMIALS

The Legendre polynomials are defined as follows:

$$(1) \qquad P_n(x) = \frac{1}{2^n n!} \frac{d^n}{dx^n} (x^2 - 1)^n \qquad (n = 0, 1, 2, \cdots).$$

(This is called Rodrigues' formula.) Alternative definitions will be given later. The first few Legendre polynomials are

$$P_0(x) = 1, \qquad P_1(x) = x, \qquad P_2(x) = \tfrac{1}{2}(3x^2 - 1),$$
$$P_3(x) = \tfrac{1}{2}(5x^3 - 3x), \qquad P_4(x) = \tfrac{1}{8}(35x^4 - 30x^2 + 3),$$
$$P_5(x) = \tfrac{1}{8}(63x^5 - 70x^3 + 15x).$$

It will be noticed that, for each fixed $n = 0, 1, 2, \cdots, P_n(x)$ is a polynomial of degree n, $P_n(x)$ contains only odd powers of x when n is odd and only even powers when n is even, $P_n(1) = 1$ for all n (see Exercise 1), and hence that $P_n(-1) = (-1)^n$.

In most applications where Legendre polynomials arise, the x in $P_n(x)$ is the cosine of some angle, and therefore $P_n(x)$ is of interest only for x in the range $-1 \leqq x \leqq 1$.

We frequently need to compute integrals of the form

$$\int_{-1}^{1} f(x) P_n(x)\, dx.$$

If f and its first n derivatives are continuous throughout this interval, this is most easily evaluated through integration by parts. We write

$$(2) \qquad \int_{-1}^{1} f(x) P_n(x)\, dx = \frac{1}{2^n n!} \int_{-1}^{1} f(x) \frac{d^n}{dx^n} (x^2 - 1)^n\, dx$$

and integrate by parts n times. Since the first $n - 1$ derivatives of $(x^2 - 1)^n$ vanish at the endpoints $x = \pm 1$, the term outside the integral sign vanishes at each step and we are left with

$$(3) \qquad \int_{-1}^{1} f(x) P_n(x)\, dx = \frac{(-1)^n}{2^n n!} \int_{-1}^{1} (x^2 - 1)^n \frac{d^n f(x)}{dx^n}\, dx$$

provided that $d^n f/dx^n$ exists and is continuous throughout the interval. This is a very useful formula.

In particular, if we take $f(x) = x^m$ where m is an integer, $0 \leqq m < n$, then $d^n f/dx^n$ is identically zero, and we have

$$(4) \qquad \int_{-1}^{1} x^m P_n(x)\, dx = 0 \qquad (m = 0, 1, \cdots, n - 1).$$

Since $P_m(x)$ is a polynomial of order m, it follows from (4) that

$$\int_{-1}^{1} P_m(x)P_n(x)\,dx = 0 \qquad (m < n). \tag{5}$$

This proves that $P_0, P_1, P_2, P_3, \cdots$ *is an orthogonal sequence* relative to the interval $-1 \leqq x \leqq 1$.

It will be observed that

$$\frac{d^nP_n}{dx^n} = \frac{1}{2^n n!}\frac{d^{2n}}{dx^{2n}}(x^2 - 1)^n = \frac{(2n)!}{2^n n!} \tag{6}$$

since the only term in $(x^2 - 1)^n = x^{2n} - nx^{2n-2} + \cdots$ having non-zero $(2n)$th derivative is x^{2n}. Therefore if we take $f(x) = P_n(x)$ in (3) we obtain

$$\int_{-1}^{1} [P_n(x)]^2\,dx = \frac{(-1)^n}{2^n n!}\frac{(2n)!}{2^n n!}\int_{-1}^{1}(x^2 - 1)^n\,dx, \tag{7}$$

which reduces to (see Exercises 2 and 3)

$$\int_{-1}^{1} [P_n(x)]^2\,dx = \frac{2}{2n + 1}. \tag{8}$$

Theorem 1. *Every polynomial of degree n that is orthogonal to all of the functions $1, x, x^2, \cdots, x^{n-1}$, relative to the interval $-1 \leqq x \leqq 1$, is a scalar multiple of $P_n(x)$.*

Proof: Suppose $p_n(x)$ is a polynomial of degree n which is orthogonal to $1, x, x^2, \cdots, x^{n-1}$. Then for some choice of the scalar C, $CP_n(x)$ will have the same leading coefficient as $p_n(x)$, so that $CP_n(x) - p_n(x)$ is a polynomial of degree at most $n - 1$ (or it is the zero polynomial). By hypothesis, and using (4), it follows that $CP_n(x) - p_n(x)$ is orthogonal to $1, x, \cdots, x^{n-1}$, and hence is orthogonal to any linear combination of $1, x, \cdots, x^{n-1}$. Therefore, $CP_n(x) - p_n(x)$ is orthogonal to itself! That is,

$$\int_{-1}^{1} [CP_n(x) - p_n(x)]^2\,dx = 0,$$

which implies $CP_n(x) - p_n(x) = 0$ and hence $p_n(x) = CP_n(x)$.

Theorem 2. *If the Gram-Schmidt process is applied to the sequence $1, x, x^2, \cdots$, the resulting orthonormal sequence is C_0P_0, C_1P_1, C_2P_2, $C_3P_3, \cdots$ where $C_n = \sqrt{(2n + 1)/2}$.*

Proof: The reader is advised to review the Gram-Schmidt orthogonalization process if he does not feel this theorem follows immediately from Theorem 1 and (8).

Theorem 3. *If f is integrable over $-1 \leqq x \leqq 1$, the polynomial*

$$q_n(x) = A_0P_0(x) + A_1P_1(x) + \cdots + A_nP_n(x) \tag{9}$$

with coefficients

$$A_k = \frac{2k+1}{2}\int_{-1}^{1} f(x)P_k(x)\,dx \tag{10}$$

($k = 0, 1, \cdots, n$) provides the best least-squares approximation to f, in the interval $-1 \leqq x \leqq 1$, of any polynomial of the same degree. In other words, if $r_n(x)$ is any other polynomial of degree n,

$$\int_{-1}^{1} [f(x) - q_n(x)]^2\,dx \leqq \int_{-1}^{1} [f(x) - r_n(x)]^2\,dx. \tag{11}$$

Proof: See Section 2.3, where an even more general result was proved.

Theorem 4. *Every function f, defined and piecewise continuous on $-1 \leqq x \leqq 1$, can be expanded in a series of Legendre polynomials which converges in the mean to f in the interval. That is, we have formally*

$$f(x) = A_0P_0(x) + A_1P_1(x) + A_2P_2(x) + \cdots \tag{12}$$

with coefficients A_k given by (10) [but now for every value of k], and

(13)

$$\int_{-1}^{1} [f(x) - A_0P_0(x) - A_1P_1(x) - \cdots - A_nP_n(x)]^2\,dx \to 0$$

as $n \to \infty$.

Outline of Proof: By the same argument given in the proof of Theorem 4, Section 3.8, we can approximate f as closely as we like, in the mean-square sense, by a continuous function. By Theorem 13, Section 3.7, we can approximate any continuous function as closely as we like uniformly, and *a fortiori* in the mean, by a polynomial (in the interval $-1 \leqq x \leqq 1$). Therefore, given any positive ϵ, there exists a polynomial $r_N(x)$ of degree N such that $\int_{-1}^{1} [f(x) - r_N(x)]^2\,dx < \epsilon$. By Theorem 3, the $N + 1$st partial sum of (12) will provide an approximation that is at least as good as this. Therefore

$$\int_{-1}^{1} [f(x) - A_0P_0(x) - \cdots - A_nP_n(x)]^2\,dx < \epsilon$$

when $n = N$, and as shown in Section 2.2 this inequality is also valid for all $n > N$. Since ϵ can be taken arbitrarily small, this proves (13).

Theorem 5. *If f is piecewise smooth in the interval $-1 \leqq x \leqq 1$, the series (12) will converge to $[f(x+) - f(x-)]/2$ for every value of x in the range $-1 < x < 1$.*

The proof of this theorem is beyond the scope of this book.

In applications, Legendre polynomials do not arise directly from the definition (1). They arise as solutions of a differential equation. This equation is easily derived from (1). In the derivation, we will use the following formula

$$[fg]^{(n)} = \sum_{r=0}^{n} \binom{n}{r} f^{(n-r)} g^{(r)}, \tag{14}$$

which is reminiscent of the binomial theorem, except that $g^{(r)}$ refers to the rth *derivative* of g and similarly for $f^{(n-r)}$. The coefficients $\binom{n}{r}$ are the binomial coefficients $n!/r!(n-r)!$ This formula is derived by the product rule for differentiation, and reduces to the product rule when $n = 1$. Keep in mind that $g^{(0)} = g$, i.e. we differentiate g "zero times" to get g.

Let $f(x) = (x^2 - 1)^n$. By the following calculation,

$$(x^2 - 1) \frac{d}{dx} (x^2 - 1)^n = (x^2 - 1)n(x^2 - 1)^{n-1}(2x) = 2nx(x^2 - 1)^n, \tag{15}$$

we see that

$$(x^2 - 1)f' = 2nxf. \tag{16}$$

Now we differentiate both sides of (16) $n + 1$ times, making use of (14). Since the derivatives of $(x^2 - 1)$ of order three and greater are zero, and derivatives of $2nx$ beyond the second also vanish, we obtain

$$\begin{aligned} (x^2 - 1)f^{(n+2)} + 2x(n+1)f^{(n+1)} + \frac{2(n+1)nf^{(n)}}{2} \\ = 2nxf^{(n+1)} + 2n(n+1)f^{(n)}. \end{aligned} \tag{17}$$

This simplifies to

$$(x^2 - 1)f^{(n+2)} + 2xf^{(n+1)} - n(n+1)f^{(n)} = 0. \tag{18}$$

Since $P_n(x) = \dfrac{1}{2^n n!} f^{(n)}$, by the definition (1), we have

$$(x^2 - 1)P_n''(x) + 2xP_n'(x) - n(n+1)P_n(x) = 0. \tag{19}$$

Letting $y = P_n$, and rewriting the equation slightly, we see that the nth Legendre polynomial is a solution of the differential equation

$$(20) \qquad [(x^2 - 1)y']' - n(n+1)y = 0.$$

This is called the *Legendre equation.*

In applications, this equation sometimes arises in the slightly disguised form

$$(21) \qquad [(x^2 - 1)y']' - \lambda y = 0$$

and it is useful to know that this equation has solutions which are *bounded* in the interval $-1 \leqq x \leqq 1$ *only* if $\lambda = n(n+1)$, for some value of $n = 0, 1, 2, 3, \cdots$ and the Legendre polynomials are (to within a scalar factor) the *only* bounded solutions. We will now proceed to prove this fact.

Part of this proof duplicates the discussion in Section 2.4. First we show that two functions must be orthogonal over $-1 \leqq x \leqq 1$ if they are bounded solutions of (21) corresponding to different values of λ. Indeed, if u and v are such functions, i.e.,

$$(22) \qquad [(x^2 - 1)u']' = \lambda_1 u, \qquad [(x^2 - 1)v']' = \lambda_2 v,$$

then multiplying the first of these equations by v, the second by u, subtracting the two equations, and integrating by parts, we obtain (see Section 2.4 if this seems too brief)

$$(23) \qquad [(x^2 - 1)(u'v - v'u)\Big|_{-1}^{1} = \int_{-1}^{1} (\lambda_1 - \lambda_2)uv\,dx.$$

Since $(x^2 - 1)$ is zero at $x = \pm 1$, the left side vanishes, and if $\lambda_1 \neq \lambda_2$ it follows that $\int_{-1}^{1} u(x)v(x)\,dx = 0$.

Now let us suppose that there is a solution w of (21) that is not a scalar multiple of one of the Legendre polynomials $P_n(x)$. There are two possibilities: either w is a solution for some λ not equal to one of the numbers $n(n+1)$, $n = 0, 1, 2, \cdots$, or it is a second solution of (20) for some n. In the first instance, the function w will be orthogonal to $P_n(x)$ for *every* $n = 0, 1, 2, \cdots$, and the second instance w is orthogonal to every $P_n(x)$ *except* perhaps for one value of n, say $n = n_0$. In the second case, we can apply the Gram-Schmidt process to the two functions P_{n_0}, w to produce a nontrivial function f that is orthogonal to P_{n_0} and is a linear combination of P_{n_0} and w. This function f is also a nontrivial solution of (21), for $n = n_0$, and is orthogonal to $P_n(x)$ for *every* $n = 0, 1, 2, \cdots$.

In either case, we have a nontrivial function f that satisfies (21) for some λ, and is orthogonal to every P_n. Since it is orthogonal to every P_n, and every polynomial can be written as a linear combination of Legendre polynomials, it must also be orthogonal to every polynomial. We have

$$\int_{-1}^{1} f(x)p_n(x)\,dx = 0 \tag{24}$$

for every polynomial $p_n(x)$. Since f is bounded and continuous, there exists a sequence of polynomials converging uniformly to f, by Theorem 13, Section 3.7. Taking this sequence to be $p_n(x)$, and passing to the limit inside the integral (24), which is justified since the sequence is uniformly convergent, we obtain $\int_{-1}^{1} [f(x)]^2\,dx = 0$. Since f is continuous, this implies $f(x) = 0$ identically in $-1 \leqq x \leqq 1$. We have proved:

Theorem 6. *Legendre's equation* (21) *has solutions that are bounded in the interval* $-1 \leqq x \leqq 1$ *if and only if* $\lambda = n(n+1)$. *If* $\lambda = n(n+1)$, *Legendre's equation has, to within a scalar factor, only one solution that is bounded in this interval, namely* $P_n(x)$.

The reader may feel that this contradicts an existence theorem quoted earlier, which states that a second-order linear differential equation, with coefficients continuous in an interval, must have two linearly independent solutions in that interval. Thus, we expect (20) to have, for every value of n, a general solution of the form

$$y(x) = C_1P_n(x) + C_2Q_n(x) \tag{25}$$

where P_n and Q_n are linearly independent. It will be recalled, however, that the existence theorem in question is valid only if the leading coefficient of the differential equation is always nonzero.

The leading coefficient in (20) is nonzero when $-1 < x < 1$, but is zero at the endpoints. Therefore the existence theorem guarantees a solution of the form (25) valid in the *open interval* $-1 < x < 1$, but not necessarily in the *closed interval* $-1 \leqq x \leqq 1$. It turns out, in this case, that every solution of (21) for $\lambda \neq n(n+1)$ is unbounded at one of the endpoints, and if $\lambda = n(n+1)$ the "other" solution $Q_n(x)$ is unbounded at *both* endpoints. Thus, Theorem 6 does not contradict the existence theorem. This also explains why we can prove orthogonality more easily here than in most Sturm-Liouville equations; the only "boundary condition" needed is boundedness at the endpoints (compare Example 6, Section 2.4).

Let us return to the expansion (12). If a function f is even, $f(x) = f(-x)$, we will have

$$\begin{aligned} A_k &= (2k+1)\int_0^1 f(x)P_k(x)\,dx \qquad (k = 0, 2, 4, \cdots) \\ A_k &= 0 \qquad\qquad\qquad\qquad\qquad\quad (k = 1, 3, 5, \cdots). \end{aligned} \tag{26}$$

If f is *not* even, and we form the expansion using these coefficients, it will represent the function f only in the interval (0,1), and will represent the *even extension* of f in the full interval (−1,1). (Compare this with Fourier cosine series.) Similarly, if we define the coefficients by

$$\begin{aligned} A_k &= (2k+1)\int_0^1 f(x)P_k(x)\,dx \qquad (k = 1, 3, 5, \cdots) \\ A_k &= 0 \qquad (k = 0, 2, 4, \cdots) \end{aligned} \tag{27}$$

we obtain the *odd extension* of f, by analogy with Fourier sine series. An example will be found at the end of Section 6.4.

EXAMPLE 1: Determine the Legendre expansion in $-1 \leqq x \leqq 1$ of the function $f(x) = 0(-1 \leqq x < 0)$, $f(x) = 1(0 \leqq x \leqq 1)$.

Solution: Since f is not continuous, we cannot make use of (3). Since $f(x) = 0$ in the interval $-1 < x < 0$, and is unity in the rest of the interval, we can write (10) as

$$A_k = \frac{2k+1}{2}\int_0^1 f(x)P_k(x)\,dx = \frac{2k+1}{2}\int_0^1 P_k(x)\,dx. \tag{28}$$

To evaluate $\int_0^1 P_k(x)$ we integrate (20) over the interval (0,1), obtaining (if we set $n = k$ in that formula)

$$k(k+1)\int_0^1 P_k(x)\,dx = \left[(x^2-1)\frac{dP_k}{dx}\right]\Bigg|_0^1 = P_k'(0). \tag{29}$$

By a slightly tedious calculation (see Exercise 4), we find that

$$\begin{aligned} P_k'(0) &= (-1)^{(k-1)/2}\frac{1\cdot3\cdot5\cdots k}{2\cdot4\cdot6\cdots(k-1)} \qquad (k = 3, 5, 7, \cdots) \\ &= 1 \qquad (\text{when } k = 1) \\ &= 0 \qquad (\text{when } k = 0, 2, 4, 6, \cdots) \end{aligned} \tag{30}$$

and therefore

$$\begin{aligned} A_k &= (-1)^{(k-1)/2}\frac{2k+1}{2k+2}\,\frac{1\cdot3\cdot5\cdots(k-2)}{2\cdot4\cdot6\cdots(k-1)} \qquad (k = 3, 5, 7, \cdots) \\ A_k &= 3/4 \qquad (\text{when } k = 1) \\ A_k &= 1/2 \qquad (\text{when } k = 0) \\ A_k &= 0 \qquad (\text{when } k = 2, 4, 6, \cdots). \end{aligned} \tag{31}$$

The first few terms are

$$f(x) = \tfrac{1}{2}P_0(x) + \tfrac{3}{4}P_1(x) - \tfrac{7}{16}P_3(x) + \tfrac{11}{32}P_5(x) + \cdots \qquad (-1 < x < 1).$$

EXAMPLE 2: Expand $\cos 2\phi$ in a series $A_0P_0(\cos\phi) + A_1P_1(\cos\phi) + A_2P_2(\cos\phi) + \cdots$.

Solution: Letting $x = \cos\phi$ in (10) and writing $g(\phi)$ instead of $f(\cos\phi)$, we obtain

$$A_k = \frac{2k+1}{2}\int_0^\pi g(\phi)P_k(\cos\phi)\sin\phi\,d\phi. \tag{32}$$

Thus, we have $A_0 = \frac{1}{2}\int_0^\pi \cos 2\phi\sin\phi\,d\phi = \frac{2}{3}$,

$A_1 = \frac{3}{2}\int_0^\pi \cos 2\phi\cos\phi\sin\phi\,d\phi = 0$,

$A_2 = \frac{5}{2}\int_0^\pi \cos 2\phi[\frac{1}{2}(3\cos^2\phi - 1)]\sin\phi\,d\phi = \frac{4}{3}$, etc.

There is an easier way, however. Keeping one eye on the list of Legendre polynomials, and letting $x = \cos\phi$, we have

$$\begin{aligned}\cos 2\phi &= 2\cos^2\phi - 1 = 2x^2 - 1\\ &= \tfrac{4}{3}[\tfrac{1}{2}(3x^2 - 1)] + \tfrac{2}{3} = \tfrac{4}{3}P_2(x) + \tfrac{2}{3}P_0(x).\end{aligned}$$

Hence,

$$\cos 2\phi = \tfrac{2}{3}P_0(\cos\phi) + \tfrac{4}{3}P_2(\cos\phi).$$

The "series" has only two nonvanishing terms.

• EXERCISES

1. Directly from (1), show that $P_n(1) = 1$ for all n. Hint: let $z = 1 - \epsilon$. Then

$$\begin{aligned}\frac{1}{2^n n!}\frac{d^n}{dx^n}(x^2 - 1)^n &= \frac{1}{2^n n!}\frac{d^n}{d\epsilon^n}(2\epsilon - \epsilon^2)^n\\ &= \frac{1}{2^n n!}\frac{d^n}{d\epsilon^n}[(2\epsilon)^n + (\text{higher order terms in } \epsilon)].\end{aligned}$$

2. (a) Show that $\int_{-1}^1 (1 - x^2)^n\,dx = \int_0^\pi \sin^{2n+1}\phi\,d\phi$.

(b) Show that $\displaystyle\int_0^\pi \sin^{2n+1}\phi\,d\phi = \frac{2n}{2n+1}\int_0^\pi \sin^{2n-1}\phi\,d\phi$.

(c) Hence, derive $\displaystyle\int_0^\pi \sin^{2n+1}\phi\,d\phi = 2\left(\frac{2\cdot 4\cdot 6\cdots 2n}{3\cdot 5\cdot 7\cdots(2n+1)}\right)$, $n = 1, 2, 3, \cdots$.

(d) Show that $\displaystyle\frac{2\cdot 4\cdot 6\cdots 2n}{3\cdot 5\cdot 7\cdots(2n+1)} = \frac{2^{2n}(n!)^2}{(2n+1)!}$.

(e) Hence, derive (8) from (7). Another method is given in the next exercise.

3. Fill in the missing steps:

$$\int_{-1}^{1} (1 - x)^n(1 + x)^n \, dx = \frac{n}{n+1} \int_{-1}^{1} (1 - x)^{n-1}(1 + x)^{n+1} \, dx$$

$$= \frac{(n!)^2}{(2n)!} \int_{-1}^{1} (1 + x)^{2n} \, dx$$

$$= \frac{(n!)^2}{(2n)!(2n + 1)} 2^{2n+1}.$$

(Compare the preceding exercise and Exercise 22.)

4. **(a)** Show that

$$\frac{d^{n+1}}{dx^{n+1}} (x^2 - 1)^n = \sum_{r=0}^{n} (-1)^r \binom{n}{r} (2n - 2r)$$

$$(2n - 2r - 1) \cdots (2n - 2r - n)x^{n-2r-1}$$

where $\binom{n}{r} = \dfrac{n!}{r!(n - r)!}$.

(b) Hence, show that

$$P_n'(0) = \frac{1}{2^n n!} (-1)^{(n-1)/2} \frac{n!(n + 1)!}{\left(\frac{n - 1}{2}\right)! \left(\frac{n + 1}{2}\right)!}$$

and state for what values of n this is valid.

(c) Derive (30) from this expression.

***5.** Show that, if $P_n(x)$ is multiplied by the reciprocal of the coefficient of x^n, we obtain a polynomial having the following property: it has the smallest mean-square norm (relative to $-1 \leqq x \leqq 1$) of any polynomial of the nth degree with leading coefficient 1.

6. Find the first three coefficients in the expansion of the function

$$f(x) = \begin{cases} 0 & (-1 \leqq x < 0) \\ x & (\ 0 \leqq x \leqq 1) \end{cases}$$

in a series of Legendre polynomials.

7. Determine $\int_{-1}^{1} x^n P_n(x) \, dx$.

8. Write $\cos^3 \phi$ as a linear combination of functions of the form $P_n(\cos \phi)$.

9. Show that $f(r,\phi) = r^n P_n(\cos\phi)$ satisfies the partial differential equation

$$\frac{1}{r^2}\frac{\partial}{\partial r}\left(r^2\frac{\partial f}{\partial r}\right) + \frac{1}{r^2\sin\phi}\left[\frac{\partial}{\partial\phi}\left(\sin\phi\,\frac{\partial f}{\partial\phi}\right)\right] = 0$$

independently of n.

10. **(a)** Show that the sequence of functions $P_0\left(\frac{x}{L}\right)$, $P_1\left(\frac{x}{L}\right)$, $P_2\left(\frac{x}{L}\right)$, $\cdots$ is orthogonal relative to the interval $-L \leqq x \leqq L$.

(b) Write a formula similar to (10) that would be useful in expanding a function in a sequence of functions $P_n\left(\frac{x}{L}\right)$ relative to the interval $-L \leqq x \leqq L$.

11. **(a)** Is $P_n(\cos\phi)$ a periodic function of ϕ for all n?

(b) For what values of n is $P_n(\cos\phi)$ an *even* function of ϕ? (The despair of every teacher is the student who automatically answers "for even values of n" without *thinking*.)

(c) Can every suitable smooth function $f(\phi)$ of period 2π be expanded in a series $\sum_{n=0}^{\infty} A_n P_n(\cos\phi)$?

12. Using (14), or otherwise, show that

$$\frac{d^{n+1}(xP_n)}{dx^{n+1}} = (n+1)\frac{d^nP_n}{dx^n} = (n+1)\frac{(2n)!}{2^n n!}.$$

13. Using (3) and the results of Exercises 3 and 12, show that

$$\frac{2n+3}{2}\int_{-1}^{1} xP_n(x)P_{n+1}(x)\,dx = \frac{n+1}{2n+1}.$$

14. Directly from the result in Exercise 13, show that

$$\frac{2n-1}{2}\int_{-1}^{1} xP_n(x)P_{n-1}(x)\,dx = \frac{n}{2n+1}.$$

15. Show that $\int_{-1}^{1} x[P_n(x)]^2\,dx = 0$. (No calculation is required.)

***16.** Show that, if $|n - m| > 1$,

$$\int_{-1}^{1} xP_n(x)P_m(x)\,dx = 0.$$

(No calculation is needed.)

17. Deduce from the preceding exercises that

$$(2n + 1)xP_n(x) = (n + 1)P_{n+1}(x) + nP_{n-1}(x).$$

Discuss what this becomes if $n = 0$. (This is an important "recurrence relation," and is related to the selection rules of quantum mechanics.)

18. With most of the simpler second-order differential equations, one can specify the values of $f(x)$ and $f'(x)$ quite arbitrarily at a point, and find a solution satisfying these conditions. Explain why this is not possible at $x = 1$ or $x = -1$ for the differential equation (19). (Absolutely no knowledge of the theory discussed in this section is needed to answer this question. Just look at the differential equation.)

19. Verify the equation $P_n'(1) = n(n + 1)/2$ for a few of the Legendre polynomials listed in the first paragraph of this section, and prove that it is valid quite generally.

20. Write $\int_{-1}^{1} (1 - x^2)^n e^x \, dx$ as an integral involving Legendre polynomials.

21. Let $I_n(x) = \int_a^x (x - t)^{n-1} f(t) \, dt$, where f is a continuous function.

(a) Show that $dI_n/dx = (n - 1)I_{n-1} \quad (n \geqq 1)$.

(b) Show that $d^nI_n/dx^n = (n - 1)!f(x)$.

(c) Show that $I_n(x)$ and its first $n - 1$ derivatives vanish when $x = a$.

(d) Hence, show that $I_n(x)/(n - 1)!$ is equal to the function obtained by integrating $f(x)$ n times from a to x. With a slight abuse in notation, this can be written in the form

$$\int_a^x \int_a^x \cdots \int_a^x f(x) \, dx \cdots dx = \frac{1}{(n - 1)!} \int_a^x (x - t)^{n-1} f(t) \, dt. \tag{33}$$

(This relation can also be deduced through integration by parts.) This relation is frequently used in solving integral equations.

22. Making use of (33), show that $\int_{-1}^{1} (1 - x^2)^n \, dx$ can be obtained by integrating $n!(1 + x)^n$, from -1 to x, $n + 1$ times, to give $n!(1 + x)^{2n+1}/(n + 1)(n + 2) \cdots (2n + 1)$, and then evaluating at $x = 1$. Compare your answer with that obtained in Exercises 2 and 3.

The following exercises are intended for readers who have trouble with factorials and binomial expansions.

23. Show how the following identity is obtained:

$$1\cdot 3\cdot 5\cdot 7 \cdots n = \frac{(n+1)!}{2^{(n+1)/2}\left(\frac{n+1}{2}\right)!}. \tag{34}$$

24. Deduce the relation $\binom{n}{r} + \binom{n}{r+1} = \binom{n+1}{r+1}$ for the binomial coefficients. Note that

$$\binom{k}{s} = \frac{k!}{s!(k-s)!}.$$

25. Schoolboys use "Pascal's triangle" to construct the binomial coefficients $\binom{n}{r}$. Explain why it works. The triangle is formed as follows, with each interior entry the sum of the two entries above it, taking boundary entries to be unity:

$$\begin{array}{ccccccccc} & & & & 1 & & & & \\ & & & 1 & & 1 & & & \\ & & 1 & & 2 & & 1 & & \\ & 1 & & 3 & & 3 & & 1 & \\ 1 & & 4 & & 6 & & 4 & & 1 \\ & & & & \cdots\cdots & & & & \end{array}$$

26. By induction, prove

$$(a+b)^n = \sum_{k=0}^{n} \binom{n}{k} a^{n-k} b^k$$

for integer values of n, $n > 0$.

27. Show that $2\cdot 4\cdot 6 \cdots n = 2^{n/2}\left(\frac{n}{2}\right)!$

4.4 • LAPLACE'S EQUATION IN SPHERICAL COORDINATES

In Section 5.1, we will show that the steady-state temperature distribution in a homogeneous, isotropic material, where there are no sources or sinks, satisfies Laplace's equation $\nabla^2 f = 0$. (A heuristic derivation was given in Section 4.2.) In spherical coordinates, Laplace's equation is

$$\frac{\partial^2 f}{\partial r^2} + \frac{2}{r}\frac{\partial f}{\partial r} + \frac{1}{r^2}\frac{\partial^2 f}{\partial \phi^2} + \frac{\cot\phi}{r^2}\frac{\partial f}{\partial \phi} + \frac{1}{r^2 \sin^2\phi}\frac{\partial^2 f}{\partial \theta^2} = 0. \tag{1}$$

Solutions of Laplace's equation are called *harmonic functions.* (A more precise definition of harmonic function will be given in Section 5.4.) Our choice of spherical coordinates is illustrated in Figure 4.1.

We will make no attempt to find a "general solution" of (1). Instead, we will obtain particular solutions by using Bernoulli's method of *separation of variables.*

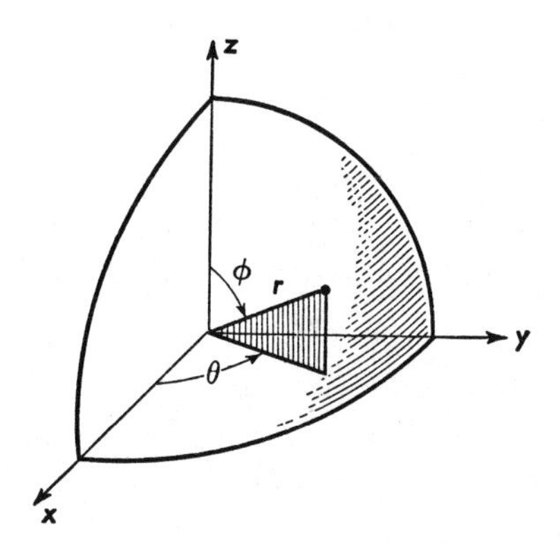

Figure 4.1

In this case, the procedure begins by assuming that there exist solutions of Laplace's equation having the form

$$u(r,\phi,\theta) = R(r)\Phi(\phi)\Theta(\theta) \tag{2}$$

where each of the functions R, Φ, and Θ is a function of only a single variable.

Substituting (2) into (1) we find that u satisfies Laplace's equation if and only if

$$R''\Phi\Theta + \frac{2}{r}R'\Phi\Theta + \frac{1}{r^2}R\Phi''\Theta + \frac{R\Phi'\Theta}{r^2 \tan\phi} + \frac{R\Phi\Theta''}{r^2\sin^2\phi} = 0. \tag{3}$$

Since each function depends only on one variable, there is no ambiguity in using primes to denote differentiation.

Multiplying by $(r^2 \sin^2 \phi)/R\Phi\Theta$ and rearranging, we have

$$\sin^2\phi\left[\frac{r^2R''}{R} + 2r\frac{R'}{R} + \frac{\Phi''}{\Phi} + \cot\left(\frac{\Phi'}{\Phi}\right)\right] = -\frac{\Theta''}{\Theta} \tag{4}$$

where the limit convention is understood to give meaning to these terms whenever denominators vanish.

Now we are ready to apply the key idea of the method: *nonconstant functions of independent coordinates are linearly independent.* In particular, a function of θ can be identically equal to a function of r and ϕ only if it is constant, since otherwise we could contradict the equality by holding r and ϕ fixed and varying θ. It follows that both sides of (4) must be constant functions. We denote this constant value by m^2 (this will be explained below) and obtain two differential equations,

$$-\frac{\Theta''}{\Theta} = m^2, \tag{5}$$

and (after a little rearrangement),

$$\frac{r^2R''}{R} + 2r\frac{R'}{R} = -\left[\frac{\Phi''}{\Phi} + \cot\phi\left(\frac{\Phi'}{\Phi}\right) - m^2\csc^2\phi\right]. \tag{6}$$

We met (5) before, in Section 2.4. If $m \neq 0$, every solution of (5) is a linear combination of the functions $\sin m\theta$ and $\cos m\theta$. To avoid the inconvenience of writing $\sqrt{m}$ rather than m, we chose the "separation constant" to be m^2 rather than m. Had we chosen $-m^2$ instead, we would obtain $e^{m\theta}$ and $e^{-m\theta}$ as real solutions of (5), instead of the sinusoidal functions $\sin m\theta$ and $\cos m\theta$. Since θ denotes an angle, and we want single-valued solutions, we reject $e^{m\theta}$ and $e^{-m\theta}$ as inacceptable. (If m were allowed to be pure imaginary, these would be acceptable *complex* solutions.)

Now let us look at (6). For the same reasons as before, both sides of this equation must be constant functions. Anticipating later developments, we take the separation constant to be $n(n + 1)$ and obtain two differential equations,

$$r^2R'' + 2rR' - n(n+1)R = 0 \tag{7}$$

$$\Phi'' + (\cot\phi)\Phi' + [n(n+1) - m^2\csc^2\phi]\Phi = 0 \tag{8}$$

First let us solve (7). This is of a type treated in books on differential equations, and such books give a trick substitution to use (see Exercise 1) but this is hardly necessary here. In the semi-infinite interval $r > 0$ (negative values of r do not interest us here) the leading coefficient is nonzero, and according to the theory of such equations we can write down the general solution provided we can, by some method, find two linearly independent solutions. The best method for finding such solutions, and one that should always be tried first, is to *guess* them. (For example, to solve (5) one need only guess two functions that, except for a scalar factor, are the negatives of their own second derivatives.) Anyone who has ever differentiated a polynomial should guess that (7) has a polynomial solution, since the second derivative of an nth order polynomial is of order $n - 2$, the first derivative is of order $n - 1$, and the multiplicative coefficients in (7) restore these derivatives to give a polynomial of order n.

If we substitute $R = r^k$ into (7) we obtain $k(k - 1)r^k + 2kr^k - n(n + 1)r^k = 0$, which is identically satisfied if $n = k$ or $n = -(k + 1)$. Since the functions r^n and r^{-n-1} are linearly independent, the general solution of (7) for $r > 0$ is an arbitrary linear combination of these two functions.

Now let us look at (8). Here, guessing is not much help, nor are the special tricks given in elementary courses in differential equations. If one is lucky, one might notice that any constant function is a solution when $m = n = 0$ and that $\cos \phi$ is a solution when $m = 0$ and $n = 1$, but this isn't much help in general.

When other methods fail, one usually seeks a series solution. (This will be demonstrated later on.) To avoid having to expand $\cot \phi$ and $\csc^2 \phi$ in powers of ϕ (which is, in fact, impossible), it is convenient to attempt to change variables to eliminate the trigonometric coefficients. We will now show that the substitution $z = \cos \phi$ is successful.

If $z = \cos \phi$ then

$$\frac{d}{d\phi} = \frac{dz}{d\phi}\frac{d}{dz} = -\sin \phi \frac{d}{dz} \quad \text{and} \quad \frac{d^2}{d\phi^2} = -\cos \phi \frac{d}{dz} + \sin^2 \phi \frac{d^2}{dz^2}.$$

(See Exercise 5.) Replacing $\sin^2 \phi$ by $1 - z^2$ and $\cos \phi$ by z, (8) becomes

$$(9) \qquad (1 - z^2)\frac{d^2\Phi}{dz^2} - 2z\frac{d\Phi}{dz} + \left(n(n + 1) - \frac{m^2}{1 - z^2}\right)\Phi = 0.$$

This equation, and its equivalent (8), is called the *associated Legendre equation.*

When $m = 0$, (9) reduces to *Legendre's equation,* introduced in the preceding section.

Since $-1 \leqq \cos \phi \leqq 1$ for every ϕ, we are concerned with solutions of (9) only in the interval $-1 \leqq z \leqq 1$. For fixed n and m, let us denote linearly independent solutions of (9) by $P_n^{(m)}(z)$ and $Q_n^{(m)}(z)$. (Notice that (m) here does *not* refer to m-fold differentiation.)

Summarizing, we see that there are many solutions of Laplace's equation having the special form (2). For fixed m and n we have solutions

$$u_n^{(m)}(r,\phi,\theta) = \begin{Bmatrix} r^n \\ r^{-n-1} \end{Bmatrix} \begin{Bmatrix} P_n^{(m)}(\cos\phi) \\ Q_n^{(m)}(\cos\phi) \end{Bmatrix} \begin{Bmatrix} \cos m\theta \\ \sin m\theta \end{Bmatrix} \tag{10}$$

where we are free to choose any combination of the three factors.

We have not yet determined the linearly independent solutions $P_k^{(m)}$ and $Q_k^{(m)}$. Discussion of the case $m \neq 0$ will be deferred to the next section. When $m = 0$, the superscript is customarily omitted. We have already proved that Legendre's equation

$$(1 - z^2)\frac{d^2\Phi}{dz^2} - 2z\frac{d\Phi}{dz} + n(n + 1)\Phi = 0 \tag{11}$$

has bounded solutions in the interval $-1 \leqq z \leqq 1$ when $n = 0, 1, 2, \cdots$, and that these solutions $P_n(z)$ are the only solutions bounded in this interval. The reader will notice that the same solutions are obtained when $n = -1, -2, -3, \cdots$ (Exercise 6).

However, we "backed into" these solutions in the preceding section. It may be well to see how they could have been obtained directly. Standard theorems on differential equations assure us that we can obtain all solutions of (11) in the form

$$\Phi(z) = a_0 + a_1 z + a_2 z^2 + \cdots \qquad (-1 < z < 1). \tag{12}$$

If we substitute (12) into (11), the coefficient of z^j on the left side will be

$$(j + 2)(j + 1)a_{j+2} - j(j - 1)a_j - 2ja_j + n(n + 1)a_j. \tag{13}$$

Since the right side of (11) is identically zero, and all of the coefficients of the power series expansion of the zero function are zero, we set (13) equal to zero, which yields the recursion relation

$$a_{j+2} = \frac{j(j + 1) - n(n + 1)}{(j + 2)(j + 1)}a_j = -\frac{(n - j)(n + j + 1)}{(j + 2)(j + 1)}a_j. \tag{14}$$

If we choose a_0 and a_1 arbitrarily, the other coefficients will be determined by (14). Assuming that the resulting series converges, we write the solution in the form

$$\Phi(z) = a_0 u_n(z) + a_1 v_n(z). \tag{15}$$

The expansion of u_n contains only even powers of z, and v_n contains only odd powers of z.

If n is a positive integer, we see from (14) that $a_{j+2} = 0 \cdot a_j = 0$ when $j = n$. Therefore, if n is even, u_n has only a finite number of nonzero terms; it is a polynomial of degree n. Similarly, if n is odd, v_n breaks off, and is a polynomial of degree n. No problem of convergence arises in these instances. Letting $P_n(z)$ denote the polynomial solutions thus obtained, and choosing the arbitrary constant a_0 or a_1 in such a manner that $P_n(1) = 1$, we obtain

$$
\begin{aligned}
P_0(z) &= a_0u_0(z) = 1 \\
P_1(z) &= a_1v_1(z) = z \\
P_2(z) &= a_0u_2(z) = -\tfrac{1}{2} + \tfrac{3}{2}z^2 \\
P_3(z) &= a_1v_3(z) = -\tfrac{3}{2}z + \tfrac{5}{2}z^3 \\
&\cdots
\end{aligned}
\tag{16}
$$

These are the Legendre polynomials, and now we have found them by a direct procedure.

If n is even, the power series expansion of v_n does not break off, and if n is odd, the series for u_n is likewise nontrivial. Let us denote these $Q_n(z)$. Thus, if $n = 0$, we obtain

$$
\begin{aligned}
Q_0(z) = a_1v_0(z) &= a_1\left(z + \frac{1\cdot 2}{2\cdot 3}z^3 + \frac{1\cdot 2\cdot 3\cdot 4}{2\cdot 3\cdot 4\cdot 5}z^5 + \cdots\right) \\
&= a_1(z + \tfrac{1}{3}z^3 + \tfrac{1}{5}z^5 + \cdots)
\end{aligned}
\tag{17}
$$

and if $n = 1$, we obtain

$$
\begin{aligned}
Q_1(z) &= a_0u_1(z) \\
&= a_0\left(1 - z^2 - \frac{(2\cdot 3 - 2)}{3\cdot 4}z^4 - \frac{(2\cdot 3 - 2)(4\cdot 5 - 2)}{3\cdot 4\cdot 5\cdot 6}z^6 - \cdots\right) \\
&= a_0(1 - z^2 - \tfrac{1}{3}z^4 - \tfrac{1}{5}z^6 - \tfrac{1}{7}z^8 - \cdots).
\end{aligned}
\tag{18}
$$

By the ratio test, these series converge in the interval $-1 < z < 1$. The resulting functions are not bounded but "tend to infinity" at both endpoints. They cannot be normalized in the same manner as the Legendre polynomials were normalized, which is why we have left the arbitrary constants a_0 and a_1 in the expression given above. These functions are called *Legendre functions of the second kind.* They will not be discussed further in this book.

It is possible to find infinite series solutions of (11) involving descending negative exponents of z, which converge when $|z| > 1$, but they are of no interest to us, since $z = \cos\phi$ in the problems that arise later.

If we ignore the solutions $Q_n(z)$, then when $m = 0$ (10) becomes

$$
u_n(r,\phi,\theta) = r^nP_n(\cos\phi) \quad \text{or} \quad r^{-n-1}P_n(\cos\phi).
\tag{19}
$$

since θ does not appear in these expressions, they are harmonic functions having *cylindrical symmetry* about the z axis. When $n = 0$, the second of these solutions is simply $1/r$, and can be identified with the electric potential due to a point charge located at the origin.

When $n = 1$, the second expression is $(\cos \phi)/r^2$, which some readers will recognize as the electric potential due to a point *dipole* at the origin. Higher values of n correspond to potentials due to higher order multipoles at the origin (i.e., $n = 2$ is the potential due to a linear quadrupole, i.e., a "dipole consisting of two dipoles"). When $m \neq 0$,

$$u_n^{(m)}(r,\phi,\theta) = r^{-n-1}P_n^{(m)}(\cos \phi) \begin{cases} \cos m\theta \\ \sin m\theta \end{cases}$$

has a similar interpretation, in terms of multipoles at the origin that do not possess cylindrical symmetry about the z axis. Interested readers will find further details in Morse and Feshbach, *Methods of Theoretical Physics* (McGraw-Hill, 1953, page 1281).

If the reader has read the earlier chapters, he is probably beginning to suspect that the basic *purpose* of the Bernoulli method is to provide a collection of functions that will be an *approximating basis* for a linear space of functions. This is essentially true, but the situation is more complicated than anything we have considered so far. To avoid burdening the reader with a lot of theory, we will consider only special cases.

First, let us suppose that a function is continuous and satisfies Laplace's equation throughout the interior of a sphere of $r = 1 + \epsilon$, where $\epsilon > 0$. This sphere is centered at the origin, and has a radius greater than unity; this ensures that the function is quite smooth, both in the interior and on the surface of the unit sphere $r = 1$. It can be shown that such a function must be the sum of a series of the form

$$(20) \quad f(r,\phi,\theta) = \sum_{n=0}^{\infty} \sum_{m=0}^{n} r^n P_n^{(m)}(\cos \phi)[A_{n,m} \cos m\theta + B_{n,m} \sin m\theta]$$

which will be uniformly convergent for $0 \leqq r \leqq 1$. Since f is continuous, we do not need to make use of terms having the factor r^{-n-1} or $Q_n^{(m)}$. The situation here is quite similar to expanding a function in a Fourier series, except that the integral relating $f(r,\phi,\theta)$ to its coefficients $A_{n,m}$ and $B_{n,m}$ is more complicated. The reader is asked for the moment not to question why the sum over m runs from 0 to n rather than from $-\infty$ to ∞ or from 0 to ∞; this will be indicated in the next section.

On the other hand, suppose that f is a continuous solution of Laplace's equation for $r > 1 - \epsilon$, with $\epsilon > 0$, and that f tends to zero as $r \to \infty$ (independently of ϕ and θ). Then a similar expansion is valid, with r^n replaced by r^{-n-1}. In view of the remarks made above, this has an immediate physical interpretation: the potential outside a sphere centered at the origin due to any distribution of charge contained entirely in the interior of the sphere is equivalent to the potential that would result from a point charge at the origin, superimposed with various point dipoles and other multipoles located at the origin. Thus, one can never determine exactly what the distribution of charge is, within the sphere, from measurements of the resulting potential field outside the sphere.

We cannot, and do not want to, discuss either of these special cases in detail, although some closely related applications will be taken up in later chapters. We conclude this section by mentioning one extremely special case, because it plays a fundamental role in the theory of Legendre polynomials.

Let us consider a point charge located at the point $x = 0$, $y = 0$, $z = 1$. Let R denote the distance between this point and an arbitrary point (x,y,z), as shown in Figure 4.2. If (r,ϕ,θ) denotes

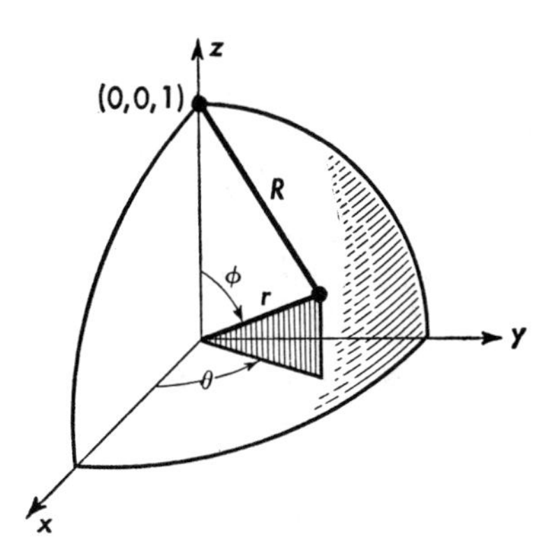

Figure 4.2

the spherical coordinates of this generic point, a simple application of the cosine law of trigonometry (the "generalized Pythagorean theorem") yields

$$R = (1 - 2r\cos\phi + r^2)^{1/2}. \tag{21}$$

We can interpret $1/R$ as the electrostatic potential due to this point charge. Since this potential is cylindrically symmetric about the

z axis, we expect to be able to expand it in a series of terms of the form $r^n P_n(\cos\phi)$ or $r^{-n-1}P_n(\cos\phi)$.

In fact, each of the following expansions is valid. (To relate them to the example just given, let $t = \cos\phi$.)

$$(22)\quad (1 - 2rt + r^2)^{-1/2} = \sum_{n=0}^{\infty} r^n P_n(t) \qquad \text{when } |r| < 1,\ -1 \leqq t \leqq 1.$$

$$(23)\quad (1 - 2rt + r^2)^{-1/2} = \sum_{n=0}^{\infty} r^{-n-1} P_n(t) \qquad \text{when } |r| > 1,\ -1 \leqq t \leqq 1.$$

Because of these expansions, the function $(1 - 2rt + r^2)^{-1/2}$ is called the *generating function* for the Legendre polynomials. As will be indicated in the exercises, it is possible to base the entire theory of Legendre polynomials on these expansions.

• EXERCISES

1. **(a)** By changing variables, from r to $t = e^r$, show that (7) becomes an equation with constant coefficients. Solve this equation.

(b) For what differential equations would you expect this procedure to work?

2. Explain why the separation constant in (7) was taken to be $n(n + 1)$.

3. What is wrong with the following argument: If (7) is rewritten in the form $R'' + (2/r)R' - n(n + 1)R/r^2 = 0$, and r is sufficiently large, the second two terms are negligible compared to the first, so we have $R'' = 0$, hence $R = Ar + B$. It follows that every solution of (7) must be similar to the function $Ar + B$ for large values of r. (Note: both the reasoning and the conclusion are false.)

4. Find solutions of (5) for $m = 0$. Why were these ignored in the text?

5. Derive (9), starting with (8), showing all steps in detail.

6. What is the bounded solution of (11) when n is a negative integer?

7. Find a harmonic function $f(r,\phi,\theta)$ taking the values $\frac{1}{2}(3\cos^2\phi - 1)$ when $r = 1$. Is this the only such harmonic function?

8. The temperature at the surface of a sphere of radius $r = 3$ centered at the origin is $-\frac{3}{2}\cos\phi + \frac{5}{2}\cos^3\phi$. Find the temperature, as a function of r and ϕ, at points within the sphere, assuming that the temperature is a harmonic function and is bounded within the sphere.

9. Find a solution of Laplace's equation that is bounded in the *exterior* of the sphere $r = 1$, taking the value $\cos 2\phi$ on the surface of the sphere.

10. Taking $\beta = -2rt + r^2$ in the binomial expansion of $(1 + \beta) - \frac{1}{2}$, show that the coefficients of r^0, r^1, and r^2 in (22) are $P_0(t)$, $P_1(t)$, and $P_2(t)$ respectively.

11. Let $g(r,t) = (1 - 2rt + r^2) - 1/2 = \sum_{n=0}^{\infty} r^n u_n(t)$.

(a) Differentiating $g(r,t)$ with respect to r, show that

$$(1 + r^2 - 2rt)\frac{\partial g}{\partial r} + (r - t)g = 0. \tag{24}$$

(b) Substituting the series expression into this equation, and equating the coefficients of r^n to zero, show that

$$(2n + 1)tu_n(t) = (n + 1)u_{n+1}(t) + nu_{n-1}(t) \qquad (n = 1, 2, 3, \cdots). \tag{25}$$

(c) By using the results of Exercise 17, Section 4.3, and the preceding exercise, show that $u_n(t) = P_n(t)$. [This does not establish (22) rigorously; for example, it does not show (22) is valid whenever $|r| < 1$.]

12. Assuming the validity of (22), establish the validity of (23).

13. Taking the natural logarithm of (22), differentiating with respect to r, and equating corresponding coefficients of r^n, obtain a recurrence relation for the Legendre polynomials.

14. Discuss the Abel summability of $\sum_{n=0}^{\infty} P_n(z)$.

15. Find a "generating function" for the functions $\cos nx$.

In the following sequence of exercises, it is shown that (22) is the Taylor series expansion of $(1 - 2rt + r^2)^{-1/2}$ considered as a function of r for fixed t. This is done by direct calculation, showing that the coefficient of r^n is

$$\frac{1}{2^n n!}\frac{d^n}{dt^n}(t^2 - 1)^n.$$

These exercises are recommended to readers having considerable interest in the techniques of analysis.

Consider the equation

$$y = a + xg(y). \tag{26}$$

Assume that g has been so chosen that this equation determines y implicitly as a function of x, and that y can be represented for sufficiently small x by

$$y = a + a_1x + a_2x^2 + a_3x^3 + \cdots \tag{27}$$

so that, if $f(y)$ is a function that can be expanded in a series

$$f(y) = f(a) + B_1(y - a) + B_2(y - a)^2 + \cdots \tag{28}$$

then we can also expand f in the Taylor series

$$f(y) = f(a) + A_1x + A_2x^2 + A_3x^3 + \cdots. \tag{29}$$

Our problem is to obtain the coefficients $A_1, A_2, A_3, \cdots$ in terms of the functions g and f and the constant a. This would be a forbidding project if attempted by brute-force methods. The answer turns out to be

$$A_n = \frac{1}{n!}\frac{d^{n-1}}{da^{n-1}}[g(a)^n f'(a)]. \tag{30}$$

Note that we are writing $g(a)^n$ rather than $[g(a)]^n$ to avoid the confusion of additional parentheses, and that a is treated as a variable in (30). This brilliant formula, due to Lagrange, is elegantly derived in advanced books on complex-variable theory. The following elementary derivation is also of considerable interest.

16. By differentiating (26) with respect to x and a, show that

$$[1 - xg'(y)]\frac{\partial y}{\partial x} = g(y), \qquad [1 - xg'(y)]\frac{\partial y}{\partial a} = 1. \tag{31}$$

17. Let $u = f(y)$. Keeping in mind that y is a function of x and a, and therefore u depends on x and a, show from (31) that

$$\frac{\partial u}{\partial x} = g(y)\frac{\partial u}{\partial a}. \tag{32}$$

18. By actually performing the indicated differentiations, show that

$$\frac{\partial}{\partial a}\left[F(y)\frac{\partial u}{\partial x}\right] = \frac{\partial}{\partial x}\left[F(y)\frac{\partial u}{\partial a}\right] \tag{33}$$

for a continuously differentiable function F.

19. Prove by induction, using (33) and (32), that

$$\frac{\partial^n u}{\partial x^n} = \frac{\partial^{n-1}}{\partial a^{n-1}}\left[g(y)^n \frac{\partial u}{\partial a}\right]. \tag{34}$$

20. Letting $x = 0$, obtain

$$\left(\frac{\partial^n u}{\partial x^n}\right)_{x=0} = \frac{d^{n-1}}{da^{n-1}}\,[g(a)^n f'(a)] \tag{35}$$

and hence derive (30) by Taylor's theorem.

21. Now apply the above result to the special case $g(y) = \frac{1}{2}(y^2 - 1)$ to deduce that the root of

$$y = a + \frac{x}{2}\,(y^2 - 1) \tag{36}$$

which equals a when $x = 0$ is given by

$$y = a + \sum_{n=1}^{\infty} \frac{1}{n!}\left(\frac{x}{2}\right)^n \frac{d^{n-1}}{da^{n-1}}\,(a^2 - 1)^n. \tag{37}$$

22. Show that $y = \frac{1}{x} - \frac{1}{x}\sqrt{1 - 2ax + x^2}$ is the root of (36) which tends to a as $x \to 0$.

23. Differentiating (37) with respect to a, obtain

$$(1 - 2ax + x^2)^{-1/2} = \sum_{n=0}^{\infty} [P_n(a)]x^n. \tag{38}$$

[With a change of notation, this is the same as (22).]

4.5 • SPHERICAL HARMONICS

This section can be omitted without loss in continuity. Our purpose is to derive expressions for the associated Legendre function $P_n^{(m)}$, introduced in the preceding section, and to show how to obtain a solution of Laplace's equation taking prescribed values on the surface of a sphere. As an example of the type of application that is possible, we may be given the temperature on the surface of a sphere (assumed independent of time) and asked to find the steady-state temperature at points inside the sphere. We have already seen how to do this in certain special cases (see Exercise 8 of the preceding section).

In this theory an especially important part is played by polynomials in x, y, and z. A monomial $x^j y^p z^s$ is said to have *degree* n if $j + p + s = n$. Any nontrivial linear combination of monomials of degree n is called a *homogeneous polynomial of degree n*. The class of all homogeneous polynomials of degree n, together with the zero polynomial, forms a linear space, denoted V_n.

For example, if $n = 2$, we have a linear space V_2 with basis $x^2, y^2, z^2, xy, xz, yz$. Thus, $y^2 - 3z^2 + 12xy + yz$ is a homogeneous polynomial of the second degree, an element of V_2.

In courses in modern algebra, polynomials are considered as purely formal objects. We treat them here as functions defined throughout space. Therefore we may refer to $x^3y + 3x^2y^2 - z^4$ as "a function of class V_4." The zero polynomial, which technically has no degree, is identified with the zero function and considered to be of class V_n for every $n = 0, 1, 2, \cdots$.

Theorem 1. *The dimension of V_n is*

$$(n + 1) + n + \cdots + 1 = \frac{(n + 2)(n + 1)}{2}. \tag{1}$$

Proof: The theorem is easily verified when $n = 0$, $n = 1$, and $n = 2$. For instance, V_0 is the class of all constant functions, and its dimension is unity. We have already shown that a basis for V_2 has six elements, so V_2 is 6-dimensional. To prove that (1) gives the dimension of V_n for arbitrary n, we proceed as follows. Assume it is true for V_{n-1}, so the dimension of V_{n-1} is

$$n + (n - 1) + \cdots + 1.$$

If we multiply the elements of V_{n-1} by (say) z, we obtain all elements of V_n except those in x and y alone, and there are $n + 1$ linearly independent monomials $x^j y^p$ with $j + p = n$. This accounts for the first term on the left side of (1). The result now follows by induction.

Now we will explore the functional dependence of functions of class V_n on the spherical coordinates r, ϕ, and θ. The first step in this direction is provided by the following theorem.

Theorem 2. *If the values taken by a function of class V_n on the surface of the unit sphere $r = 1$ are given by $g(\phi,\theta)$, then $r^n g(\phi,\theta)$ gives the values of the function throughout space.*

Proof: Since $z = r \cos \phi$, $x = r \sin \phi \cos \theta$, and $y = r \sin \phi \sin \theta$, a function of class V_n can be written (in spherical coordinates) as

$r^n h(\phi,\theta)$. (Recall that each term $x^j y^p z^s$ has the property $j + p + s = n$.) If this equals $g(\phi,\theta)$ identically when $r = 1$, then $h(\phi,\theta) = g(\phi,\theta)$ and the function must be $r^n g(\phi,\theta)$.

Now we wish to analyze the dependence on ϕ and θ. The substitutions given in the preceding proof yield a complicated expression involving terms of the form $r^n \sin^{j+p} \phi \cos^s \phi \cos^j \theta \sin^p \theta$ which are difficult to manipulate. We shall therefore write these functions in complex form. We first prove that every homogeneous polynomial of degree n can be written as a linear combination (with complex coefficients) of functions of the following form:

$$r^n e^{im\theta} \sin^{-m} \phi q_n^{(m)}(\cos \phi) \qquad (m = -n, -n+1, \cdots, n) \tag{2}$$

where $q_n^{(m)}$ is a polynomial of degree at most $n + m$.

Let us introduce the new variables

$$\alpha = x + iy, \quad \beta = x - iy \tag{3}$$

and observe at once that

$$\alpha\beta = x^2 + y^2 = r^2 - z^2 \tag{4}$$

and

$$\beta^m = (x - iy)^m = r^m \sin^m \phi \, e^{-im\theta} \tag{5}$$

[since $r \sin \phi \, e^{-i\theta} = r \sin \phi(\cos \theta - i \sin \theta) = x - iy$].

Obviously, any homogeneous polynomial of degree n in x, y, and z can be written also as a homogeneous polynomial of degree n in α, β, and z. When thus written, we can express the polynomial as a sum of polynomials $u_n^{(m)}$, $m = -n, -n+1, \cdots, n$, where $u_n^{(m)}$ consists entirely of terms in which the exponents of α and β have the fixed difference m. Then $\beta^m u_n^{(m)}$ will be a polynomial in $\alpha\beta$ and z, and by virtue of (4) can be written as a polynomial in r^2 and z.

On the surface of the unit sphere $r = 1$, we have $z = \cos \phi$, and $\beta^m u_n^{(m)}$ will be a polynomial of degree at most $n + m$ in $\cos \phi$. Write this $q_n^{(m)}(\cos \phi)$. We now have

$$\beta^m u_n^{(m)} = q_n^{(m)}(\cos \phi) \qquad (\text{for } r = 1 \text{ only}) \tag{6}$$

and hence, by (5)

$$u_n^{(m)} = e^{im\theta} \sin^{-m} \phi \, q_n^{(m)}(\cos \phi) \qquad (\text{for } r = 1 \text{ only}). \tag{7}$$

The expression (2) now follows at once from Theorem 2.

The functions $q_n^{(m)}$ are not intended to bear any relation to the $Q_n^{(m)}$ discussed earlier. However, we will show that, if $u_n^{(m)}$ satisfies Laplace's equation, they are *related* to the functions $P_n^{(m)}$ introduced in the preceding section. Indeed, if (2) satisfies Laplace's equation, then (since it is a function of the form $R\Phi\Theta$) we expect it to be of the form of (10) Section 4.4, except perhaps with the factor Θ in complex form $e^{im\theta}$ rather than as $\cos m\theta$ or $\sin m\theta$.

Having completed these preliminaries, we are now ready to determine the *associated Legendre function* $P_n^{(m)}(z)$ introduced before. They are required to be solutions of (9), Section 4.4, bounded in the interval $-1 \leqq z \leqq 1$. On comparing (7) above with (10), Section 4.4, we are quite naturally led to substitute $\Phi = (1 - z^2)^{-m/2}q$ into that differential equation. (Why? Because

$$\sin^{-m} \phi = (1 - z^2)^{-m/2}$$

when $z = \cos \phi$, and by the above remarks we expect q to be a polynomial in z, which should be pretty easy to work with.) The preceding analysis was mainly intended to motivate this substitution, which would otherwise seem quite artificial.

If we do this, we obtain (after a slightly tedious calculation) the following differential equation:

$$(8) \quad (1 - z^2)\frac{d^2q}{dz^2} + 2(m - 1)z\frac{dq}{dz} + [n(n + 1) - m(m - 1)]q = 0.$$

It is easy to show that this equation has, to within a scalar factor, at most one solution $q = q_n^{(m)}$ that is a polynomial in z (Exercise 25). If we differentiate both sides of (8), we find that dq/dz satisfies the same differential equation with m replaced by $m - 1$. It follows that, when suitably normalized,

$$(9) \qquad q_n^{(m-1)}(z) = dq_n^{(m)}/dz.$$

When $m = 0$, (8) is identical to Legendre's equation, with the polynomial solution

$$(10) \qquad q_n^{(0)} = P_n(z) = \frac{1}{2^n n!}\frac{d^n}{dz^n}(z^2 - 1)^n.$$

Because of (9), we are led to

$$(11) \qquad q_n^{(m)}(z) = \frac{1}{2^n n!}\frac{d^{n-m}}{dz^{n-m}}(z^2 - 1)^n.$$

By direct substitution one can verify that (11) is indeed a solution of (8). From this it follows that the functions

$$(12) \qquad (1 - z^2)^{-m/2}\frac{1}{2^n n!}\frac{d^{n-m}}{dz^{n-m}}(z^2 - 1)^n$$

are solutions of the associated Legendre equation [(9), Section 4.4]. We state that, for $n = 0, 1, 2, 3, \cdots$ and (for each fixed n) $m = -n,$

$-n+1, \cdots, n$ these are the only *bounded* solutions in the interval $-1 \leqq z \leqq 1$. (This is proved, in essentially the same way that the corresponding fact for Legendre's equation was proved, by using Theorem 3, given later in this section.)

Since m occurs in the associated Legendre equation as m^2, and that equation has (to within a scalar factor) only one bounded solution for fixed values of n and m, we obtain essentially the same solution if we replace m by $-m$ in (12). We therefore *define* the associated Legendre functions by

$$P_n^{(m)} = CP_n^{(-m)} = \frac{(1-z^2)^{m/2}}{2^n n!} \frac{d^{n+m}}{dz^{n+m}} (z^2-1)^n \tag{13}$$

where the constant C is easily evaluated by comparing the terms of highest order in (12) and (13). One obtains

$$C = (-1)^m \frac{(n+m)!}{(n-m)!}. \tag{14}$$

It is seen from (13) that $P_n^{(m)}$ for even m is a polynomial of degree n, and for odd m is $(1-z^2)^{1/2}$ times a polynomial of degree $n-1$. These functions are nontrivial only if $m = -n, -n+1, \cdots, n$, and vanish identically whenever $|m| > n$.

To summarize, we have found solutions of Laplace's equation, bounded within the unit sphere, of the form

$$r^n e^{im\theta} P_n^{(m)}(\cos \phi),$$

where $P_n^{(m)}(\cos \phi) = \sin^m \phi \, q_n^{(-m)}(\cos \phi)$, the functions $q_n^{(-m)}(\cos \phi) = q_n^{(-m)}(z)$ being polynomials in $z = \cos \phi$.

To evaluate $\int_{-1}^{1} [P_n^{(m)}(z)]^2 \, dz$, which will be needed later, we write it as

$$C \int_{-1}^{1} P_n^{(m)}(z) P_n^{(m)}(z) \, dz$$

which can be integrated by parts m times to yield

$$(-1)^m C \int_{-1}^{1} P_n^{(0)}(z) P_n^{(0)}(z) \, dz$$

where $P_n^{(0)} = P_n$ are the familiar Legendre polynomials. Therefore

$$\int_{-1}^{1} [P_n^{(m)}(z)]^2 \, dz = \frac{2}{2n+1} \frac{(n+m)!}{(n-m)!}. \tag{15}$$

Now we are ready to derive the expansion formula that is the central object of this section. We assume that we are given the values $g(\phi,\theta)$ of a harmonic function on the surface of the sphere $r = 1$. It is desired to obtain the values of the function within the sphere. The formal solution is

(16) $$f(r,\phi,\theta) = \sum_{n=0}^{\infty} \sum_{m=-n}^{n} C_{n,m} r^n e^{im\theta} P_n^{(m)}(\cos \phi)$$

where the coefficients $C_{n,m}$ are given by

(17) $$C_{n,m} = \frac{(2n+1)}{4\pi} \frac{(n-m)!}{(n+m)!}.$$

$$\int_0^{\pi} \int_0^{2\pi} g(\phi,\theta) e^{-im\theta} P_n^{(m)}(\cos \phi) \sin \phi \, d\theta \, d\phi.$$

This formula is derived formally just like that for the coefficients in a Fourier series. Letting $r = 1$ in (16), we multiply both sides of (16) by $e^{-im\theta} P_n^{(m)}(\cos \phi) \sin \phi$ and integrate over the unit sphere $r = 1$. Notice that $e^{-im\theta}$ cancels the $e^{im\theta}$ term in (16), wherever it occurs, and the integral over $(0,2\pi)$ yields a factor of 2π. This explains the factor of 4π in the denominator, the rest of the factor coming from (15). All that is required to complete the formal derivation is assurance that

(18) $$\int_0^{\pi} \int_0^{2\pi} e^{im\theta} P_n^{(m)}(\cos \phi) e^{-im'\theta} P_{n'}^{(m')}(\cos \phi) \sin \phi \, d\theta \, d\phi = 0$$

whenever *either* $m \neq m'$ *or* $n \neq n'$.

When $m \neq m'$ there is nothing new to prove; the integration over θ gives zero at once.

In Section 5.4, we will show that, if u and v are harmonic functions, and S is a smooth closed surface,

(19) $$\iint_S \left(v \frac{\partial u}{\partial n} - u \frac{\partial v}{\partial n} \right) dS = 0$$

where $\partial/\partial n$ denotes differentiation in the direction of the outer normal to the closed surface S. Taking S to be the unit sphere $r = 1$, we have $\partial/\partial n = \partial/\partial r$. We let $u = r^n e^{im\theta} P_n^{(m)}(\cos \phi)$ and $v = r^{n'} e^{-im\theta} P_{n'}^{(-m)}(\cos \phi)$. Notice that $dS = \sin \phi \, d\theta \, d\phi$ in this case, since $r = 1$. Since both u and v are harmonic, we can apply (19) to obtain

(20) $$C \int_0^{\pi} \int_0^{2\pi} (n - n') r^{n+n'-1} P_n^{(m)}(\cos \phi) P_{n'}^{(m)}(\cos \phi) \sin \phi \, d\theta \, d\phi = 0$$

[for the meaning of C, see (14)]. Since $n \neq n'$ and $C \neq 0$, the value of the integral

(21) $$\int_0^{\pi} \int_0^{2\pi} P_n^{(m)}(\cos \phi) P_{n'}^{(m)}(\cos \phi) \sin \phi \, d\theta \, d\phi$$

[obtained by taking $r = 1$ in (20)] is zero. This completes the formal derivation of (17), and even provides a rigorous derivation if we know that (16) converges uniformly to $f(r,\phi,\theta)$ when $r \leqq 1$, since term-by-term integration is then justified.

In many applications, the function $f(r,\phi,\theta)$ satisfies Laplace's equation in the interior of the sphere, but does not actually satisfy Laplace's equation on the surface. In these cases, we do not expect the series to be uniformly convergent, but it is still possible to prove that the series converges in a mean-square sense. This would be the case, for instance, if the prescribed temperature $g(\phi,\theta)$ had a discontinuity along the "equator" $\phi = 90°$. If g is bounded and continuous elsewhere on the surface of the sphere, the integral (17) will still define coefficients, and it can be shown the series will converge within the sphere to a harmonic function. However, it will not converge uniformly for $r \leqq 1$.

Incidentally, some readers may feel that the assumption of a temperature discontinuity on the surface would lead to an infinite temperature gradient and therefore an infinite heat flow within the sphere, which would be absurd. However absurd the assumption may be from a microscopic viewpoint, it does no violence to the macroscopic theory via Laplace's equation. It is perfectly possible to find a function satisfying Laplace's equation at every point within a sphere, tending to a discontinuous function as $r \to 1$. Since the temperature discontinuity is only a surface phenomenon, it cannot produce an "infinite" rate of heat transport.

So far we have not defined the term *spherical harmonic.* We have not done so because the term is used with different meaning by different authors. We call any function of the form $r^n e^{im\theta} P_n^{(m)}(\cos \phi)$ a "spherical harmonic" and the functions

$$Y_n^{(m)} = e^{im\theta} P_n^{(m)}(\cos \phi) \tag{22}$$

the *surface harmonics* of degree n. (See Exercise 19.)

We shall merely outline the proof of the following theorem. The interested reader will then have no trouble proving a theorem for functions integrable over a sphere that is analogous to Theorem 3, Section 4.3, and (after studying Chapter 5) somewhat deeper theorems concerning the validity of the expansion (16) at points *within* the sphere. This theorem is concerned entirely with points on the *surface* of the sphere.

Theorem 3. *Every continuous function $f(\phi,\theta)$ defined on the surface of the unit sphere $r = 1$ can be uniformly approximated by linear combinations of the surface harmonics $Y_n^{(m)}$.*

Outline of Proof: From the orthogonality relation (18) it follows that the functions $r^n Y_n^{(m)}$, $r^n Y_{n-2}^{(m)}$, $r^n Y_{n-4}^{(m)}$, $\cdots$ are linearly independent, since they are mutually orthogonal over the unit sphere. For any k, $r^k Y_k^{(m)}$ can be shown to be a homogeneous polynomial of degree k in α, β, and z. Therefore each of the functions just listed is a homogeneous polynomial of degree n. Those of the form $r^n Y_n^{(m)}$, for $m = -n, -n+1, \cdots, n$, provide $2n + 1$ such polynomials; those of the form $r^n Y_{n-2}^{(m)}$ provide $2(n-2)+1$ more, and so on. In all, we have

$$(23) \qquad [(n+1)+n] + [(n-1)+(n-2)] + \cdots + 1$$

homogeneous polynomials of degree n. By Theorem 1, they must provide a basis for V_n. Therefore every homogeneous polynomial of degree n must be a linear combination of these. Only the first $2n+1$ of them, those of the form $r^n Y_n^{(m)}$, are actually spherical harmonics. However, on the sphere $r = 1$, each of them equals a surface harmonic.

It follows, therefore, that every polynomial has values on the unit sphere $r = 1$ that are identical to the values taken by some linear combination of surface harmonics. Therefore, the proof is reduced to showing that every continuous function $f(\theta,\phi)$ can be uniformly approximated by polynomials over the surface of the unit sphere. This, however, is a well-known theorem of analysis (see Exercise 21).

• EXERCISES

1. Determine the degree of each of the following polynomials.

(a) $x^3 + xy^5 - 3x^2y^2 + z^4 - 8$.

(b) $x^2y^2 + 3xy^3 - z^4$.

(c) $\dfrac{d^2}{dz^2}(z^2-1)^5$.

(d) $(z^2-1)^{-2}\dfrac{d^4}{dz^4}(z^2-1)^7$.

[Check to see that (d) is really a polynomial.]

2. **(a)** What is the dimension of V_5?

(b) Is it true that the product of two functions, one of class V_2 and the other of class V_3, is always a function of class V_5?

3. With the notation of this section, write r^4 as a polynomial in x, y, and z, and also as a polynomial in α, β, and z.

4. **(a)** Can r^3 be written as a polynomial in x, y, and z?
(b) Can r^3 be written as a polynomial in α, β, and z?

5. Find a homogeneous polynomial of degree 2 in α, β, and z, whose values for $r = 1$ are given by

$$g(\phi,\theta) = 1 - \cos^2 \phi + \sin^2 \phi e^{-i2\theta}.$$

6. Show that, in the coordinates α, β, and z, Laplace's equation becomes

$$4 \frac{\partial^2 f}{\partial\alpha\partial\beta} + \frac{\partial^2 f}{\partial z^2} = 0. \tag{24}$$

7. Find a set of product solutions of (24), of the form

$$f(\alpha,\beta,z) = A(\alpha)B(\beta)Z(z).$$

8. Substitute a polynomial of the form

$$v_k = \sum C_{j,p,s}\alpha^j\beta^p z^s = 0 \tag{25}$$

into (24), and show that the resulting equation

$$\sum D_{j,p,s}\alpha^j\beta^p z^s = 0 \tag{26}$$

is satisfied, provided that

$$D_{j,p,s} = 4(j+1)(p+1)C_{j+1,p+1,s} + (s+2)(s+1)C_{j,p,s+2} = 0. \tag{27}$$

9. Use the recursion relation

$$4(j+1)(p+1)C_{j+1,p+1,s} + (s+2)(s+1)C_{j,p,s+2} = 0 \tag{28}$$

to show that the linear space consisting of those homogeneous polynomials of degree n that satisfy Laplace's equation is of dimension $2n + 1$.

***10.** (For students familiar with linear algebra.) The Laplacian ∇^2 defines a linear transformation from V_n into V_{n-2}.
(a) By comparing the dimensions of these two linear spaces, show that the null space (the elements of V_n satisfying Laplace's equation) is of dimension at least $2n + 1$.
(b) Show that this null space is exactly of dimension $2n + 1$.

11. Actually construct a polynomial of the type $u_5^{(1)}$ satisfying Laplace's equation—in other words, a polynomial in α, β, and z, homogeneous of degree 5, each term having exponent of α one greater than the exponent of β, and satisfying (24).

***12.** By directly substituting (2) into Laplace's equation, find the differential equation that must be satisfied by $q_n^{(m)}$ if (2) defines a harmonic function. Hence, show that $r^n e^{im\theta} v_n^{(m)}$ is a harmonic function, bounded for $r \leqq 1$, if and only if $v_n^{(m)}$ is a scalar multiple of $P_n^{(m)}$.

13. What happens to (13) if $m > n$?

14. Show that, except for a scalar factor, $P_n^{(n)}(\cos\phi)$ equals $\sin^n\phi$ for $n = 0, 1, 2, \cdots$.

15. **(a)** Determine $P_2^{(1)}(\cos\phi)$ directly from (13).
(b) Using (14) and your answer to (a), determine $P_2^{(-1)}(\cos\phi)$.

16. Is this formula valid?

$$\int_0^\pi P_n^{(m)}(\cos\phi)P_{n'}^{(m')}(\cos\phi)\sin\phi\, d\phi = 0$$

$$(\text{whenever} \quad n \neq n' \quad \text{or} \quad m \neq m').$$

If not, find conditions under which it is valid.

17. The temperature at the surface of the unit sphere $r = 1$ is given by $g(\phi,\theta) = \cos\theta\sin 2\phi$. Assuming the temperature $f(r,\phi,\theta)$ is a harmonic function, find its values within the sphere.

18. Show that, except for a scalar factor, the first few associated Legendre functions are as follows, and determine the factor in each case.
$P_0^{(0)}(\cos\phi) = 1$, $P_1^{(0)}(\cos\phi) = \cos\phi$, $P_1^{(1)}(\cos\phi) = \sin\phi$,
$P_2^{(0)}(\cos\phi) = 3\cos^2\phi - 1$, $P_2^{(1)}(\cos\phi) = \sin\phi\cos\phi$,
$P_2^{(2)}(\cos\phi) = \sin^2\phi$, $P_3^{(0)}(\cos\phi) = \frac{5}{3}\cos^3\phi - \cos\phi$,
$P_3^{(1)}(\cos\phi) = \sin\phi\,(5\cos^2\phi - 1)$, $P_3^{(2)}(\cos\phi) = \sin^2\phi\cos\phi$,
$P_3^{(3)}(\cos\phi) = \sin^3\phi$.
(In other words, revise this list, making it correct.)

19. Show that, under a rotation of the coordinate axes, a function that was a spherical harmonic of degree n becomes, in the new coordinates, a linear combination of at most $2n + 1$ spherical harmonics. Hence, criticize the definition of "spherical harmonic" given in the text. Give a definition of "surface harmonic of degree n" that will be invariant under rotations of the coordinate system.

20. **(a)** Directly from (13), show that

$$P_3^{(-1)}(z) = (1 - z^2)^{-1/2}[\tfrac{1}{2}z^2(z^2 - 1) + \tfrac{1}{8}(z^2 - 1)^2].$$

(b) Letting $z = \cos\phi$, show that

$$P_3^{(-1)}(\cos\phi) = \tfrac{1}{8}\sin\phi(1 - 5\cos^2\phi).$$

(c) Show that, in rectangular coordinates,

$$r^3 e^{i\theta} P_3^{(-1)}(\cos\phi) \quad \text{becomes} \quad \tfrac{1}{8}(x + iy)(x^2 + y^2 - 4z^2).$$

(d) Similarly, show that

$$r^3 e^{-i\theta} P_3^{(1)}(\cos\phi) \quad \text{becomes} \quad \tfrac{3}{2}(x - iy)(4z^2 - x^2 - y^2).$$

(e) Do the functions $4z^2x - x^3 - xy^2$ and $4z^2y - x^2y - y^3$ span the same two-dimensional (complex) linear space as the two functions of parts (c) and (d)?

21. Can Theorem 14, Section 3.7, be used to prove the assertion made in the last paragraph of this section?

22. (a) Find a harmonic function whose values on the surface of the unit sphere are given by $g(\phi,\theta) = 3\cos^2\phi - 1$.

(b) Find a harmonic function, as in (a), with

$$g(\phi,\theta) = [3\cos^2\theta - 1]\sin^2\phi.$$

23. Does Theorem 2 imply that any polynomial function that is identically zero on the surface of the sphere $r = 1$ is identically zero everywhere?

24. Show that $\iint_S zY_k^{(m)}Y_j^{(m)}\,dS = 0$ unless $j - k = \pm 1$. (The integral is over the surface S of the unit sphere $r = 1$.) [Hint: show that $zr^kY_k^{(m)}$ is a homogeneous polynomial of degree $k + 1$, and hence is a sum $C_1r^{k+1}Y_{k+1}^{(m)} + C_2r^{k+1}Y_{k-1}^{(m)} + \cdots$ and use the orthogonality relations.]
(Compare Exercises 13-17, Section 4.3.)

25. Show that, to within a scalar factor, (8) has only one polynomial solution.

4.6 • BESSEL FUNCTIONS

In cylindrical coordinates Laplace's equation is

$$\frac{1}{r}\frac{\partial}{\partial r}\left(r\frac{\partial f}{\partial r}\right) + \frac{1}{r^2}\frac{\partial^2 f}{\partial \theta^2} + \frac{\partial^2 f}{\partial z^2} = 0. \tag{1}$$

The functions

$$\begin{aligned} &e^{az}\sin n\theta\, J_n(ar), \qquad e^{-az}\sin n\theta\, J_n(ar) \\ &e^{az}\cos n\theta\, J_n(ar), \qquad e^{-az}\cos n\theta\, J_n(ar) \end{aligned} \tag{2}$$

satisfy this equation if $J_n(r)$ satisfies Bessel's equation below. This is shown by substituting any one of these four expressions into (1). As in our earlier work, it is convenient to use the complex form

$$(3) \qquad e^{az}e^{in\theta}J_n(ar) \quad \text{or} \quad e^{-az}e^{in\theta}J_n(ar)$$

and consider either the real or the imaginary part as the desirable "real" solution. After simplifying, we obtain on substituting into (1)

$$(4) \qquad \left[r^2 \frac{d^2}{dr^2} + r\frac{d}{dr} + (r^2 - n^2)\right] J_n(r) = 0.$$

Notice that in (4) we have eliminated the arbitrary constant a, which is easy to do since (by the chain rule for differentiation)

$$\frac{d}{dr}[J_n(ar)] = aJ_n'(ar) \quad \text{and} \quad \frac{d^2}{dr^2}[J_n(ar)] = a^2J_n''(ar).$$

Equation (4) also arises when the method of separation of variables is applied to (1) (see Exercise 1).

If (4) is written in "standard form," dividing through by the leading coefficient r^2, its coefficients are continuous except at $r = 0$, and therefore there are two linearly independent solutions of (4) valid in the half-line $r > 0$. Every solution valid in the interval $r > 0$ is a linear combination of these. According to the fundamental existence theorem (Section 1.4) such solutions exist whether or not n is an integer. Since n appears only as n^2, solutions for $-n$ are identical to those for n.

On the other hand, if n is not an integer, for example if $n = \frac{1}{2}$, then the functions (2) are not single-valued functions of the space coordinate θ, since (for instance) $\cos \frac{1}{2}\theta$ is not periodic of period 2π. Therefore if we are interested in studying only solutions of (1) of the form (2), valid throughout space, we have no need to consider solutions of (4) for nonintegral values of n. This does not mean that such solutions do not exist, but only that they are not relevant to our present discussion.

Again, only for nonnegative r do we require solutions, although solutions of (4) exist for $r < 0$ as well as $r \geqq 0$. However, we see that (4) is symmetric with respect to replacing r by $-r$, so we actually lose no generality by restricting our attention to $r \geqq 0$.

If n is zero or a positive integer, one readily obtains a solution of (4) by substituting

$$\sum_{k=0}^{\infty} a_k x^k.$$

This yields a recursion relation that is easily satisfied, and the resulting series, when normalized in a conventional (arbitrary) manner, is

$$J_n(r) = \sum_{k=0}^{\infty} \frac{(-1)^k (r/2)^{n+2k}}{k!(n+k)!}. \tag{5}$$

By using the ratio test, one can show that (5) converges for all values of r, so termwise differentiation is valid and it follows that (5) does define a solution of (4) when n is zero or a positive integer.

We state, without proof, that to within a scalar factor, (5) is the *only* solution of (4) valid for $r \geqq 0$. Other solutions of (4) are unbounded in the vicinity of $r = 0$, and not defined at $r = 0$ itself.

These functions are called *Bessel functions of the first kind* of order n. Since we shall not have occasion to mention any other Bessel functions, we shall simply refer to J_n as the Bessel function of order n. The word "order" here refers to the order of the first nonvanishing term. If $n = 0$ we have

$$J_0(r) = 1 - \frac{r^2}{2^2} + \frac{r^4}{2^4(2!)^2} - \frac{r^6}{2^6(3!)^2} + \frac{r^8}{2^8(4!)^2} - \cdots \tag{6}$$

and if $n = 1$ we have

$$J_1(r) = \frac{r}{2} - \frac{r^3}{2^3 2!} + \frac{r^5}{2^5 2!3!} - \frac{r^7}{2^7 3!4!} + \frac{r^9}{2^9 4!5!} - \cdots. \tag{7}$$

These two Bessel functions are worth special attention since, as we shall show below, one can in principle determine $J_n(r)$ for any n and r if one has a table of values of J_0 and J_1. Graphs of these two functions are shown in Figure 4.3. It will be noted that they look some-

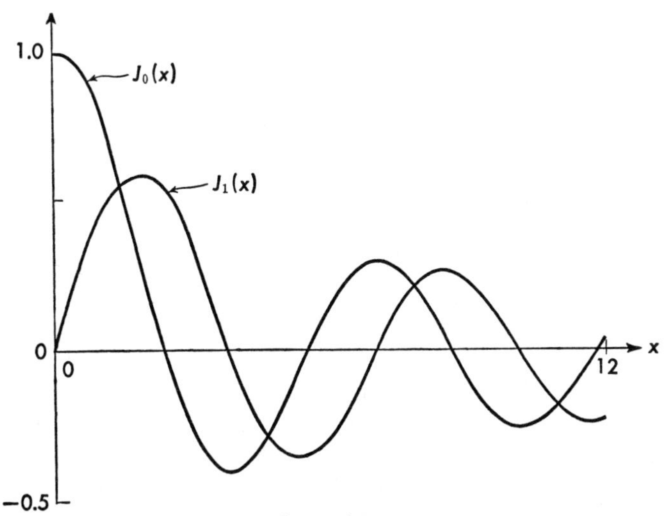

Figure 4.3

what like damped sinusoids, and for large values of r this is "almost" true. More precisely it can be shown that the ratio of $J_n(r)$ to the function

$$\sqrt{\frac{2}{\pi r}}\cos(r - \alpha_n) \qquad \left[\alpha_n = (2n+1)\frac{\pi}{4}\right] \tag{8}$$

tends to unity as $r \to \infty$. For small values of r, this expression is far from being valid. In particular, the zeros of the functions $J_n(r)$ are not evenly spaced, as is the case with damped sinusoids.

On differentiating (6) and comparing with (7), we find that

$$J_0'(r) = -J_1(r) \tag{9}$$

and it is similarly easy to verify that

$$\int_0^x rJ_0(r)\,dr = xJ_1(x). \tag{10}$$

By calculating directly from (5), one finds

$$rJ_n'(r) = nJ_n(r) - rJ_{n+1}(r) \qquad (n = 0, 1, 2, \cdots) \tag{11}$$

and also

$$rJ_n'(r) = -nJ_n(r) + rJ_{n-1}(r) \qquad (n = 1, 2, 3, \cdots) \tag{12}$$

Summing (11) and (12), we obtain

$$J_n'(r) = \tfrac{1}{2}[J_{n-1}(r) - J_{n+1}(r)] \qquad (n = 1, 2, 3, \cdots), \tag{13}$$

which is somewhat more complicated than the corresponding differentiation formulas for sinusoidal functions.

On subtracting (11) and (12), we find

$$J_{n+1}(r) = \frac{2n}{r}J_n(r) - J_{n-1}(r) \qquad (n = 1, 2, 3, \cdots), \tag{14}$$

which is a type of recursion formula. This shows that in principle one can obtain a table of values for J_n given tables for J_0 and J_1, whenever $n = 2, 3, 4, \cdots$. In practice, this is not usually necessary, as tables of Bessel functions have been extensively tabulated and are available in any good library.

In Figure 4.4 we show a *purely schematic* graph of J_n for a larger value of n than $n = 0$ or $n = 1$. It is included simply to emphasize the fact that the leading term in (5) is a scalar times r^n, so for $n = 2, 3, 4, \cdots$ not only does J_n vanish at the origin, but its first $n - 1$ derivatives do also.

We shall now derive certain orthogonality relations for Bessel functions that will be useful later. These could be derived by methods similar to those used earlier in connection with the Sturm-Liouville theorem (note Exercises 4 and 5) but the method we employ here is of interest in itself, and shows more clearly the significance of orthogonality with respect to a weighting function.

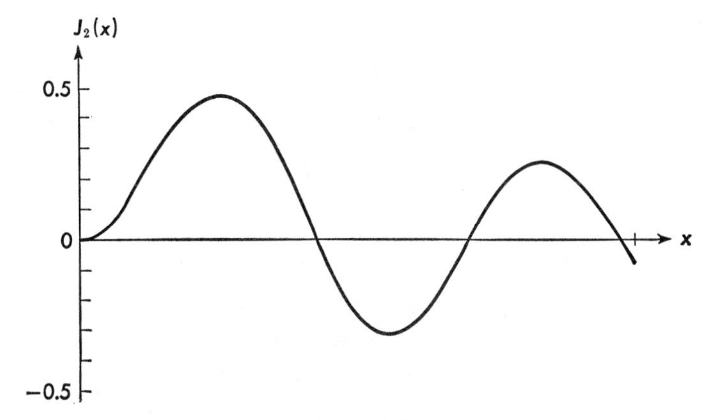

Figure 4.4

First, let us briefly review the orthogonality relations for the sinusoidal functions. For simplicity, let us only consider those of the form sin ax. According to the Sturm-Liouville theorem, we have

$$\int_0^L \sin a_1x \sin a_2x \, dx = 0 \tag{15}$$

whenever $a_1^2 \neq a_2^2$, provided that

$$\sin a_1L = 0 \qquad \text{and} \qquad \sin a_2L = 0. \tag{16}$$

Of course, the Sturm-Liouville theorem is not needed to prove this: (16) implies that a_1L and a_2L are integral multiples of π and therefore (15) is equivalent to

$$\int_0^L \sin\frac{n\pi x}{L} \sin\frac{m\pi x}{L} \, dx = 0 \qquad (n^2 \neq m^2), \tag{17}$$

which is easily verified by direct integration, if n and m are integers. The reason we can rewrite (15) and (16) in the simpler form (17) is that the zeros of the function sin x are evenly spaced. Unfortunately, the zeros of the Bessel functions are not evenly spaced, so there is no formulation as simple as (17). Our formula will look more like (15), and to make practical use of it we need a table giving the values of r

for which the Bessel function vanishes. This will become clear as we proceed.

The relation we desire to prove is

$$\int_0^c rJ_n(a_1r)J_n(a_2r)\,dr = 0, \tag{18}$$

which is valid whenever a_1 and a_2 are chosen so that

$$J_n(a_1c) = 0, \quad J_n(a_2c) = 0, \qquad (a_1 \neq a_2). \tag{19}$$

This is similar to (15) and (16), except that the functions are orthogonal with respect to the weight function r in this instance. (We assume a_1, a_2, and c are positive.)

We will also prove (18) is valid if

$$J_n'(a_1c) = 0, \quad J_n'(a_2c) = 0, \qquad (a_1 \neq a_2). \tag{20}$$

More general conditions suffice—for instance, requiring that a_1 and a_2 be distinct roots of the equation $J_n(ac) + hJ_n'(ac) = 0$ for some fixed h—but we have no need for this degree of generality.

We now proceed to derive (18). We assume a_1 and a_2 chosen so that (19) is satisfied (or, alternatively, so that (20) is satisfied). We denote by u and v the functions

$$u = e^{a_1z}\cos n\theta\, J_n(a_1r), \quad v = e^{a_2z}\cos n\theta\, J_n(a_2r). \tag{21}$$

As already noted, these are harmonic functions. We make use of Green's formula (for a derivation see Section 5.4) which asserts, for harmonic functions u and v, that

$$\iint_S \left(u\frac{\partial v}{\partial n} - v\frac{\partial u}{\partial n}\right) dS = 0 \tag{22}$$

where the integral is over a closed bounded surface; here, $\partial/\partial n$ denotes the derivative in the direction of the outward normal. We let S be the surface of a cylinder of radius c concentric with the z axis. The curved part of the cylinder has equation $r = c$ (in cylindrical coordinates). Let the top of the cylinder lie in the plane $z = 0$ and the bottom lie in the plane $z = z_0$ for some negative value of z_0 (see Figure 4.5).

With this choice of u and v, there will be no contribution to (22) from the curved part of the cylinder, since both u and v are zero on this surface. (If the alternative conditions (20) apply, both $\partial u/\partial n$ and $\partial v/\partial n$ will be zero.) By taking the absolute value of z_0 sufficiently large, we can make the contribution to the integral from the bottom surface as nearly zero as we like. It follows that the integral over the top surface must be zero, for otherwise (22) would not be valid.

On the top surface, $z = 0$ and $\partial/\partial n = \partial/\partial z$, with $dS = r\,dr\,d\theta$.

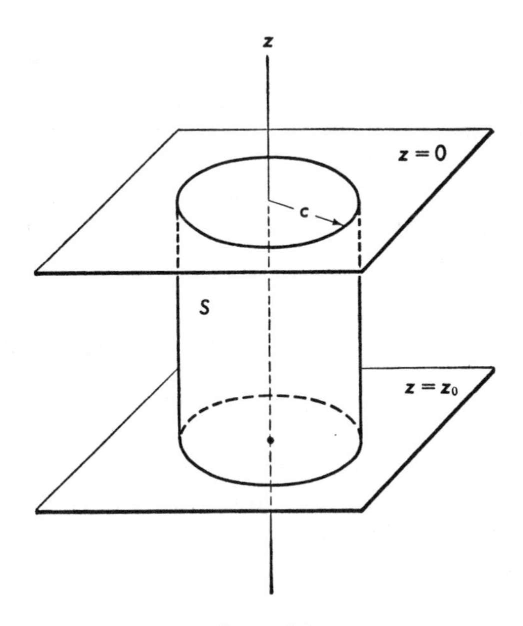

Figure 4.5

From (21), we have $\partial u/\partial n = a_1 u$ and $\partial v/\partial n = a_2 v$ and the integral over this surface becomes

(23) $$\int_0^{2\pi}\int_0^c [u(a_2 v) - v(a_1 u)]r\,dr\,d\theta = 0.$$

Since $a_1 \neq a_2$, this implies

(24) $$\int_0^{2\pi}\int_0^c uvr\,dr\,d\theta = 0,$$

which becomes, on substituting,

(25) $$\int_0^{2\pi}\int_0^c \cos^2 n\theta\, J_n(a_1 r)J_n(a_2 r)r\,dr\,d\theta = 0.$$

Integrating first over θ, we obtain

(26) $$\pi\int_0^c J_n(a_1 r)J_n(a_2 r)r\,dr = 0,$$

which implies (18).

Had we not chosen a_1 and a_2 in a very special way, we could not have neglected the contribution to the integral (22) from the curved surface; letting $z_0 \to -\infty$, the integral over the bottom face would tend to zero and (22) would become

$$(27)\quad (a_2 - a_1) \int_0^{2\pi} \int_0^c uvr\, dr\, d\theta + \int_0^{2\pi} \int_{-\infty}^0 \left(u \frac{\partial v}{\partial r} - v \frac{\partial u}{\partial r} \right) c\, dz\, d\theta = 0.$$

Notice that the element of area on the curved surface is $c\, dz\, d\theta$ since c is the radius of the cylinder. The integral over θ can be explicitly carried out in both terms; and the integral over z can also be carried out in the second term. (As an elementary exercise, the reader is urged to do this.) We obtain (after cancellation of π)

$$(28)\quad \begin{aligned} &(a_2 - a_1) \int_0^c J_n(a_1 r) J_n(a_2 r) r\, dr \\ &\qquad + \frac{c}{a_1 + a_2} [a_2 J_n(a_1 c) J_n'(a_2 c) - a_1 J_n'(a_1 c) J_n(a_2 c)] = 0. \end{aligned}$$

Using the differentiation formulas given above, one can evaluate the second term explicitly, if one has tables of Bessel functions. We point this out, not because (28) is an especially useful formula, but simply to emphasize that such formulas exist. Indeed, they exist by the hundreds, and can be found in more advanced treatises.

If $J_n(a_1 c) = 0$, (28) reduces to

$$(29)\quad \int_0^c J_n(a_1 r) J_n(a_2 r) r\, dr = \frac{c a_1}{a_2^2 - a_1^2} J_n'(a_1 c) J_n(a_2 c),$$

which unfortunately is not valid when $a_1 = a_2$ since the denominator on the right side is then zero. However, its numerator is also zero, and using l'Hôpital's rule (in this case, differentiating numerator and denominator with respect to a_2 and letting a_2 tend to a_1) we obtain

$$(30)\quad \int_0^c [J_n(a_1 r)]^2 r\, dr = \frac{c^2}{2} [J_n'(a_1 c)]^2$$

valid when $J_n(a_1 c) = 0$.

If one desires to expand an "arbitrary" function in a series of terms in $J_n(a_m r)$, where $J_n(a_m r) = 0$ for $m = 1, 2, 3, \cdots$, one quite clearly needs a formula like (30). Notice that according to (11) we have

$$(31)\quad J_n'(a_1 c) = \frac{n}{a_1 c} J_n(a_1 c) - J_{n+1}(a_1 c)$$

and since $J_n(a_1 c) = 0$ we can rewrite (30) in the form

(32) $$\int_0^c r[J_n(ar)]^2\,dr = \frac{c^2}{2}\,[J_{n+1}(ac)]^2$$

valid whenever $J_n(ac) = 0$.

The same technique yields a different formula when a is so chosen that $J_n'(ac) = 0$. This formula is given in Exercise 10.

We shall now derive *Bessel's integral form* for $J_n(r)$, which is

(33) $$J_n(r) = \frac{1}{\pi}\int_0^\pi \cos\,(r\sin\phi - n\phi)\,d\phi.$$

Before giving the derivation, we note that $\sin\,(r\sin\phi)$ for any fixed r is symmetric about $\phi = \pi/2$ (for the very simple reason that $\sin\phi$ possesses this symmetry) and we are led to

(34) $$\int_0^\pi \sin n\phi\,\sin\,(r\sin\phi)\,d\phi = 0 \qquad (n = 0, \pm 2, \pm 4, \cdots).$$

Similarly, we have when n is odd,

(35) $$\int_0^\pi \cos n\phi\,\cos\,(r\sin\phi)\,d\phi = 0 \qquad (n = \pm 1, \pm 3, \pm 5, \cdots).$$

If the integrand is expanded thus:

(36) $$\cos\,(r\sin\phi - n\phi) = \cos\,(r\sin\phi)\cos n\phi + \sin\,(r\sin\phi)\sin n\phi,$$

it follows that if n is even, we can replace (33) by

(37) $$J_n(r) = \frac{1}{\pi}\int_0^\pi \cos n\phi\,\cos\,(r\sin\phi)\,d\phi \quad (n = 0, \pm 2, \pm 4, \cdots)$$

and if n is odd

(38) $$J_n(r) = \frac{1}{\pi}\int_0^\pi \sin n\phi\,\sin\,(r\sin\phi)\,d\phi \quad (n = \pm 1, \pm 3, \pm 5, \cdots).$$

The derivation of (33) is not complicated. Since the integrand in

(39) $$y(r) = \int_0^\pi \cos\,(r\sin\phi - n\phi)\,d\phi$$

has a continuous partial derivative with respect to r, it is permissible to interchange the order of differentiation and integration, to obtain

(40) $$y'(r) = -\int_0^\pi [\sin\,(r\sin\phi - n\phi)]\sin\phi\,d\phi.$$

Integrating by parts yields

(41) $$\begin{aligned} y'(r) &= [-\cos\phi\sin\,(r\sin\phi - n\phi)]_0^\pi \\ &\quad - \int_0^\pi (r\cos\phi - n)\cos\phi\cos\,(r\sin\phi - n\phi)\,d\phi. \end{aligned}$$

The first term vanishes at $\phi = 0$ and $\phi = \pi$ and hence

$$y'(r) = -\int_0^\pi (r\cos\phi - n)\cos\phi\cos(r\sin\phi - n\phi)\,d\phi. \tag{42}$$

Differentiating again with respect to r, we obtain

$$\begin{aligned} y''(r) &= -\int_0^\pi \cos^2\phi\cos(r\sin\phi - n\phi)\,d\phi \\ &\quad + \int_0^\pi \sin\phi\cos\phi\,[(r\cos\phi - n)\sin(r\sin\phi - n\phi)]\,d\phi. \end{aligned} \tag{43}$$

In the second integral, the factor in brackets is the derivative of $-\cos(r\sin\phi - n\phi)$ with respect to ϕ, so it is not difficult to integrate it by parts. If the reader does this, he finds on combining the result with the first integral on the right side of (43) that

$$y''(r) = -\int_0^\pi \cos(r\sin\phi - n\phi)\sin^2\phi\,d\phi. \tag{44}$$

Now we substitute (39), (42), and (44) into the left side of Bessel's equation; the reason for integrating by parts above was to ensure that all three integrands would have a common factor $\cos(r\sin\phi - n\phi)$. We obtain

$$\begin{aligned} r^2y'' &+ ry' + (r^2 - n^2)y \\ &= \int_0^\pi [\cos(r\sin\phi - n\phi)](-r^2\sin^2\phi - r^2\cos^2\phi \\ &\qquad + rn\cos\phi + r^2 - n^2)\,d\phi \\ &= n\int_0^\pi (r\cos\phi - n)\cos(r\sin\phi - n\phi)\,d\phi \\ &= n\sin(r\sin\phi - n\phi)\Big|_0^\pi = 0, \end{aligned} \tag{45}$$

from which it follows that y satisfies Bessel's equation. Directly from (39) we see that, if $r = 0$, we have $y(0) = \pi$ when $n = 0$, and $y(0) = 0$ if $n = 1, 2, 3, \cdots$, and therefore

$$J_0(r) = \frac{1}{\pi}\int_0^\pi \cos(r\sin\phi)\,d\phi \tag{46}$$

and to within a scalar factor

$$J_n(r) = \frac{1}{\pi}\int_0^\pi \cos(r\sin\phi - n\phi)\,d\phi, \qquad (n = 1, 2, 3, \cdots). \tag{47}$$

That this "scalar factor" is, indeed, unity, is seen by comparing the first n derivatives of (47) with respect to r with those obtained from (5). The first $n - 1$ derivatives in each instance are zero, and the nth derivative is 2^{-n} as computed both from (47) and from (5) (see Exercise 12). Here again, we are making use of the fact that the functions $J_n(r)$ are, to within a scalar factor, the only solutions of Bessel's equation that are bounded at $r = 0$.

Another way of writing (46) is given in Exercise 13.

As a consequence of (35) and (37), the Fourier cosine expansion of $\cos(r \sin \phi)$ is

$$\cos(r \sin \phi) = J_0(r) + 2[J_2(r) \cos 2\phi + J_4(r) \cos 4\phi + \cdots] \tag{48}$$

and by virtue of (34) and (38) we have

$$\sin(r \sin \phi) = 2[J_1(r) \sin \phi + J_3(r) \sin 3\phi + \cdots] \tag{49}$$

as the Fourier sine series representing $\sin(r \sin \phi)$.

As an elementary application, we now briefly discuss the notion of *frequency modulation*. We recall from elementary physics that in amplitude modulation, the transmitted radio signal is of the form $f(t) = A \cos \omega t$, where ω is a constant and A varies. The carrier frequency is $\omega/2\pi$ and is much larger than the maximum possible frequency with which A varies, the latter being in the audio range. Note that this can be written $f(t) = A \cos \phi(t)$ where $\phi(t) = \omega t$, and $\phi'(t) = \omega$, a constant.

In frequency modulation, we have $f(t) = C \cos \phi(t)$ where $\phi'(t) = \omega(t)$, which varies. If the frequency of the carrier wave is denoted p, we have $\omega(t) = p + A \cos qt$; for simplicity here, we are assuming the audio signal to be transmitted has a constant amplitude A and a frequency $q/2\pi$. It follows then that $\phi(t) = pt + (A/q) \sin qt$. The ratio A/q is called the *deviation ratio*.

We have for the transmitted radio signal

$$\begin{aligned} f(t) &= \cos\left(pt + \frac{A}{q} \sin qt\right) \\ &= \cos pt \cos\left(\frac{A}{q} \sin qt\right) - \sin pt \sin\left(\frac{A}{q} \sin qt\right) \\ &= \cos pt\left[J_0\left(\frac{A}{q}\right) + 2J_2\left(\frac{A}{q}\right) \cos 2qt + \cdots\right] \\ &\quad - \sin pt\left[2J_1\left(\frac{A}{q}\right) \sin qt + 2J_3\left(\frac{A}{q}\right) \sin 3qt + \cdots\right] \\ &= J_0\left(\frac{A}{q}\right) \cos pt + J_1\left(\frac{A}{q}\right) [\cos(p+q)t + \cos(p-q)t] \\ &\quad + J_2\left(\frac{A}{q}\right) [\cos(p+2q)t + \cos(p-2q)t] + \cdots. \end{aligned} \tag{50}$$

In the notes to Section 1.2, we briefly mention the sideband phenomenon in radio engineering, whereby if the carrier is modulated (amplitude modulation) by an audio signal of (angular) frequency q, there are two transmitted signals, one of frequency $p + q$ and the other $p - q$. (See page 366.) We see from the above analysis, that with frequency modulation there are an *infinite* number of sidebands. Fortunately, however, $J_n(A/q)$ tends to zero with increasing n (Exercise 17), so for practical purposes only a finite number of sidebands are of importance. It will be noted that the relative strength of the received carrier varies as $J_0(A/q)$. For maximum q, the deviation ratio A/q is about 5 in good practice, which radio enthusiasts may understand by looking at a graph of $J_0(r)$ in the vicinity of $r = 5$.

At this point, we have derived most of the relations that will be useful to us later. The rest of the book is devoted to applying these relations to specific problems, and broadening our understanding of the ideas already introduced.

• EXERCISES

1. Show that expressions of the form (2) arise by the method of separation of variables.

2. Show that one can find four linearly independent solutions of (4) continuous for all real r except $r = 0$.

3. Verify (11) by direct calculation.

4. Why don't the orthogonality relations for Bessel functions follow trivially from those proved earlier for self-adjoint differential operators?

5. Derive the orthogonality relations for Bessel functions by methods similar to those of Section 2.4.

6. If (16) is replaced by the analog of (20), is (15) still valid?

7. Prove (18) for a more general set of conditions than (19) or (20).

8. Verify (28).

9. **(a)** Show by differentiating (11) with respect to r, and "simplifying," using (11) and (12), that

$$J_n'(r) + rJ_n''(r) = \frac{n^2}{r} J_n(r) - rJ_n(r). \tag{51}$$

(b) Do you recognize this equation?

10. **(a)** Show by the method suggested in the text that

$$\int_0^c [J_n(ar)]^2 r\,dr = -\frac{cJ_n(ac)}{2a}[J_n'(ac) + acJ_n''(ac)] \tag{52}$$

if $J_n'(ac) = 0$.

(b) Combining this with the results of Exercise 9, show that

$$\int_0^c [J_n(ar)]^2 r\,dr = \frac{a^2c^2 - n^2}{2a^2}[J_n(ac)]^2 \tag{53}$$

when $J_n'(ac) = 0$.

11. Using $\sin\alpha\sin\beta = \frac{1}{2}\cos(\alpha - \beta) - \frac{1}{2}\cos(\alpha + \beta)$ and $\sin\alpha \cos\beta = \frac{1}{2}\sin(\alpha + \beta) + \frac{1}{2}\sin(\alpha - \beta)$, show that

(a) $\sin^k\phi = (-1)^{k/2}(\frac{1}{2})^{k-1}\cos k\phi + \cdots \qquad (k = 2, 4, 6, \cdots)$

(b) $\sin^k\phi = (-1)^{(k-1)/2}(\frac{1}{2})^{k-1}\sin k\phi + \cdots \qquad (k = 3, 5, 7, \cdots)$

where the omitted terms are scalar multiples of $\sin j\phi, j < k$.

12. Show [by differentiating (47)], that

(a) if k is even, $J_n^{(k)}(0) = \dfrac{1}{\pi}\displaystyle\int_0^\pi (-1)^{k/2}\cos(-n\phi)\sin^k\phi\,d\phi$,

(b) if k is odd, $J_n^{(k)}(0) = \dfrac{1}{\pi}\displaystyle\int_0^\pi (-1)^{(k+1)/2}\sin(-n\phi)\sin^k\phi\,d\phi$.

(c) Using the orthogonality relations for trigonometric functions and the results of Exercise 11, show that $J_n^{(k)}(0) = 0$ for $k < n$, and $J_n^{(n)}(0) = 2^{-n}$.

(d) Obtain (c) directly from (5).

13. Show that $J_0(r) = \dfrac{1}{\pi}\displaystyle\int_0^\pi \cos(r\cos\phi)\,d\phi$.

14. Is the function $r^3\sin^3[J_3(r\cos^2\phi)]$ a periodic function of ϕ for every fixed positive r?

15. Evaluate $\displaystyle\int_0^1 \frac{\cos xt}{\sqrt{1 - t^2}}\,dt$.

***16.** Show that $x = 2[J_1(x) + 3J_3(x) + 5J_5(x) + \cdots]$ in some sense, and explain the sense in which the equation is valid.

17. Prove that $\lim_{n\to\infty} J_n(r) = 0$ for every fixed positive r.

18. Show formally how to obtain the coefficients B_n in the expansion of a function $f(r)$ in a series

$$f(r) = \sum_{n=1}^{\infty} B_n J_0(a_n r)$$

where $J_0(a_n c) = 0$, $(n = 1, 2, 3, \cdots)$.

Note: Additional exercises on Bessel functions will be found at the end of Chapter 6, beginning on page 345.

CHAPTER 5

HEAT AND TEMPERATURE

5.1 • THEORY OF HEAT CONDUCTION

Under certain idealized conditions (see any advanced physics text) the rate of flow of heat across a surface S is given by the surface integral

$$\iint_S K \frac{\partial T}{\partial n} \, dS \tag{1}$$

where K is the thermal conductivity of the media and $\partial T/\partial n$ refers to the derivative of the temperature in a direction normal to the surface. If S is the boundary of a region of space, it is customary to take $\partial/\partial n$ to be the derivative in the direction of the outward normal (i.e., away from the region enclosed by S). Since K is a positive quantity, this convention ensures that the integral gives the heat flow *into* the region bounded by S, since heat flows in a direction opposite to that of the temperature gradient.

Again under suitable conditions (see Section 4.2), the effect of this heat flow is to change the temperature within the region bounded by S. If s denotes the specific heat and ρ the mass density, the rate of flow of heat into the region is equal to

$$\iiint_V s\rho \frac{\partial T}{\partial t} \, dV \tag{2}$$

where the integral is taken throughout the region V enclosed by S.

By making use of the divergence theorem of vector analysis (see Section 5.4), one can rewrite (1) as an integral taken throughout the region V, provided one assumes K is a constant throughout V. The integral becomes

$$\iiint_V K\nabla^2 T\, dV \tag{3}$$

where $\nabla^2 T$ denotes the Laplacian of T, discussed in Section 4.2.

Since these integrals are equal, we can combine them into a single integral,

$$\iiint_V \left[K\nabla^2 T - s\rho \frac{\partial T}{\partial t}\right] dV = 0. \tag{4}$$

If the integrand in (4) is a continuous function of the space coordinates, it must be identically zero. Otherwise, one could choose a small region V in which it is positive throughout, or one in which it is negative throughout, and (4) would not be valid for such a region. This line of reasoning leads to the *heat equation*.

$$\nabla^2 T = \frac{1}{\alpha^2}\frac{\partial T}{\partial t} \qquad (\alpha^2 = K/s\rho > 0). \tag{5}$$

The quantity α^2 is called the *thermal diffusivity* of the media.

By definition, the *steady-state* condition is that in which the temperature is not a function of the time t, but varies only with the position in space. Then $\partial T/\partial t = 0$ and (5) reduces to *Laplace's equation*

$$\nabla^2 T = 0. \tag{6}$$

The heat equation (5), sometimes called the *equation of heat conduction*, is also satisfied by the concentration of a substance in a porous media, provided (as usual) that certain ideal conditions are assumed. Laplace's equation (6) is satisfied by the electrostatic potential in a region that is charge-free. A detailed knowledge of the physical situation is not required in the applications we give later.

A standard boundary-value problem of mathematical physics is the *Dirichlet problem*. In this problem, which should more properly be called a class of problems, we are given two pieces of information: (i) the knowledge that a function T satisfies a certain partial differential equation, and (ii) the values of T on the boundary of a prescribed region.

The Dirichlet problem consists in finding the values of T within the region.

In connection with the Dirichlet problem, it is of fundamental importance to know whether or not a solution exists, whether a solution (if it exists) is *unique* and *stable* (we will explain these terms below), whether or not a constructive means exists for finding the solution, and whether there are efficient numerical procedures for determining desired values in any given special case.

The question of existence will be taken up briefly in Section 5.5. This is of little interest to engineers or physicists, since physical considerations usually suffice to convince them on this matter. However, existence proofs are of great interest to mathematicians, and they can be divided roughly into two classes, those that are constructive and those that are not. Some of the most elegant proofs (not discussed in this book) are completely nonconstructive, in that they do not suggest a procedure for finding the solution. On the other hand, a constructive existence proof shows in principle how a solution can be found. In practice, this may be too laborious to even contemplate using to obtain numerical results. Even in some of the simplest cases, there do not exist at the present time efficient procedures for obtaining numerical solutions.

Roughly speaking, a solution of the Dirichlet problem is said to be *stable* if "small" changes in the given values of T on the boundary result in "small" changes in the solution. It is *unstable* if a small change in the given conditions results in an enormous change in the solution. From a practical viewpoint, an unstable solution would be of no value, since there is always some lack of precision in experimental data. Stability of solutions will be discussed briefly later. (See Section 5.5.)

Insofar as *uniqueness* is concerned, this is quite easily proved for Laplace's equation, provided that the surface S is sufficiently regular that the usual integral theorems of vector analysis apply. In Section 5.4 we give the standard proof.

Closely related to the Dirichlet problem is the *Neumann problem*, in which (ii) is replaced by:

(iii) the values of $\partial T/\partial n$, the normal derivative, are given on the boundary of prescribed region.

A slightly more general differential equation than Laplace's equation is obtained by considering the steady-state temperature distribution in a region V where there is an endothermal chemical process taking place. By this, we mean there is a chemical process that is "using up" the heat available without resulting in an increase in temperature. As a matter of convenience, let us assume that this process takes place at a rate proportional to the (local) temperature T. Then (2) is replaced by

$$\iiint_V KpT\,dV \tag{7}$$

where p is a positive function depending on the local density of the reacting material (we shall take it to be a constant). Instead of obtaining Laplace's equation in the steady-state case, we obtain the equation

$$\nabla^2 T = pT \qquad (p > 0). \tag{8}$$

In the next chapter, we will consider the same equation with *negative* p, and will find it presents an altogether different situation.

Before proceeding with any further theoretical material, we shall digress in the next two sections to discuss several special cases in which these equations appear.

• EXERCISE

In spherical coordinates, the element of area on the surface of the sphere $r = a$ is $dS = a^2 \sin\phi\, d\phi\, d\theta$ and (1) becomes

$$\int_0^{2\pi}\int_0^{\pi} K\frac{\partial T}{\partial r}\, a^2 \sin\phi\, d\phi\, d\theta$$

when S is the surface of a sphere of radius a centered at the origin. Write a similar expression in cylindrical coordinates if S is the curved surface of the cylinder of radius a concentric with the z axis, contained between the planes $z = -1$ and $z = 1$.

5.2 • TEMPERATURE OF PLATES

Laplace's equation in two dimensions, in rectangular coordinates, is

$$\frac{\partial^2 T}{\partial x^2} + \frac{\partial^2 T}{\partial y^2} = 0. \tag{1}$$

This equation applies in the three-dimensional case as well, provided T does not depend on the z coordinate.

By the method of separation of variables, introduced in Section 4.4, we can find *product solutions* of the form

$$T(x,y) = X(x)Y(y). \tag{2}$$

We do this by substituting (2) into (1) and separating the variables, which yields

$$-\frac{1}{X}\frac{d^2X}{dx^2} = \frac{1}{Y}\frac{d^2Y}{dy^2} \tag{3}$$

where the limit convention is understood to give meaning to the expressions when the denominators are zero for isolated values of x or y. As noted previously (Section 4.4), this implies both sides of (3) are independent of both x and y. We set both sides equal to k^2 (or $-k^2$), anticipating that this will eliminate some awkward square roots later. We obtain two ordinary differential equations,

$$X''(x) + k^2X(x) = 0, \qquad Y''(y) - k^2Y(y) = 0 \tag{4}$$

or, if we take the constant to be $-k^2$,

$$X''(x) - k^2X(x) = 0, \qquad Y''(y) + k^2Y(y) = 0. \tag{4'}$$

For simplicity, we consider only (4) in the sequel, since clearly (4′) yields the same solutions with roles of X and Y interchanged. Solutions of (4) for X are of the form

$$X(x) = C_1 \cos kx + C_2 \sin kx \qquad (k \neq 0) \tag{5}$$

$$X(x) = C_1 + C_2x \qquad (k = 0) \tag{5'}$$

and solutions of (4) for Y are of the form

$$Y(y) = C_3 \cosh ky + C_4 \sinh ky \qquad (k \neq 0) \tag{6}$$

$$Y(y) = C_3 + C_4y \qquad (k = 0). \tag{6'}$$

In combining these to obtain product solutions, it is important to keep in mind that we must use the same k in both, since otherwise (3) will not be valid.

The simple solutions obtained by combining (5′) and (6′) are not needed in many applications. Ignoring them for the moment, we combine (5) and (6) in (2) and obtain product solutions of the form

$$T_k(x,y) = (C_1 \cos kx + C_2 \sin kx)(C_3 \cosh ky + C_4 \sinh ky). \tag{7}$$

As an elementary application, let us consider a thin rectangular plate, with edges along $x = 0$, $x = L$, $y = 0$, and $y = H$. Let the three edges $x = 0$, $x = L$, and $y = 0$ be held at zero temperature, and suppose the temperature at the top edge to be maintained at

$$T(x,H) = f(x). \tag{8}$$

We seek the steady-state temperature within the plate.

We first observe that the collection of all functions $T(x,y)$ satisfying (1) is a linear space. Moreover, the functions of this class that are zero along the three lines $x = 0$, $x = L$, $y = 0$ is a subspace of this linear space, since any linear combination of functions vanishing on these three lines will itself vanish on the three lines (Figure 5.1).

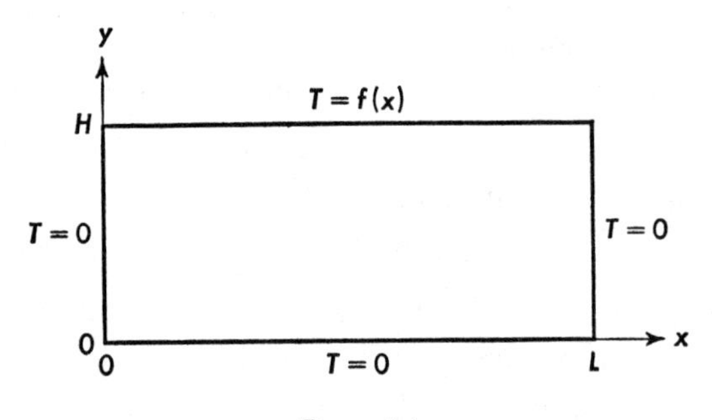

Figure 5.1

Next, we select from (7) those functions belonging to this subspace. To ensure that $T_k(0,y)$ be zero, we take $C_1 = 0$. For $T_k(L,y)$ to be zero, we must select k so that $\sin kL = 0$, so k must be an integral multiple of π/L. To ensure that $T_k(x,0)$ be zero for all x, we take $C_3 = 0$. We obtain the following sequence of functions:

$$T_n(x,y) = A_n \sin \frac{n\pi x}{L} \sinh \frac{n\pi y}{L}, \qquad (n = 1, 2, 3, \cdots). \tag{9}$$

Now assuming that this sequence of functions provides an approximating basis for the subspace, we attempt to find an infinite series of such functions whose sum will have the desired property $T(x,H) = f(x)$. That is, we determine A_n so that

$$T(x,H) = f(x) = \sum_{n=1}^{\infty} \left(A_n \sinh \frac{n\pi H}{L} \right) \sin \frac{n\pi x}{L}, \qquad (0 \leqq x \leqq L). \tag{10}$$

We do this quite simply by comparing (10) with the Fourier sine series expansion of $f(x)$, in the interval $0 \leqq x \leqq L$,

$$f(x) = \sum_{n=1}^{\infty} B_n \sin \frac{n\pi x}{L}, \tag{11}$$

where

$$B_n = \frac{2}{L} \int_0^L f(x) \sin \frac{n\pi x}{L}\, dx. \tag{12}$$

The desired solution is seen (at least formally) to be

$$T(x,y) = \sum_{n=1}^{\infty} B_n \frac{(\sin n\pi x/L)\,(\sinh n\pi y/L)}{\sinh n\pi H/L} \tag{13}$$

where the coefficients B_n are given by (12).

In polar coordinates, Laplace's equation in two dimensions is

$$\frac{\partial^2 T}{\partial r^2} + \frac{1}{r}\frac{\partial T}{\partial r} + \frac{1}{r^2}\frac{\partial^2 T}{\partial \theta^2} = 0. \tag{14}$$

In this case (Exercise 1) separation of variables leads to product solutions of the form

$$(C_1 + C_2 \log r)(C_3 + C_4\theta) \tag{15}$$

and

$$(C_1 r^k + C_2 r^{-k})(C_3 \cos k\theta + C_4 \sin k\theta), \qquad (k \neq 0) \tag{16}$$

which may be superimposed as in the preceding example.

As an application, suppose we are given the temperature distribution, as a function of θ, along the inner and outer rim of a circular annulus (Figure 5.2). Our problem is to determine the temperature at points within the annulus.

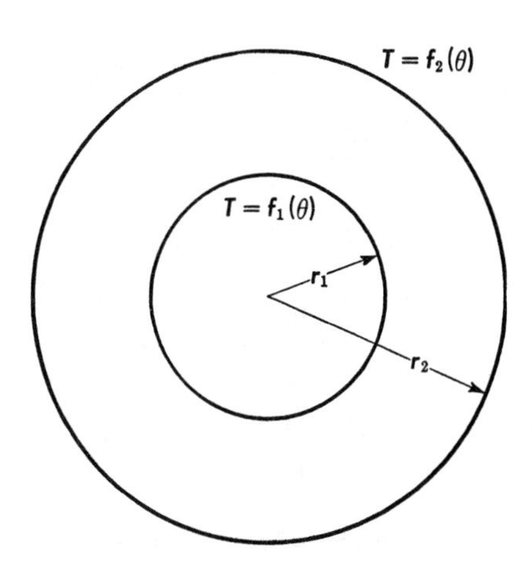

Figure 5.2

In this case, we are given

$$T(r_1,\theta) = f_1(\theta), \qquad T(r_2,\theta) = f_2(\theta) \tag{17}$$

and we desire a solution $T(r,\theta)$ of (14) satisfying the boundary conditions (17).

Again, we consider the linear space of all solutions of (14) that make sense as possible temperature distributions for such a circular annulus. Such functions must clearly be bounded within the annulus and, as functions of θ, must have period 2π. In this linear space, we seek to find an approximating basis consisting of functions of the type given above.

It is clear that solutions of type (16) will be periodic of period 2π only if k is an integer. Any solution of type (15) must have $C_4 = 0$ for the same reason. It might appear at first that we should take C_2 in (15) and (16) to be zero, but this is not so. Although $\log r$ and r^{-k} are not bounded as r tends to zero, they are perfectly well-behaved within the circular annulus and therefore are elements of the linear space.

We therefore try to determine the coefficients in the infinite series

$$\begin{aligned} T(r,\theta) &= (A_0 + B_0 \log r) \\ &\quad + \sum_{n=1}^{\infty} [(A_n r^n + B_n r^{-n}) \cos n\theta + (C_n r^n + D_n r^{-n}) \sin n\theta] \end{aligned} \tag{18}$$

in order to satisfy (17). We do this by finding the Fourier series expansions of $f_1(\theta)$ and $f_2(\theta)$ and comparing these with (18). Each of the coefficients of (18) must match the coefficients in the expansion of $f_1(\theta)$ when $r = r_1$, and those of $f_2(\theta)$ when $r = r_2$. Thus we have

$$A_0 + B_0 \log r_j = \frac{1}{2\pi} \int_0^{2\pi} f_j(\theta)\, d\theta, \qquad (j = 1,2) \tag{19}$$

$$A_n r_j^n + B_n r_j^{-n} = \frac{1}{\pi} \int_0^{2\pi} f_j(\theta) \cos n\theta\, d\theta, \qquad (j = 1,2) \tag{20}$$

$$C_n r_j^n + D_n r_j^{-n} = \frac{1}{\pi} \int_0^{2\pi} f_j(\theta) \sin n\theta\, d\theta, \qquad (j = 1,2). \tag{21}$$

Each of the equations (20) and (21) represents two equations in two unknown coefficients and they are solved to obtain the coefficients needed in (18).

A somewhat simpler problem is that of determining the temperature distribution in a circular plate, given the temperature $f(\theta)$

around its outer rim. The above considerations apply, with r_1 set equal to zero. In this case we must reject the product solutions that are not bounded as r tends to zero, and the resulting solution can be written in the form

$$(22)\quad T(r,\theta) = a_0 + \sum_{n=1}^{\infty} \left(\frac{r}{r_2}\right)^n (a_n \cos n\theta + c_n \sin n\theta), \qquad (r < r_2)$$

where the a's and c's are the Fourier coefficients in the Fourier expansion of $f(\theta)$.

In this connection, it is of interest to note that

$$(23)\qquad T(0,\theta) = a_0 = \frac{1}{2\pi} \int_0^{2\pi} f(\theta)\, d\theta$$

is the temperature at the center of the plate. Since these considerations are valid generally for all bounded harmonic functions of two variables, whether interpreted as temperatures or otherwise, we see formally that the *value of such a function at the center of a circle is equal to its average value on the circumference of the circle.* (A rigorous proof will be given in Section 5.5.)

The reader can easily verify that if we are given values $f(\theta)$ on the circumference of a circle of radius r_1, and we seek a harmonic function of two variables in the region *exterior* to the circle, matching these values when $r = r_1$ and tending to zero with increasing r, we formally obtain

$$(24)\quad T(r,\theta) = a_0 + \sum_{n=0}^{\infty} \left(\frac{r_1}{r}\right)^n (b_n \cos n\theta + d_n \sin n\theta), \qquad (r_1 < r)$$

where the coefficients a_0, b_n, and d_n are the Fourier coefficients of the function f.

The striking similarity between (22) and (24) above and (22) and (23) of Section 4.4 suggests that these may also be related to a "generating function." This is, in fact, the case. If one expands the function

$$\frac{1}{2}\left(\frac{1 - r^2}{1 - 2r\cos\theta + r^2}\right)$$

in powers of r, one obtains $\frac{1}{2} + r\cos\theta + r^2 \cos 2\theta + \cdots$, so this function may in some sense be considered a generating functon for the cosines.

Let us explore this idea further. If we substitute the integrals defining the Fourier coefficients into (22), and assume the legitimacy of interchanging differentiation and summation, we obtain (letting $r_2 = a$)

$$T(r,\theta) = \frac{1}{\pi}\int_0^{2\pi} f(\phi)$$

$$(25) \qquad \left[\frac{1}{2} + \sum_{n=1}^{\infty} \left(\frac{r}{a}\right)^n (\cos n\theta \cos n\phi + \sin n\theta \sin n\phi)\right] d\phi$$

$$= \frac{1}{\pi}\int_0^{2\pi} f(\phi)\left[\frac{1}{2} + \sum_{n=1}^{\infty}\left(\frac{r}{a}\right)^n (\cos n(\theta - \phi))\right] d\phi \qquad (r < a).$$

The reader will notice the similarity between this and the discussion in Section 2.6 which preceded the introduction of the Dirichlet kernel. By (36), Section 3.9, we have

$$(26) \quad \frac{1}{2} + \sum_{n=1}^{\infty}\left(\frac{r}{a}\right)^n \cos n\theta = \frac{1}{2}\frac{a^2 - r^2}{a^2 - 2ar\cos\theta + r^2} \qquad (r < a)$$

and introducing this into (25), we obtain

$$(27) \quad T(r,\theta) = \frac{1}{2\pi}\int_0^{2\pi} \frac{a^2 - r^2}{a^2 - 2ar\cos(\theta - \phi) + r^2} T(a,\phi)\, d\phi \qquad (r < a).$$

It will be noted that (27), called *Poisson's integral formula*, expresses the value of a harmonic function T at points within a circle of radius a in terms of its values on the circumference. It will also be noted that this formula is of convolution type (Section 2.6).

• EXERCISES

1. By "separating variables" in (14), obtain solutions of the form (15) and (16).

2. The temperature along three edges of a square plate is maintained at $T = 0°$, and along the fourth edge at $T = 100°$. These temperatures are maintained until steady-state conditions are reached. (Strictly speaking, this would involve an infinite temperature gradient at each corner; for a comment on this, see the Answers and Notes.)

(a) Without doing any calculating at all (except mental arithmetic), determine the temperature at the center of the square.

(b) Find a series expansion for the temperature as a function of position throughout the plate.

(c) Taking the corners of the square to be the points (0,0), (1,0), (0,1), and (1,1), show by symmetry that $T(\frac{1}{3},\frac{1}{3}) = T(\frac{2}{3},\frac{1}{3})$ and determine the numerical value of T at these two points.

3. A function $T(x,y)$ satisfying (1) has the known values $T(0,y) = 0$, $T(x,0) = 0$, $T(\pi,y) = 0$, $T(x,\pi) = 3\sin x - 4\sin 2x$. Determine $T(x,y)$.

4. A function $T(x,y)$ satisfying (1) has known values $T(0,y) = 0$, $T(x,0) = 0$, $T(\pi,y) = \sin y$, $T(x,\pi) = \sin 2x$. Determine $T(x,y)$.

5. A function $T(r,\theta)$ satisfying (14) is bounded within the circle $r = 4$ and has values $T(4,\theta) = \sin\theta - 3\cos\theta + 5\sin 4\theta$. Determine $T(r,\theta)$.

6. A function $T(r,\theta)$ satisfying (14) except at the origin is bounded in the *exterior* of the circle $r = 3$ and has values $T(r,\theta) = \theta(2\pi - \theta)$, $0 \leqq \theta \leqq 2\pi$. Find an infinite series representation of $T(r,\theta)$ for $r > 3$.

7. Given a circular annulus with radii 3 and 6 and temperatures $T(3,\theta) = 5\cos\theta$, $T(6,\theta) = 10\sin\theta$, find $T(r,\theta)$ within the annulus.

8. Given that a function is bounded within the unit circle $r = 1$, that $T(1,\theta) = 1$ when $0 < \theta < \pi$, $T(1,\theta) = -1$ when $\pi < \theta < 2\pi$, and that the function satisfies Laplace's equation when $r < 1$,
(a) find formally a solution $T(r,\theta)$ valid when $r < 1$;
(b) what is the value of T along the diameter $y = 0$?

***9.** All points on the circumference of a large circular disk of radius R are maintained at 0° except for one small region near $(R,0)$ where the temperature is maintained by a steady blow-torch at a very high temperature. Find an approximate steady-state temperature distribution $T(r,\theta)$ valid when $r < R$ (necessarily to within a scalar factor).

***10.** A function f is continuous and periodic with period 2π. Its Fourier series expansion is

$$f(\theta) = a_0 + \sum_{n=1}^{\infty} (a_n \cos n\theta + b_n \sin n\theta).$$

A new function is defined by

$$g(\theta) = a_0 + \sum_{n=1}^{\infty} \frac{1}{2^n} (a_n \cos n\theta + b_n \sin n\theta).$$

Determine a function $h(\theta)$ whose convolution product with f gives g. (Despite appearances, this is related to the material in this section.)

11. The temperature along the straight edges of a circular sector $0 \leqq \theta \leqq \alpha$, $0 \leqq r \leqq a$ is maintained at zero degrees, and along the curved edge we have $T(a,\theta) = 3 \sin (\pi\theta/\alpha)$. Find the temperature within the sector.

12. The temperature along the straight edge $\theta = 0$ of the sector (Exercise 11) is maintained at zero degrees, but along the other straight edge $\theta = \alpha$ the sector is insulated so that $\partial T/\partial\theta = 0$ along this edge. The temperature $T(a,\theta)$ is arbitrarily prescribed. What product solutions would you use in setting up a series solution for $T(r,\theta)$ within the sector?

5.3 • TEMPERATURE OF SOLIDS

In three dimensions, Laplace's equation in rectangular coordinates is

$$\frac{\partial^2 T}{\partial x^2} + \frac{\partial^2 T}{\partial y^2} + \frac{\partial^2 T}{\partial z^2} = 0. \tag{1}$$

As before, we can obtain product solutions of the form

$$T(x,y,z) = X(x)Y(y)Z(z) \tag{2}$$

by the method of separation of variables. The reader can readily verify that a large number of possibilities arise. One possible set of such solutions is the following:

$$X(x) = C_1 \sin k_1 x + D_1 \cos k_1 x, \tag{3}$$

$$Y(y) = C_2 \sin k_2 y + D_2 \cos k_2 y, \tag{4}$$

$$Z(z) = C_3 \sinh k_3 z + D_3 \cosh k_3 y \tag{5}$$

provided that

$$k_1^2 + k_2^2 = k_3^2. \tag{6}$$

By interchanging the roles of x, y, and z, other possibilities arise; our choice of Z as the factor containing the hyperbolic functions was purely arbitrary. One can also obtain factors of the form $X(x) = A_1 + B_1 x$, $Y(y) = A_2 + B_2 y$, and the like (obviously the function xyz is harmonic).

We will consider only product solutions of the above type, and relegate discussion of others to the exercises.

It is of crucial importance that the condition (6) not be neglected. Except for this, which introduces a slight degree of complication, the only new idea involved in superimposing product solutions to match prescribed boundary conditions is the notion of a *double Fourier series.* In this book, we shall restrict attention to the double Fourier *sine* series.

Given real-valued functions $f(x,y)$ and $g(x,y)$ continuous on the rectangle $0 \leqq x \leqq L_1$, $0 \leqq y \leqq L_2$, we define their inner product by

$$(f|g) = \int_0^{L_2} \int_0^{L_1} f(x,y)g(x,y)\,dx\,dy. \tag{7}$$

Relative to this inner product, one readily verifies that the double sequence of functions

$$f_{mn}(x,y) = \sin\frac{m\pi x}{L_1}\sin\frac{n\pi y}{L_2} \tag{8}$$

are orthogonal, which in this case means

$$(f_{mn}|f_{ij}) = 0 \qquad (\text{if } m \neq i \text{ or } n \neq j). \tag{9}$$

On the other hand, it is easy to verify that

$$(f_{mn}|f_{mn}) = L_1L_2/4. \tag{10}$$

This suggests the possibility that an "arbitrary" continuous function $f(x,y)$ can be expanded in a double series of the form

$$f(x,y) = \sum_{m=1}^{\infty}\sum_{n=1}^{\infty} C_{mn}\sin\frac{m\pi x}{L_1}\sin\frac{n\pi y}{L_2} \tag{11}$$

where

$$C_{mn} = \frac{4}{L_1L_2}\int_0^{L_2}\int_0^{L_1} f(x,y)\sin\frac{m\pi x}{L_1}\sin\frac{n\pi y}{L_2}\,dx\,dy \tag{12}$$

in complete analogy with ordinary Fourier sine series.

Remarks similar to those made in connection with ordinary sine series can be made here. For instance, we do not expect this series to be convergent to $f(x,y)$ *outside* the rectangle unless $f(x,y)$ satisfies some rather stringent conditions. Even in the rectangle, the series cannot converge uniformly to $f(x,y)$ unless $f(x,y)$ vanishes on the boundaries of the rectangle and satisfies other conditions more stringent than mere continuity, although it is true that $\lim_{\substack{n\to\infty \\ m\to\infty}} \|f - S_{mn}\| = 0$, where the meaning of the partial sum S_{mn} and the norm are defined in obvious ways. (See Section 2.1.)

Now suppose that we are told that the temperature on five faces of a rectangular parallelepiped has been maintained at zero, and its sixth face maintained at a temperature distribution $f(x,y)$, so that

$$T(0,y,x) = T(L_1,y,x) = T(x,0,z) = T(x,L_2,z) = T(x,y,0) = 0, \quad \text{and } T(x,y,H) = f(x,y), \tag{13}$$

and held at these values until a steady-state temperature distribution prevails throughout the region within the parallelepiped. Then choosing appropriate product solutions and reasoning in an entirely similar way to that leading to (13), Section 5.2, we obtain a formal solution for $T(x,y,z)$ as follows:

$$T(x,y,z) = \sum_{m=1}^{\infty} \sum_{n=1}^{\infty} C_{mn} \sin \frac{m\pi x}{L_1} \sin \frac{n\pi x}{L_2} \frac{\sinh k_{mn}z}{\sinh k_{mn}H} \tag{14}$$

where

$$k_{mn} = \pi \sqrt{\frac{m^2}{L_1^2} + \frac{n^2}{L_2^2}}$$

and the C_{mn} are determined by (12).

Laplace's equation in *spherical coordinates* was discussed in Section 4.4. The product solutions of greatest value in practice are those of the form

$$r^k \cos m\theta P_k^{(m)}(\cos \phi) \quad \text{or} \quad r^k \sin m\theta P_k^{(m)}(\cos \phi) \tag{15}$$

and

$$r^{-k-1} \cos m\theta P_k^{(m)}(\cos \phi) \quad \text{or} \quad r^{-k-1} \sin m\theta P_k^{(m)}(\cos \phi) \tag{16}$$

where the functions $P_k^{(m)}$ are the associated Legendre functions, already discussed in some detail.

If we are given a continuous function $g(\theta,\phi)$ as the values of a harmonic function $T(r,\theta,\phi)$ on the surface of the sphere $r = r_0$, then the solution to the Dirichlet problem within the sphere is given by

$$T(r,\theta,\phi) = \sum_{k=0}^{\infty} \sum_{m=0}^{k} \left(\frac{r}{r_0}\right)^k (A_{km} \cos m\theta + B_{km} \sin m\theta) P_k^{(m)}(\cos \phi) \tag{17}$$

where, making use of a modification of (17), Section 4.5, the coefficients B_{km} are given by

$$B_{km} = \frac{\int_0^{\pi} \int_0^{2\pi} g(\theta,\phi) \sin m\theta P_k^{(m)}(\cos \phi) \sin \phi \, d\theta \, d\phi}{\pi \left(\frac{2}{2k+1}\right) \frac{(k+m)!}{(k-m)!}}. \tag{18}$$

The expression for A_{km} is the same with $\sin m\theta$ replaced by $\cos m\theta$ in the integrand, except of course that the denominator must be doubled when $m = 0$. (See Exercise 8.)

Here we have not used product solutions of the form (16) since they are not bounded as r tends to zero. If, however, we were given the values of T on the outer and inner surfaces of a hollow spherical shell, then we would need to include solutions of the form (16) as well.

• EXERCISES

1. Show that (2) is a product solution of (1) if we take

$$X(x) = A_1 + B_1x, \quad Y(y) = A_2 + B_2y, \quad \text{and} \quad Z(z) = A_3 + B_3z.$$

2. Find a set of product solutions of (1) other than those mentioned in the text or in Exercise 1.

3. Write simplified versions of (17) and (18) valid when $g(\theta,\phi)$ has cylindrical symmetry, i.e., does not depend on θ.

4. **(a)** Determine all solutions of Laplace's equation that depend only on the spherical coordinate r.

(b) The temperature at the outer surface $r = 6$ of a hollow spherical shell is maintained at 100° and the temperature at its inner surface $r = 2$ is maintained at 50° until steady-state conditions prevail. Find T as a function of r, and determine T at $r = 4$.

5. Show formally, using (17) and (18), that the value of a harmonic function at the center of a sphere is the average of its values on the surface.

6. How must (17) and (18) be modified to produce solutions $T(r,\theta,\phi)$ bounded in the region *exterior* to the sphere?

7. **(a)** Find product solutions of Laplace's equation in cylindrical coordinates that do not depend on z or θ.

(b) The temperature at the outer surface $r = 6$ of a long hollow cylinder is maintained at 100° and the temperature at its inner surface $r = 2$ is maintained at 50° until steady-state conditions prevail. Find the temperature T as a function of the cylindrical coordinate r, and determine T at $r = 4$.

***8.** Show how (18) was obtained. [Compare (17), Section 4.5.]

5.4 • HARMONIC FUNCTIONS

Preparatory to proving some of the fundamental theorems concerning harmonic functions, we shall briefly review some basic ideas of vector analysis.

We recall that, if $f(x,y,z)$ is a continuously differentiable real-valued function of position in space, the *gradient* of f, denoted grad f or ∇f, is a vector-valued function of position that gives, at each point, the direction in which f increases most rapidly and the magnitude of this rate of increase. More generally, if n is a unit vector,

$$\frac{\partial f}{\partial s} = \nabla f \cdot n, \tag{1}$$

which means that, if we wish to know the directional derivative $\partial f/\partial s$ (rate of change per unit distance) of f in the direction of n, we take the scalar product of grad f with n.

If F is a vector-valued function of position in space, which we assume to have scalar components that are continuously differentiable, then the *curl* of F, denoted curl F or $\nabla \times F$, is a vector field which can be defined by the requirement that the surface integral of the normal component of curl F over a smooth oriented surface S in space, bounded by a piecewise smooth closed curve C, equals the line integral of the tangential component of F about C. That is, we have

$$\iint_S (\nabla \times F) \cdot n \, dS = \int_C F \cdot T \, ds. \tag{2}$$

In this equation, n denotes (at each point of S) a unit vector normal to S, dS denotes the element of surface area, T denotes a unit tangent vector to the curve C (which forms the boundary of S), and ds is the element of arc length on C. There are two possible choices for n, and similarly for T, and these must be chosen consistently, so that $n \times T$ at any point on the boundary of S points in the general direction of the interior of S.

The *divergence* of a vector-valued function of position in space, denoted div F or $\nabla \cdot F$, is a scalar-valued function. Assuming that the components of F are continuously differentiable, div F can be defined by the requirement that the integral of div F throughout any connected region of space D bounded by a closed suitably smooth surface S is equal to the surface integral of the normal component of F over the bounding surface S. That is, we have

$$\iiint_D \operatorname{div} F \, dV = \iint_S F \cdot n \, dS \tag{3}$$

where dV is the element of volume and n is a unit normal which, at any point on S, is directed away from the interior of S.

At any point P, one obtains (in principle) the value of div F from (3) by a limiting process, whereby one considers increasingly small domains containing P. A similar limiting process would be needed to obtain curl F at a point from the definition (3). We shall have no occasion to use the analytical expressions for the gradient, divergence, and curl, in various coordinate systems, and therefore they are omitted here. The reader will find an extensive discussion of the intuitive ideas behind these concepts in an elementary book by the present author, *Introduction to Vector Analysis* (Allyn and Bacon, 1961).

The Laplacian of a real-valued function f of position in space is defined to be the divergence of its gradient, and is denoted $\nabla^2 f$. Analytical expressions for the Laplacian have already been given in Sections 4.2 and 4.4. If f is twice continuously differentiable, then its Laplacian is a continuous function of position in space, and gives a measure of the "smoothness" of f, as discussed heuristically in Section 4.2.

It can be shown that, if f is a real-valued function and F is a vector-valued function, then

$$\nabla \cdot (fF) = \nabla f \cdot F + f \nabla \cdot F. \tag{4}$$

Suppose that $F = \nabla g$, and f and g are twice continuously differentiable functions. Then by (4) we have

$$\nabla \cdot (f \nabla g) = \nabla f \cdot \nabla g + f \nabla^2 g \tag{5}$$

and it then follows from (3) that

$$\iiint_D (\nabla f \cdot \nabla g + f \nabla^2 g) \, dV = \iint_S (f \nabla g) \cdot n \, dS. \tag{6}$$

From (1) it then follows that

$$\iiint_D (\nabla f \cdot \nabla g + f \nabla^2 g) \, dV = \iint_S f \frac{\partial g}{\partial n} \, dS \tag{7}$$

where $\partial g / \partial n = \nabla g \cdot n$ is the derivative of g in the direction of the outward normal to the closed surface S. This equation is sometimes called the *nonsymmetric Green's theorem.* (It should not be confused with a more elementary theorem which appears in some books and is also called Green's theorem.)

If we interchange the position of f and g in (7) and subtract the resulting equation from (7) we obtain

$$\iiint_D (f\nabla^2 g - g\nabla^2 f)\, dV = \iint_S \left(f\frac{\partial g}{\partial n} - g\frac{\partial f}{\partial n}\right) dS. \tag{8}$$

This is called the *symmetric form of Green's theorem.*

If we take f to be identically equal to unity in (8), we obtain

$$\iiint_D \nabla^2 g\, dV = \iint_S \frac{\partial g}{\partial n}\, dS \tag{9}$$

which provides some justification for the intuitive notion that $\nabla^2 g$ provides a measure of the smoothness of g. To see this, suppose that $\nabla^2 g$ is a constant, $\nabla^2 g = K$, and let us apply (9) to a spherical region. Taking the origin to be the center of a sphere of radius r, (9) then becomes $\frac{4}{3}\pi r^3 K = \iint_S \partial g/\partial r\, dS$. If we integrate both sides of this expression from $r = 0$ to $r = R$, we obtain

$$\tfrac{1}{3}\pi R^4 K = \tfrac{1}{2}\iint_S [g(R,\phi,\theta) - g(0,\phi,\theta)]\, dS$$

and the integral can be rewritten $4\pi R^2\bar{g}(R) - 4\pi R^2 g(0)$ where $\bar{g}(R)$ denotes the mean value of g over the surface of a sphere of radius R, and $g(0)$ denotes its value at the center. This yields

$$\bar{g}(R) - g(0) = \tfrac{1}{6}R^2\nabla^2 g \tag{10}$$

provided that $\nabla^2 g$ is a constant. Multiplying this by $4\pi R^2$ and integrating again, one obtains

$$\bar{g} - g(0) = \tfrac{1}{10}R^2\nabla^2 g \tag{11}$$

where $\bar{g}$ denotes the mean value of g within a sphere of radius R, $g(0)$ its value at the center, and $\nabla^2 g$ is assumed a constant. These expressions are consistent with the heuristic remarks in Section 4.2, but it must be noted that they are valid only when $\nabla^2 g$ is constant. Of course, if $\nabla^2 g$ is continuous, then it will be "approximately" constant in a sufficiently small sphere, and this is what we had in mind in Section 4.2.

Now let us give a more precise definition of harmonic function. A function f is said to be harmonic at a point if it is continuous and satisfies Laplace's equation throughout some neighborhood of the point. It is harmonic in a bounded region D if it is harmonic at every point within D and continuous both within and on the boundary of D. It is harmonic in an infinite (i.e., unbounded) region D if it is harmonic in every bounded region within D and if both rf and $r^2\nabla f$ are bounded outside a sphere of sufficiently large radius.

EXAMPLE 1: The function $1/r$ (r denotes the spherical coordinate) is harmonic except at the origin. Letting D denote the region consisting of all points except the origin, then D is an unbounded region, and the above requirements are satisfied.

EXAMPLE 2: The function $e^z J_0(r)$, where r here is the cylindrical coordinate, is not harmonic throughout space, although it satisfies Laplace's equation (see Section 4.6) and is continuous everywhere. This is because it does not satisfy the boundedness condition demanded above. However, it is harmonic at every point, and is harmonic in every bounded region D. It is also harmonic in every infinite region that is bounded above by any plane z = constant.

EXAMPLE 3: It is possible for a function to satisfy Laplace's equation

$$\frac{\partial^2 f}{\partial x^2} + \frac{\partial^2 f}{\partial y^2} + \frac{\partial^2 f}{\partial z^2} = 0$$

and not even be continuous. Of course, for the necessary derivatives to exist, the function must be continuous as a function of x, y, or z separately, but this does not ensure continuity in these variables simultaneously. In particular, on changing to spherical coordinates, the function may not be continuous as a function of r, θ, and ϕ. If, however, a function satisfies Laplace's equation in rectangular coordinates, and is known to be continuous, then it can be shown that it will satisfy Laplace's equation in any other system of coordinates as well, which is the reason for the requirement above that f be continuous. (See Exercise 3.)

In view of this definition of harmonic function, some of the examples given in preceding sections were rather improper, since (for instance) one problem called for finding a harmonic function with prescribed boundary conditions that were discontinuous. (Indeed, it is common in applications to ignore this definition, and to consider *generalized harmonic functions*, which need not be continuous or differentiable but which can be approximated arbitrarily closely by harmonic functions.)

Theorem 1. *If f and g are harmonic and continuously differentiable in the interior or the exterior of a closed piecewise smooth surface S, including the surface S itself, then*

$$\iint_S \left(f\,\frac{\partial g}{\partial n} - g\,\frac{\partial f}{\partial n} \right) dS = 0. \tag{12}$$

Proof: If the region is bounded, i.e., is within S, this follows from (8) on setting $\nabla^2 g$ and $\nabla^2 f$ equal to zero. If the region is exterior to S, we apply (8) to the region between S and a large sphere of radius R containing S, and then let R tend to infinity, whereby since $r^2 \nabla f$ is bounded we have $\partial f/\partial n < C/r^2$ and since rf is bounded, f tends to zero, so the integral over the outer sphere tends to zero. [The hypothesis of continuous differentiability is needed to ensure the validity of (3), on which (8) depends.]

Theorem 2. *If f is continuously differentiable and harmonic in the interior or the exterior of a closed piecewise smooth surface S, including the surface S itself, then*

$$\iint_S \frac{\partial f}{\partial n}\, dS = 0. \tag{13}$$

Proof: Let g be identically equal to unity in (12).

Theorem 3. *If a function f is harmonic in a sphere, the value of f at the center of the sphere is equal to the average of its values over the surface of the sphere.*

Proof: The requirement that f be harmonic in the sphere means, by definition, that it is harmonic at points interior to the sphere, and continuous not only at interior points but also on the boundary. This implies continuous differentiability within (but not necessarily on the surface of) the sphere. Therefore (10) is valid for smaller spheres within the given sphere, and since $\nabla^2 f = 0$, we have $\bar{g}(R) = g(0)$ for such spheres. By continuity, this is therefore true for the given sphere.

The weaker hypothesis (not requiring continuous differentiability on the boundary) could not be used in Theorem 2, since the definition of "harmonic" in a region ensures only the continuity of f as we approach the boundary, and not that of $\partial f/\partial n$.

Theorem 4. *If a function f is harmonic in a sphere, the value of f at the center of the sphere is equal to the average of its values throughout the sphere.*

Proof: The proof is the same as that for Theorem 3, making use of (11) instead of (10).

Theorem 5. *If a function f is harmonic and not constant in a region which is either the interior or the exterior of a closed piecewise smooth surface S, then it cannot have a local maximum or local minimum inside the region.*

By a local minimum we mean a point with the property that the value of the function at the point is less than its value at every point on some sphere with center at the point (contained entirely in the given region). Local maximum is defined similarly. The theorem is therefore an obvious corollary of Theorem 3.

The reader should not be at all surprised at Theorem 5, since a sufficient condition for a local minimum of a suitably smooth function to exist is that its partial derivatives $\partial f/\partial x$, $\partial f/\partial y$, and $\partial f/\partial z$ be zero and its second derivatives $\partial^2 f/\partial x^2$, $\partial^2 f/\partial y^2$, and $\partial^2 f/\partial z^2$ be positive, at the given point, which would violate the requirement that the function satisfy Laplace's equation. We have not based our proof on this, however, since this is not a *necessary* condition for a local minimum.

Theorem 6. *If a function f is harmonic in a region bounded by a closed piecewise smooth surface S on which the function is constant, then the function is constant throughout the region. (If the given region is exterior to S, and therefore unbounded, this can happen only if f is identically zero throughout the region.)*

Proof: If the region is interior to S, this follows at once from Theorem 5. The rest of the proof is left to the reader (Exercise 2).

Theorem 7. *If two functions f and g are harmonic in a region bounded by a closed piecewise smooth surface S, and if $f = g$ for points on S, then $f = g$ identically throughout the region.*

Proof: Apply Theorem 6 to the function $f - g$.

Theorem 8. *If f is harmonic and continuously differentiable in a connected region bounded by a closed piecewise smooth surface S, and $\partial f/\partial n = 0$ on S, then f is constant throughout the region.*

Proof: We take $f = g$ in (7). Since $\nabla^2 f = 0$ this yields

$$\iiint_D |\nabla f|^2 \, dV = 0. \tag{14}$$

Clearly $|\nabla f|^2$ is not negative and, by the hypothesis, is continuous. If it were positive at a point, then by continuity it would be positive throughout a sufficiently small sphere centered at that point, and this would imply that the left side of (14) is positive. This contradiction shows that $|\nabla f|^2$ is identically zero. Since D is connected, any two points can be joined by a smooth path, and the difference between the values of f at the two points is obtained by integrating (1) along this path. It follows that this difference must be zero, since ∇f has zero magnitude, and therefore the function is the same at any two points, which proves the theorem.

Theorem 9. *If f and g are harmonic and continuously differentiable in a connected region bounded by a closed piecewise smooth surface S, and if $\partial f/\partial n = \partial g/\partial n$ at every point on S, then f and g must differ at most by a constant.*

Proof: Apply Theorem 8 to $f - g$.

Theorem 10. *If D is an unbounded region, the exterior of a piecewise smooth closed surface S, and if f is harmonic and continuously differentiable in D, then*

$$\iint_{S_1} \frac{\partial f}{\partial n} \, dS = \iint_S \frac{\partial f}{\partial n} \, dS \tag{15}$$

where S_1 is any closed piecewise smooth surface containing S in its interior, the normals in both integrals being taken away from the region enclosed by S (or both towards it).

Proof: This follows from Theorem 2 applied to the region between S_1 and S_2. Notice that, in integrating over any surface which is considered the boundary of an unbounded region, as for instance the exterior of a sphere, the normal direction on the surface is away from this region, i.e., in the opposite direction from the region.

Theorem 7 is the *uniqueness* theorem for solutions of the Dirichlet problem for Laplace's equation. Theorem 9 shows that, for Laplace's equation, solutions of the Neumann problem (see page 249) are unique to within an additive constant. Both of these results are what we would expect intuitively because they both have obvious interpretations in connection with problems involving steady-state temperature distributions. For example, we obtain a special case of the Neumann problem when a body is insulated so that no heat can flow across its bounding surface; then $\partial f/\partial n = 0$ and by Theorem 9 (taking g to be identically zero) it follows that f must be constant throughout the body. This is just what we would expect: if the body is insulated, the heat will flow within the body in such a way that temperature differences are evened out, and will eventually approach a steady-state condition where a constant temperature prevails.

We note that these theorems do not constitute proofs of *existence.* We have not yet shown that solutions of either the Dirichlet problem or the Neumann problem need to exist. Indeed, we see from (13) that we cannot prescribe $\partial f/\partial n$ arbitrarily on the bounding surface; a necessary condition for a solution of the Neumann problem is that $\iint_S \partial f/\partial n \, dS = 0$. From Theorem 9, we see that even if this condition is satisfied, a solution to the Neumann problem will be uniquely determined even then only if the value of f at one point also be prescribed, so that the arbitrary additive constant can be determined.

We now consider the equation

$$\nabla^2 f = pf \qquad (p > 0), \tag{16}$$

introduced in Section 5.1. We call a function *subharmonic* in a region if it is continuous throughout a region (including its boundary) and satisfies (16). A comment on this definition of "subharmonic" will be found in the Answers and Notes section. For simplicity, we shall assume in the following that f is twice continuously differentiable throughout any region under discussion, without any further mention of this assumption.

We can rewrite (7) in a completely equivalent form

(17)

$$\iiint_D (\nabla f \cdot \nabla g + pfg) \, dV = \iint_S f \frac{\partial g}{\partial n} \, dS - \iiint_D f(\nabla^2 g - pg) \, dV.$$

Except for a little transposing, we have done nothing to (7) except add $\iiint_D pfg \, dV$ to both sides.

Now suppose that f is subharmonic, so the last integral on the right vanishes, and let us put $f = g$. We obtain

(18) $$\iiint_D (|\nabla f|^2 + pf^2)\, dV = \iint_S f \frac{\partial f}{\partial n}\, dS.$$

Keeping in mind that p is a positive function, we see that the integral on the left side can never be negative. In particular, we see that if a subharmonic function f vanishes on the boundary S, then f must vanish identically throughout the region D. This is also true, as we have seen, for harmonic functions. More remarkable, however, is that if $\partial f/\partial n$ vanishes on the boundary S, then f must vanish identically throughout D. (We recall that, if f is assumed to be harmonic, all we could conclude in this instance is that f is a constant throughout D.) This is not surprising in view of the way in which we obtained (16) (see Section 5.1). If an endothermic reaction is going inside an insulated body, we would expect it to tend to a zero steady-state temperature, at which time all the heat is used up and the reaction ceases.)

Without taking the time to give detailed proofs, we can tell even more from (18). We see that the left side of (18) is never negative, and thus the right side is not negative; therefore if a point P is a local minimum, so that $\partial f/\partial n$ is (on the average) positive on a small sphere S centered at P, then f must also be positive. Therefore

(19) a subharmonic function must be positive at any point within its region of definition where it has a local minimum,

and similarly, if $\partial f/\partial n$ is negative in (18), f must also be negative, so that

(20) a subharmonic function must be negative at any point within its region of definition where it has a local maximum.

Theorem 11. *If f is subharmonic in D, then either (a) f has no local maxima or minima in D but takes its extreme values on the bounding surface S, or (b) f has a minimum value at an interior point of D, in which case f is not negative at any point of D, or (c) f has a maximum value at an interior point of D, in which event f is never positive in D.*

Proof: This follows from (19) and (20). For instance, if f has a minimum within D, then by (19) its value at this minimum point must be positive, and since this point is a minimum point f must be positive throughout D.

Theorem 12. *If f is subharmonic in D, and is not negative on the bounding surface S, it cannot be negative within S. If f is not positive on S, it cannot be positive within S.*

Proof: This is really a corollary to Theorem 11. For example, if f is not negative on S, and were positive within S, then f would have a positive maximum within S, contradicting (c) of Theorem 11.

• EXERCISES

1. It is desired to find a function satisfying Laplace's equation in the *exterior* of a sphere taking prescribed values $f(\theta,\phi)$ on the surface of the sphere. Show that there is no unique solution to this problem.
2. Finish the proof of Theorem 6.
3. Let $f(x,y,z)$ be equal to the *real part of* $e^{-1/(x+iy)^4}$ when $x^2 + y^2 \neq 0$, and equal to zero on the z axis. Show that f satisfies Laplace's equation but is discontinuous.
4. Show that a continuous solution of (16) that is not negative throughout a region must always be less inside the region than its largest value on the boundary.
5. Show that if f and g are continuous solutions of (16) in a region D, and if $f \geqq g$ on the bounding surface S, then $f \geqq g$ everywhere in D.
6. **(a)** Show that if h is twice continuously differentiable in D, and $\nabla^2 h - ph \leqq 0$ in D, where $p > 0$, and for points on the bounding surface S we have $h = f$ where f is a solution of (16), then $h \geqq f$ throughout D.
   (b) Hence, explain why the word "subharmonic" is appropriate for nonnegative solutions of (16).
7. Assuming $p = c^2$ where c is a nonzero constant, and using the method of separation of variables, find product solutions of (16) of the form $X(x)Y(y)Z(z)$ that vanish whenever $x = 0$, $x = L$, and $y = 0$.
8. Was the positivity of p used in deriving (18)?

5.5 • EXISTENCE THEOREMS

To give existence theorems for solutions of partial differential equations is beyond the scope of this book. Even to prove the existence of solutions of Laplace's equation satisfying prescribed boundary conditions is not an elementary task. We shall be content with proving *rigorously* the existence of a harmonic function of two variables taking prescribed values on the circumference of a circle. This is closely related to the theory of Fourier series, is of considerable interest without being terribly complicated, and indicates the main problems that arise in more general contexts.

We let A denote a region in the xy plane. We assume this region is *open*, by which we mean that given any point in A one can find a circle with center at this point that is entirely contained in A. We also assume that A is *bounded*, by which we mean that the least upper bound of distances between pairs of points in A is a (finite) number, which we denote d and call the *diameter* of A. (If A is the set of points interior to a circle, d is the diameter of the circle.) We let C denote the *boundary* of A, by which we mean those points p with the property that every circle having center at p contains at least one point in A and at least one point not in A. (If A is the region interior to a circle, C is the set of points on the circle itself.) We let $\overline{A}$ denote the set of points in either A or C. (If A is the interior of a circle, $\overline{A}$ is the set including *both* the points in the interior and on the circle.)

Let us suppose that $f(x,y)$ is a function that satisfies Laplace's equation $\partial^2 f/\partial x^2 + \partial^2 f/\partial y^2 = 0$ at all points (x,y) of $\overline{A}$ and that f is continuous at all these points as well. Let m denote the maximum value that $f(x,y)$ assumes on C. We shall prove that f does not take on any values in A greater than m.

Our proof proceeds by contradiction. If f does assume any values in A greater than m, then let M denote the maximum value of f on $\overline{A}$. Then $M > m$ and therefore the point Q at which f takes this value must be within A. Let us move the origin to the point Q; this does not change the harmonicity of f. We obtain a contradiction now by considering the following function:

$$g(x,y) = f(x,y) + \frac{M - m}{2d^2}(x^2 + y^2) \tag{1}$$

where d is the diameter of A. Clearly $g(0,0) = f(0,0) = M$, and $x^2 + y^2 \leqq d^2$ for (x,y) in $\overline{A}$, so for points in C

$$g(x,y) \leqq m + \frac{M - m}{2} = \frac{m + M}{2} < M. \tag{2}$$

Since the value of g at the origin is M and by (2) its value on C is less than M, g must also take its maximum value somewhere in A

(and not on C). This, however, leads to a contradiction, since at every point in A,

$$\frac{\partial^2 g}{\partial x^2} + \frac{\partial^2 g}{\partial y^2} = \frac{\partial^2 f}{\partial x^2} + \frac{\partial^2 f}{\partial y^2} + \frac{M - m}{d^2} = \frac{M - m}{d^2} \tag{3}$$

which is positive, but at a maximum point none of the second derivatives of a function can be positive.

Similarly, one can show that the values $f(x,y)$ takes in A are not less than its minimum value on C.

From this one proves the *uniqueness* in $\bar{A}$ of any continuous solution of Laplace's equation taking on prescribed values on C, as in Section 5.4. One also sees that, if two continuous solutions f and h of Laplace's equation differ by relatively little on C, i.e., $|f - h| < \epsilon$, then they must also differ by relatively little throughout A, i.e., $|f - h| < \epsilon$ throughout A. To prove this, one simply applies the above remarks to $f - h$. This continuous dependence (of solutions to the Dirichlet problem) on the boundary conditions will be useful to us in the following proof. By Theorem 5, Section 5.4, the same remarks apply equally well in three dimensions, of course, and therefore we have disposed of the *stability* question referred to in Section 5.1.

Now let us suppose that we are given values $f(\theta)$ of a continuous function, of period 2π. We wish to prove the existence of a solution to the Dirichlet problem within the unit circle $x^2 + y^2 = 1$, taking these values on its circumference. In polar coordinates, we have found a formal solution, namely

$$u(r,\theta) = A_0 + \sum_{n=1}^{\infty} r^n(A_n \cos n\theta + B_n \sin n\theta) \tag{4}$$

where A_0, A_n, B_n are the Fourier coefficients of f.

Let r_0 be any positive number less than 1, so that the series

$$M(1 + r_0 + r_0^2 + r_0^3 + \cdots) \tag{5}$$

converges. Let M be an upper bound for the numbers $|A_0|$, $|A_n|$, $|B_n|$, $n = 1, 2, 3, \cdots$. (This upper bound certainly exists for the Fourier coefficients of any integrable function.) The terms in (4) for $r < r_0$ are dominated in absolute value by the corresponding terms in (5), and therefore (4) converges uniformly for $r < r_0$ quite independently of θ (Theorem 6, Section 3.7). Moreover, if we differentiate (4) term by term with respect to r, the resulting series is dominated by series

$$M(1 + 2r_0 + 3r_0^2 + 4r_0^3 + \cdots) \tag{6}$$

obtained by differentiating (5) with respect to r_0. This series is also convergent (by the ratio test) and therefore the series obtained by differentiating (4) is uniformly convergent. It follows (Exercise 2) that term-by-term differentiation of (4) with respect to r is justified whenever $r < r_0$. Exactly the same reasoning shows that term-by-term differentiation with respect to θ is also justified; the resulting series is also dominated by (6).

Indeed, differentiating (6) again with respect to r_0 and repeating the above argument, we see that it is justified to differentiate (4) term by term as often as we like, and since its individual terms satisfy Laplace's equation

$$\frac{\partial^2 u}{\partial r^2} + \frac{1}{r}\frac{\partial u}{\partial r} + \frac{1}{r^2}\frac{\partial^2 u}{\partial \theta^2} = 0, \tag{7}$$

it follows that the function defined by (4) satisfies Laplace's equation. This is valid whenever $r < r_0 < 1$, hence whenever $r < 1$.

This proves that (4) defines a solution of Laplace's equation within the circle, provided only that f is integrable; it is not even necessary to require that f be continuous.

However, this does not show that the values of (4) for $r = 1$ coincide with $f(\theta)$, and indeed when $r = 1$ that this is the Fourier series representing f, and we have already noted that this series is not necessarily convergent. If, however, f is continuous and piecewise smooth, then we know that the series does converge to f uniformly; this was proved in Section 3.8.

We still do not know that (4) defines a *continuous* solution of Laplace's equation. All we know is that it satisfies Laplace's equation. Since every term of (4) is continuous, and (4) converges uniformly for $r < r_0 < 1$, we know that this solution is continuous for $r < 1$, but we do not know that it may not tend to infinity (or oscillate without having a limit) as r tends to 1, even though the series converges to f at $r = 1$. However, this is readily seen from the fact (see the Answers and Notes) that, if f is continuous and piecewise smooth, the series

$$|A_0| + \sum_{n=1}^{\infty} [|A_n| + |B_n|] \tag{8}$$

converges. This series dominates (4) when $r \leqq 1$, and therefore (4) is uniformly convergent for $r \leqq 1$, and since the sum of a uniformly convergent series of continuous function is continuous, it follows that (4) converges to a continuous function throughout $r \leqq 1$.

Summarizing, we have shown that if $f(\theta)$ is a continuous piecewise smooth function defined on the circumference of the unit circle, (4) converges uniformly to a function continuous for $r \leqq 1$. This function is a solution of Laplace's equation in the *interior* of the circle and equals $f(\theta)$ on its circumference.

Now suppose that f is continuous but is not necessarily piecewise smooth. Then we can construct a sequence f_n of piecewise smooth functions converging uniformly to f. (For instance, we can take the f_n to be Cesaro sums of the Fourier series for f, or we can use the broken-line functions described in Section 3.7.) By the above analysis, we can obtain a corresponding sequence $u_n(r,\theta)$ of solutions. By the remarks on stability presented earlier, these solutions must be uniformly convergent within the circle. The limit of the functions $u_n(r,\theta)$ is a continuous function $u(r,\theta)$ which coincides with $f(\theta)$ on the boundary; it remains only to show that $u(r,\theta)$ is harmonic for $r < 1$. For points with $r < 1$, the remarks above concerning the uniform convergence of (4) justify the derivation of Poisson's integral

$$u_n(r,\theta) = \frac{1}{2\pi} \int_0^{2\pi} f_n(\phi) \frac{1 - r^2}{1 - 2r \cos(\theta - \phi) + r^2} \, d\phi \tag{9}$$

and since $f_n(\theta)$ converges uniformly to $f(\theta)$, the right side of (9) tends at every point within $r < 1$ to a limit, which we have already denoted $u(r,\theta)$, and hence

$$u(r,\theta) = \frac{1}{2\pi} \int_0^{2\pi} f(\phi) \frac{1 - r^2}{1 - 2r \cos(\theta - \phi) + r^2} \, d\phi, \tag{10}$$

which is completely equivalent to (4) for $r < 1$, and we have already shown that $u(r,\theta)$ in (4) satisfies Laplace's equation for $r < 1$.

One can also verify directly that (10) satisfies Laplace's equation, by differentiation under the integral sign, which is justified here since repeated differentiation of the integrand with respect to r or θ yields a continuous function whenever $r < 1$.

This completes the existence proof. It will be noted that we used a number of standard facts from advanced calculus here. The reader not familiar with these theorems may find that his understanding of this section will provide a good yardstick with which to measure his progress, if he cares to take the time to learn these theorems from one of the standard texts.

The general pattern of this proof is typical of most proofs which justify calling *formal solutions*, obtained by separating variables, *solutions* of the boundary-value problem. The details, may, however, be quite different. The general idea is to prove that within the region the series converges and can be differentiated term-by-term, so that it does constitute a solution to the differential equation within the region. Then one must show the existence of a function that is equal to this solution within the region, is continuous both within the region and on its boundary, and matches the prescribed function on the boundary. The "formula," obtained as an infinite series, is usually not convergent in the ordinary sense at points on the boundary [as, for example, (4) above].

One reason we do not pursue this matter further is that this kind of existence proof depends on the possibility of obtaining a formal solution in the form of an infinite series. If the boundary of the region is not simply expressed in one of the standard systems of orthogonal coordinates, then the method of separation of variables is not applicable. The author has been told that there are only eleven sets of coordinates in which separation can be effected (to preserve his integrity, the author must admit he has never checked this), the familiar ones being rectangular, cylindrical, and spherical coordinates; the others (ellipsoidal, prolate spheroidal, etc.) are not discussed here. It is not true, as many students mistakenly think, that one can devise weird coordinates to suit the problem, if one is only sufficiently clever. For example, if the region is bean-shaped, none of the methods we have discussed so far can be applied, although the uniqueness theory of Section 5.4 is valid, and there are theorems proving the existence of solutions to the Dirichlet problem for such a region. They cannot be found by the method of separation of variables.

• EXERCISES

1. Write down a two-dimensional equivalent to (7), Section 5.4, and use it to prove uniqueness for solutions of the Dirichlet problem in two dimensions.

2. Use the fact that a uniformly convergent series can be integrated term-by-term (Section 3.7) to show that a series can be differentiated term-by-term provided the resulting series uniformly converges.

3. If f is continuous and piecewise smooth, is (10) valid for points on the circumference of the circle?

4. Show that, if $|a| < 1$,

$$\sum_{n=1}^{\infty} a^n \sin nx = \frac{a \sin x}{1 - 2a \cos x + a^2}.$$

5.6 • HEAT FLOW

The study of time-dependent temperature distributions is much more difficult than the steady-state case we have been discussing. The complications can achieve truly magnificent proportions.

For example, let us consider the temperature distribution within an insulating wall of thickness L. We imagine the wall constructed of a homogeneous material, and assume its faces to be flat, one face in the plane $x = 0$ and the other in the plane $x = L$. We imagine that it extends sufficiently far in the y and z directions so that, in the region of interest, we can ignore variations with y and z altogether.

The steady-state temperature distribution in the wall is in this instance of a completely trivial nature. If, for instance, the temperature at the face $x = 0$ is $T = 0$, and at $x = L$ is $T = 100$, then the steady-state temperature is $T = 100x/L$. We see that this is so by observing that Laplace's equation $\nabla^2 T = 0$ in this case becomes $d^2T/dx^2 = 0$, with solutions of the form $T = a + bx$; the coefficients a and b are easily determined from the values at $x = 0$ and $x = L$.

This steady-state problem is the easiest of its kind considered in an elementary course in physics. The corresponding time-dependent problem is not considered in such courses, although it is about as simple as such time-dependent problems can possibly be. Let us consider this problem: imagine the wall initially to have zero temperature throughout, and suppose that, at time $t = 0$, the temperature of the face at $x = L$ is suddenly raised to a temperature $T = 100°$. We seek an expression giving the temperature $T(x,t)$ as a function of x and t for points within the wall. We assume that the temperatures for positive values of t are *maintained* at

(1) $$T(0,t) = 0°, \qquad T(L,0) = 100°.$$

In principle, we solve this in the same manner as before. We separate variables in the heat equation, which in this case takes the simple form

(2) $$\frac{\partial^2 T}{\partial x^2} = \frac{1}{\alpha^2}\frac{\partial T}{\partial t}$$

and seek to combine the product solutions thus obtained in an infinite series that will satisfy the boundary conditions (1).

One slight complication arises here. In selecting the product solutions in earlier problems we have considered, the boundary conditions have always been homogeneous. That is, any linear combination of functions satisfying these boundary conditions also will satisfy the conditions. This is not the case with the conditions (1). The sum of two functions satisfying these conditions will satisfy the altogether different conditions $T(0,t) = 0$, $T(L,0) = 200°$. This situation is easily rectified, however, by writing the desired solution as the sum of the steady-state solution and another solution, called the *transient* solution, which *does* satisfy homogeneous boundary conditions. We have

$$T(x,t) = 100x/L + T_1(x,t) \tag{3}$$

and seek a solution $T_1(x,t)$ satisfying the homogeneous conditions

$$T_1(0,t) = 0, \qquad T_1(L,t) = 0. \tag{4}$$

The final solution to the problem is then obtained from (3).

The general situation is much the same. If we seek solutions of the heat equation

$$\nabla^2 T = \frac{1}{\alpha^2}\frac{\partial T}{\partial t} \tag{5}$$

satisfying nonhomogeneous boundary conditions, we first attempt to find a steady-state solution satisfying these conditions, and then find a time-dependent solution that satisfies homogeneous conditions. This covers up a multitude of special difficulties, however, for in general the boundary conditions may themselves vary with time.

It is easy to separate the time dependence from (5) by the method of separation of variables, or simply by assuming a solution of the form

$$T(x,y,z) = u(x,y,z)e^{kt} \tag{6}$$

which on substituting into (5) yields $(\nabla^2 u)e^{kt} = kue^{kt}/\alpha^2$, and since e^{kt} never takes zero values this is equivalent to

$$\nabla^2 u = \frac{k}{\alpha^2}\,u. \tag{7}$$

We have discussed the equation $\nabla^2 u = pu$ for $p > 0$, and noted that, for positive p, this has no *nontrivial* solutions taking the value zero on a closed surface; but in cases of interest here, the value of k on the right side of (7) is negative, and we have not discussed this situation yet. It will be discussed in greater detail in the next chapter.

In the special problem we are considering here, (7) reduces to

$$\frac{d^2u}{dx^2} = \frac{k}{\alpha^2}\,u. \tag{8}$$

For each particular solution u_k of (8) satisfying $u_k(0) = 0$ and $u_k(L) = 0$ we will have a particular solution of (2) of the form

$$T_k(x,t) = u_k(x)e^{kt} \tag{9}$$

where $T_k(0,t) = 0$, $T_k(L,t) = 0$, which we will try to superimpose in order to satisfy (3). That is, we will try to determine coefficients B_k so that

$$T(x,t) = 100x/L + \sum_{k=1}^{\infty} B_k u_k(x)e^{kt} \qquad (t > 0) \tag{10}$$

will satisfy the prescribed conditions.

Since we cannot satisfy these conditions with positive k, we write $k/\alpha^2 = -s^2$ and solutions of (8) are of the form $A \cos sx + B \sin sx$, and even these are zero at $x = 0$ only if $A = 0$. We must have $\sin sx = 0$ when $x = L$, so $s = n\pi/L$, $n = 1, 2, 3, \cdots$. It follows that $k = -\alpha^2 s^2 = -\alpha^2 n^2\pi^2/L^2$ and (10) becomes

$$T(x,t) = 100x/L + \sum_{n=1}^{\infty} B_n e^{-n^2\alpha^2\pi^2 t/L^2} \sin \frac{n\pi x}{L}. \tag{11}$$

The coefficients B_n are determined by the requirement that at $t = 0$ (11) shall be zero for $0 \leqq x < L$ so that

$$-100x/L = \sum_{n=1}^{\infty} B_n \sin \frac{n\pi x}{L}, \tag{12}$$

which is a Fourier sine series. The coefficients are therefore

$$B_n = \frac{2}{L}\int_0^L (-100x/L) \sin \frac{n\pi x}{L}\,dx = 200(-1)z/\pi n \tag{13}$$

and the formal solution to the problem is provided by

$$T(x,t) = 100x/L + \frac{200}{\pi}\sum_{n=1}^{\infty} \frac{(-1)^n e^{-n^2\alpha^2\pi^2 t/L^2}}{n} \sin \frac{n\pi x}{L}. \tag{14}$$

• EXERCISES

1. An infinitely long homogeneous cylindrical bar of radius R is initially at a steady-state temperature of zero degrees throughout.

The surface of the cylinder is brought to 100° at time $t = 0$. Find the temperature $T(r,t)$ for $t > 0$ (taking r to the cylindrical coordinate).

2. What would be the solution to the problem discussed in the text if the initial steady-state temperature of the wall had been $T = A + Bx$, and if at time $t = 0$ the temperature at $x = 0$ were suddenly changed from A to 0 and that at $x = L$ from $A + BL$ to 0, and these new values maintained afterwards?

CHAPTER 6

WAVES AND VIBRATIONS; HARMONIC ANALYSIS

6.1 • THE VIBRATING STRING

Our study of harmonic analysis begins with an investigation of the vibrations of a stretched elastic string. We have in mind a piano string, a violin string, or a long flexible chain.

We assume the tension in the string is sufficiently great that we can neglect the effects of air resistance, and (in the case of the piano string or violin string) can neglect forces due to gravity. We assume the string moves in a single plane and that the vibrations are transverse, i.e., that the points of the string move along straight lines perpendicular to the line of equilibrium of the string. We assume that the vibrations are quite small, so that the tension in the string does not depend on the displacement from equilibrium, although in the case of a vertically hanging chain the tension will presumably vary with position along the chain (being greater near the top of the chain, since the weight of the chain cannot be neglected in this case). We also assume that the vibrations are small, so the angle between the string and the line of equilibrium, at any instant of time, can be approximated by its tangent or by its cosine. (From the engineer's viewpoint, these are the small angles that one reads on the combined sine-tangent scale of a sliderule.)

It should be noted that an applied mathematician interested in the theory of elasticity would *not* make these assumptions *initially* but would set up the differential equation without them, and carefully analyze the maximum error each of these assumptions can produce under reasonable conditions. Such an analysis is far beyond the scope of this book.

We let the x axis be the line of equilibrium. By virtue of the assumptions made, we can take $y(x,t)$ to be the displacement from equilibrium, at time t, of a point on the string whose equilibrium position is the point $(x,0)$ in the xy plane. In other words, at any time t, this particle is located at the point $(x,y[x,t])$.

The differential equation satisfied by $y(x,t)$ is

$$\frac{\partial}{\partial x}\left(T\frac{\partial y}{\partial x}\right) = \rho\frac{\partial^2 y}{\partial t^2} \tag{1}$$

where T is the tension in the string and ρ is the density (mass per unit length) of the string.

This equation is "derived" in the following manner. We isolate a portion of the string, with x coordinates in an interval (x_0,x_1), and apply Newton's second law to this portion. If we assume perfect flexibility, the force acting on this portion at either endpoint is tangential to the string. The y component of this force at x_1 is $(T \sin \theta)|_{x=x_1}$ where T is the tension at the point $x = x_1$ and $\theta = \tan^{-1}(\partial y/\partial x)$ (i.e., the slope angle of the string at $x = x_1$). A similar force is exerted at the endpoint x_0, and the resultant force acting on this portion of the string, tending to accelerate it in the direction of increasing y, is

$$F = (T \sin \theta)|_{x_1} - (T \sin \theta)|_{x_0}. \tag{2}$$

According to the above assumptions, we can replace $\sin \theta$ by $\tan \theta$ with negligible effect, and since $\tan \theta = \partial y/\partial x$, we obtain

$$F = \left(T\frac{\partial y}{\partial x}\right)\bigg|_{x_1} - \left(T\frac{\partial y}{\partial x}\right)\bigg|_{x_0}. \tag{3}$$

The reader will notice that we have been tacitly assuming $y(x,t)$ to be a smooth function, with derivatives $\partial y/\partial x$ at each point, and as our analysis progresses he will observe we are assuming even more than that. For now we apply the "law of the mean" and write (3) in the form

$$F = \frac{\partial}{\partial x}\left(T\frac{\partial y}{\partial x}\right)\Big|_{x'}(x_1 - x_0) \tag{4}$$

where the derivative of $T\ \partial y/\partial x$ is evaluated at a point x' between x_0 and x_1; we are assuming that $T\ \partial y/\partial x$ is differentiable.

According to Newton's second law, this resultant force equals the mass of the portion of string times the acceleration of its center of gravity. If the mass density ρ is a continuous function of x, then by the mean value theorem for integrals the mass of the portion of string is $\rho(x_1 - x_0)$, where ρ is the value of the mass density at some point x'' between x_0 and x_1, and therefore

$$F = \rho\frac{\partial^2 y}{\partial t^2}(x_1 - x_0) \tag{5}$$

where the acceleration $\partial^2 y/\partial t^2$ is evaluated at the center of mass, whose x coordinate is yet a third value x''' in the interval (x_0,x_1). Equating (4) and (5), and letting x_1 tend to x_0, so that x', x'', and x''' also tend to x_0, we obtain (1), valid at every point $x = x_0$.

In the rest of this section, we will concentrate on the special case in which the tension T is a constant, independent of position. Then the equation becomes

$$T\frac{\partial^2 y}{\partial x^2} = \rho\frac{\partial^2 y}{\partial t^2}, \tag{6}$$

which we shall rewrite in the form

$$\frac{\partial^2 y}{\partial x^2} = \frac{1}{v^2}\frac{\partial^2 y}{\partial t^2} \tag{7}$$

in anticipation of the result we shall prove in Section 6.2, that $v = \sqrt{T/\rho}$ will turn out to be the velocity of waves transmitted by the string. One might guess this from purely dimensional considerations, however; recall that, in cgs units, tension is measured in dynes, where one dyne is the force that will accelerate a particle of mass one gram by one centimeter per second per second; therefore T/ρ has dimension

$$\frac{\text{dyne}}{\text{gm/cm}} = \frac{\text{gm-cm/sec}^2}{\text{gm/cm}} = (\text{cm/sec})^2$$

and therefore v, as defined above, has the dimensions of velocity.

Equation (7) is called the *one-dimensional wave equation*, and will be discussed in some generality in Section 6.2. Here we shall proceed at once to consider the vibrations of a string that is fixed at two points, $x = 0$ and $x = L$, where L denotes the length of the stretched string. Let us suppose that we are given the initial displacement and initial velocity of points on the string

$$y(x,0) = g(x), \qquad \frac{\partial y}{\partial x}(x,0) = h(x), \tag{8}$$

and we wish to determine the displacement $y(x,t)$ for $t > 0$, subject to the conditions

$$y(0,t) = 0, \qquad y(L,t) = 0 \tag{9}$$

and

$$\frac{\partial y}{\partial t}(0,t) = 0, \qquad \frac{\partial y}{\partial t}(L,t) = 0, \tag{10}$$

which are consequences of the assumption that the string is pinned down at $x = 0$ and $x = L$.

We begin as usual by looking for solutions of the form

$$y(x,t) = X(x)T(t). \tag{11}$$

Substituting (11) into (7), we obtain

$$\frac{X''(x)}{X(x)} = \frac{1}{v^2}\frac{T''(t)}{T(t)} \tag{12}$$

and, since the left side is a function of x alone and the right a function of t, we conclude both sides are constant. Taking the constant to be k^2, 0, or $-k^2$, we obtain respectively

$$y(x,t) = (A \sinh kx + B \cosh kx)(C \sinh kvt + D \cosh kvt), \tag{13}$$

$$y(x,t) = (A + Bx)(C + Dt), \tag{14}$$

$$y(x,t) = (A \sin kx + B \cos kx)(C \sin kvt + D \cos kvt). \tag{15}$$

We reject solutions of the form (13) and (14) since, in order to satisfy boundary conditions (9) and (10) we would have to take the coefficients to be zero, yielding a trivial solution. To satisfy these conditions any solution of form (15) must have $B = 0$ and $k = n\pi/L$ for integral n. We thus obtain an infinite number of solutions of (7) satisfying the required conditions; these solutions are of the form

$$y_n(x,t) = \sin\frac{n\pi x}{L}\left(C_n \sin\frac{n\pi vt}{L} + D_n \cos\frac{n\pi vt}{L}\right) \tag{16}$$

where we omit the constant before $\sin(n\pi x/L)$ since it can be combined with the other two constants.

Now we attempt to find a solution satisfying the initial conditions (8), in the form of an infinite series

$$y(x,t) = \sum_{n=1}^{\infty} \sin \frac{n\pi x}{L} \left(C_n \sin \frac{n\pi vt}{L} + D_n \cos \frac{n\pi vt}{L} \right). \tag{17}$$

To satisfy (8) we must formally have

$$y(x,0) = g(x) = \sum_{n=1}^{\infty} D_n \sin \frac{n\pi x}{L} \tag{18}$$

obtained by taking $t = 0$ in (17), and

$$\frac{\partial y}{\partial t}(x,0) = h(x) = \sum_{n=1}^{\infty} \frac{n\pi v}{L} C_n \sin \frac{n\pi x}{L} \tag{19}$$

obtained formally by differentiating (17) term by term with respect to t and setting t equal to zero.

In Chapter 3, we learned that (18) and (19) will be valid (at least formally) if we determine the coefficients D_n and C_n by

$$D_n = \frac{2}{L} \int_0^L f(x) \sin \frac{n\pi x}{L} \, dx, \tag{20}$$

$$\frac{n\pi v}{L} C_n = \frac{2}{L} \int_0^L h(x) \sin \frac{n\pi x}{L} \, dx. \tag{21}$$

With these coefficients, (17) constitutes a formal solution to the problem. In the next section we will establish rigorously that this series actually provides a solution if f and h are sufficiently smooth functions.

From a physical viewpoint, (17) represents perpetual motion. This is a consequence of neglecting air resistance and other damping factors. We must therefore recognize that a solution of this type can, at best, be an approximation to the actual motion of a string for short durations of time.

A special case of (17) is obtained if all the coefficients except those for $n = 1$ are equal to zero. In this mode of oscillation, the string always retains the same general shape of a single arch (shown in Figure 6.1) and the string produces a musical note called the *fundamental*. The frequency f of this note is obtained by equating $2\pi f$ and the coefficient of t, i.e., $2\pi f = \pi v/L$ so

$$f = v/2L = \frac{1}{2L}\sqrt{\frac{T}{\rho}}.$$

In a piano, this fundamental frequency is varied from string to string by varying their lengths, their mass densities (compare the appearance of the different strings in a piano) and their tensions. A piano tuner, of course, performs his fine adjustments by varying the tension alone; increasing the tension increases the pitch (i.e., frequency) of the note.

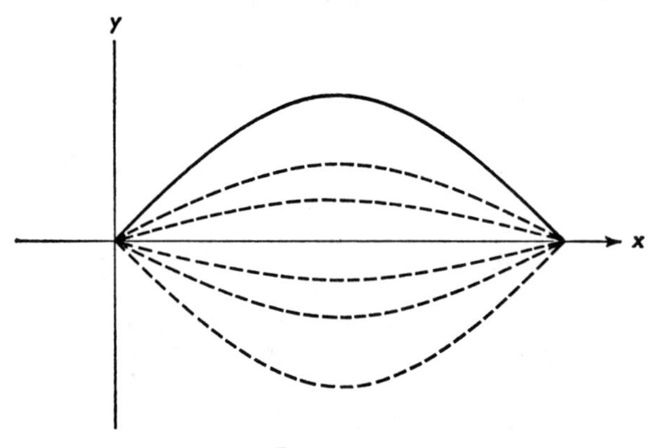

Figure 6.1

One way to obtain this fundamental mode would be to displace the string from equilibrium, as shown in the solid curve (Figure 6.1), and release the string from rest. Successive displacements would appear as shown by the dotted curves. This is not what happens when a piano key is struck, however. The string is set into motion by a felt hammer, so one obtains more than one term in (17). The term corresponding to a given n is called the nth *harmonic* or the $(n - 1)$st *overtone* of the fundamental. The frequency of the nth harmonic is precisely n times that of the fundamental. The musical quality of the piano (or violin) is due to this circumstance; we will see later that the overtones of a vibrating membrane (drum) have frequencies that are not integral multiples of the fundamental frequency.

• EXERCISES

1. At $t = 0$, y is given by $y(x,0) = 2 \sin \pi x/L$ and $\partial y/\partial t$ is given by $(\partial y/\partial t)\ (x,0) = 2 \sin 2\pi x/L$. Find $y(x,t)$ for $t > 0$, if y satisfies (7).

2. The displacement of a freely vibrating string is given by $y(x,t) = 2 \sin \pi x/L \cdot \sin 3\pi t/L$. (The length of the string is L feet, x is measured in feet and t in seconds.) Find the ratio T/ρ.

3. Show in detail why (14) and (15) must be rejected as possible solutions of the boundary-value problem.

4. The initial displacement of a string of length 2π is $y(x,0) = hx(2\pi - x)$ where h is a constant, and the initial velocity $(\partial y/\partial t)(x,0)$ is identically zero. Find $y(x,t)$ for $t > 0$.

5. Show that the kinetic energy of a vibrating string is

$$\frac{1}{2}\int_0^L \rho \left(\frac{\partial y}{\partial t}\right)^2 dx.$$

6. Show that the potential energy of the vibrating string (assuming its undisplaced potential energy to be zero) is

$$\frac{1}{2} T \int_0^L \left(\frac{\partial y}{\partial x}\right)^2 dx.$$

What assumptions do you make in this derivation?

7. A mode of oscillation of the string of the form (16) is called a *normal mode* of oscillation. Show that the total energy of a normal mode is given by $n^2\pi^2 T[C_n^2 + D_n^2]/4L$. Hence, give an energy interpretation to Parseval's equality.

8. Show that the time average, over a cycle, of the kinetic energy of a normal mode of vibration (Exercise 7) is the same as the time average, over a cycle, of the potential energy. Determine to what extent this is characteristic of normal modes (can this happen, for instance, if the mode is a sum of distinct nonzero normal modes?).

9. Show that the energy of a vibrating string is the sum of the energies of its constituent harmonics.

10. Which harmonics will be dominant if a string is set into motion by being plucked exactly at its center?

11. Show that $\sum_{n=1}^{\infty} \frac{1}{n} \sin \frac{n\pi x}{L} \cos \frac{n\pi vt}{L}$ could not possibly represent a vibration of a string,

(a) by summing the series,

(b) by energy considerations.

6.2 • THE ONE-DIMENSIONAL WAVE EQUATION

We continue the discussion of the vibrations of a string under constant tension T and constant mass density ρ. Suppose the string in this instance to be infinitely long, with the x axis as the line of equilibrium. Then we have

$$\frac{\partial^2 y}{\partial x^2} = \frac{1}{v^2}\frac{\partial y}{\partial t}, \quad v^2 = \frac{T}{\rho} \tag{1}$$

as before. Let us suppose we are given the initial displacement $g(x)$ and the initial velocity $h(x)$, so that $y(x,t)$ must satisfy the initial conditions

$$y(x,0) = g(x), \tag{2}$$

$$\frac{\partial y}{\partial t}(x,0) = h(x). \tag{3}$$

We desire to determine $y(x,t)$ for $t > 0$.

In this discussion, we assume that the given functions g and h have the properties

(4) g is twice continuously differentiable

(5) h is continuously differentiable

for simplicity, although these are more stringent than necessary to establish existence and uniqueness of the solution. In the course of this discussion, we will obtain as a by-product a proof that (17) of the preceding section constitutes a solution of the problem stated there, provided the functions g and h [which, in that section, were only defined in the interval $(0,L)$] have odd period $2L$ extensions satisfying (4) and (5).

In the preceding section we showed that v has the dimensions of velocity. One intuitively feels that if one sets up a disturbance in such a string, there will be waves traveling in both directions away from the origin of the disturbance. This suggests introducing the new variables

$$\alpha = x + vt, \qquad \beta = x - vt. \tag{6}$$

Changing to these new variables, (1) becomes

$$\frac{\partial^2 y}{\partial\alpha\partial\beta} = 0, \tag{7}$$

from which it follows that

$$\frac{\partial y}{\partial \beta} = u(\beta) \tag{8}$$

and hence

$$y(\alpha,\beta) = \int u(\beta)\, d\beta + q(\alpha) = p(\beta) + q(\alpha). \tag{9}$$

In terms of the original variables, it follows that

$$y(x,t) = p(x - vt) + q(x + vt). \tag{10}$$

From (2) we obtain, on letting $t = 0$ in (10),

$$g(x) = p(x) + q(x) \tag{11}$$

and from (3),

$$h(x) = -vp'(x) + vq'(x) \tag{12}$$

or, equivalently,

$$-p(x) + q(x) = \frac{1}{v}\int_0^x h(s)\, ds + C. \tag{13}$$

Solving (11) and (13) simultaneously, we obtain

$$q(x) = \tfrac{1}{2}g(x) + \frac{1}{2v}\int_0^x h(s)\, ds + C/2 \tag{14}$$

and

$$p(x) = \tfrac{1}{2}g(x) - \frac{1}{2v}\int_0^x h(s)\, ds - C/2 \tag{15}$$

or

$$p(x) = \tfrac{1}{2}g(x) + \frac{1}{2v}\int_x^0 h(s)\, ds - C/2 \tag{16}$$

and therefore (10) becomes

$$y(x,t) = \frac{g(x - vt) + g(x + vt)}{2} + \frac{1}{2v}\int_{x-vt}^{x+vt} h(s)\, ds. \tag{17}$$

This is called *d'Alembert's formula.*

Directly from (10) we see that the general solution of the one-dimensional wave equation consists of the superposition of two waves, one traveling to the right and the other to the left, both having velocity v. When rewritten in form (17) we immediately obtain some very interesting information. As a matter of convenience we consider two special cases.

First, suppose the string is initially at rest, so that h is identically zero and the second term on the right side of (17) vanishes. Let us suppose the initial displacement $g(x)$ is zero except in an interval near the origin $x = 0$. Then we see from (17) that at a much later time t the points at the origin will have returned to their equilibrium position. Yet far to the right of the origin there will be a wave traveling to the right, with waveform exactly duplicating the shape of the original displacement except having exactly half of the amplitude. At the same time, an identical wave exists to the left of the origin, traveling to the left with the same velocity v. In this case, the waveform is not actually *distorted* in any manner (Figure 6.2).

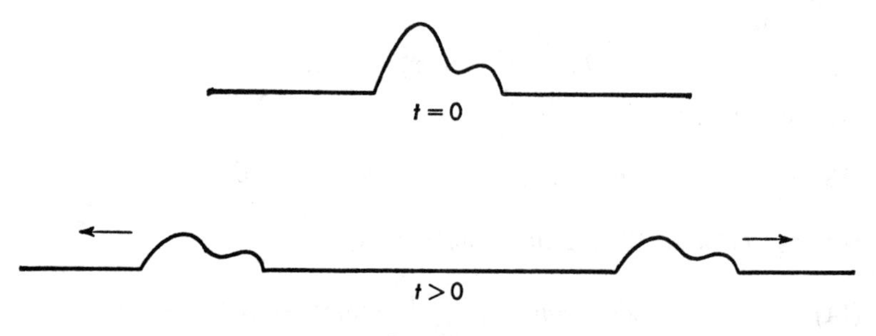

Figure 6.2

Now let us consider a second special case. We suppose that, at $t = 0$, the string is struck by a hammer aimed at $x = 0$. To fix ideas, we assume the blow to be in the direction of increasing y, so the initial velocity $h(x)$ is not negative and is zero outside a small interval $-\epsilon \leqq x \leqq \epsilon$. Let $D = (1/2v) \int_{-\epsilon}^{\epsilon} h(s)\, ds$. D is positive, and no matter how large K may be, if it is greater than ϵ we have $(1/2v) \int_{-K}^{K} h(s)\, ds = D$. Assuming the initial displacement $g(x)$ of the string is zero everywhere, so only the second term in (17) is relevant, we see that

(18) $y(x,t) = 0$ whenever the interval $(x - vt, x + vt)$ does not overlap the interval $(-\epsilon,\epsilon)$, and

(19) $y(x,t) = D$ when the interval $(x - vt, x + vt)$ is sufficiently large to contain the interval $(-\epsilon,\epsilon)$.

The net effect of the blow with the hammer is to eventually displace all points an amount D in the positive y direction. Successive displacements are shown in Figure 6.3.

In particular, we see from (17) that in this second case the displacement $y(x,t)$ at a point x depends only on values of h in the interval of length $2vt$ centered at x. This is in agreement with the observation that v is the velocity with which disturbances propagate along the string. We notice from the above that it would be more proper to call v the *maximum* velocity with which disturbances

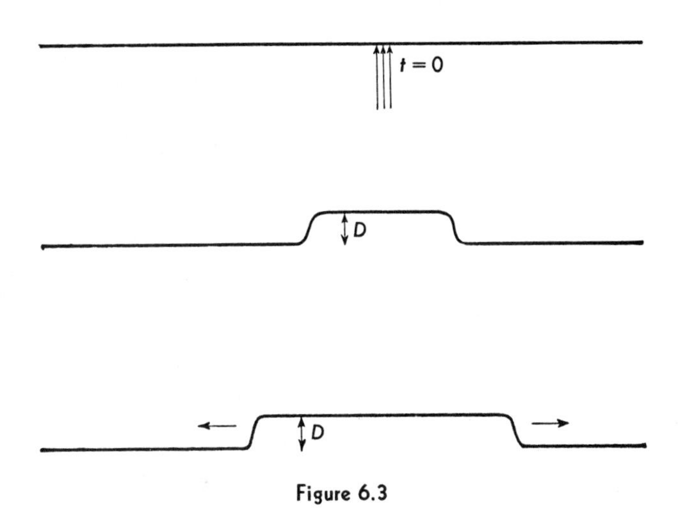

Figure 6.3

propagate. For the effect of the blow on the string is permanent; each point is displaced ultimately an amount D and does not ever return to its equilibrium position. Only in the *first case* discussed above do the disturbances truly *pass by* each point with velocity v.

Now let us return to the problem discussed in the preceding section. Or rather, let us consider the equivalent problem: investigate the vibrations of an infinite string that are periodic of period $2L$, for which both displacement and velocity are zero at $x = 0$ and $x = L$ (and hence zero at any x an integral multiple of L).

Modes of oscillation of the form

$$y_n(x,t) = \sin\frac{n\pi x}{L}\left(C_n \sin\frac{n\pi vt}{L} + D_n \cos\frac{n\pi vt}{L}\right) \tag{20}$$

are sometimes called *normal modes* or *standing waves*. The last term seems especially appropriate because these "waves" do not appear to be traveling along the string. However, if (10) is really a general solution, it *must* be possible to write (20) as a superposition of two waves, one traveling to the right and the other to the left, and this is indeed possible. Using the identities

$$\sin A \cos B = \tfrac{1}{2}[\sin (A + B) + \sin (A - B)]$$

and

$$\sin A \sin B = \tfrac{1}{2}[\cos (A - B) - \cos (A + B)]$$

we can write (20) in the form

$$(21) \qquad y_n(x,t) = \frac{1}{2}\left[D_n \sin \frac{n\pi(x + vt)}{L} + D_n \sin \frac{n\pi(x - vt)}{L}\right] + \frac{1}{2}\left[C_n \cos \frac{n\pi(x - vt)}{L} - C_n \cos \frac{n\pi(x + vt)}{L}\right],$$

which shows that even a standing wave is a superposition of "traveling" waves.

Under the hypothesis that g is a twice continuously differentiable odd function of period $2L$, its Fourier series

$$(22) \qquad g(x) = \sum_{n=1}^{\infty} D_n \sin \frac{n\pi x}{L}$$

converges uniformly and absolutely to $g(x)$ everywhere. If $h(x)$ is a continuously differentiable odd function of period $2L$, then

$$-(1/v) \int_0^x h(s)\, ds$$

is an even function that is twice continuously differentiable, and is zero at $x = 0$, so it has a Fourier series

$$(23) \qquad -\frac{1}{v} \int_0^x h(s)\, ds = \sum_{n=1}^{\infty} C_n \cos \frac{n\pi x}{L},$$

which also converges uniformly and absolutely, and can be differentiated term by term,

$$(24) \qquad \frac{h(x)}{v} = \sum_{n=1}^{\infty} \frac{n\pi}{L} C_n \sin \frac{n\pi x}{L}$$

to produce a series which also converges uniformly and absolutely (since h itself is continuously differentiable). It follows therefore that the series solution (17) of the preceding section is the sum of four convergent series:

$$
\begin{aligned}
y(x,t) &= \sum_{n=1}^{\infty} \sin\frac{n\pi x}{L}\left(C_n \sin\frac{n\pi vt}{L} + D_n \cos\frac{n\pi vt}{L}\right) \\
&= \frac{1}{2}\sum_{n=1}^{\infty} D_n \sin\frac{n\pi(x+vt)}{L} + \frac{1}{2}\sum_{n=1}^{\infty} D_n \sin\frac{n\pi(x-vt)}{L} \\
&\quad + \frac{1}{2}\sum_{n=1}^{\infty} C_n \cos\frac{n\pi(x-vt)}{L} - \frac{1}{2}\sum_{n=1}^{\infty} C_n \cos\frac{n\pi(x+vt)}{L} \\
&= \frac{1}{2}g(x+vt) + \frac{1}{2}g(x-vt) - \frac{1}{2v}\int_0^{x-vt} h(s)\,ds \\
&\quad + \frac{1}{2v}\int_0^{x+vt} h(s)\,ds \\
&= \frac{g(x+vt)+g(x-vt)}{2} + \frac{1}{2v}\int_{x-vt}^{x+vt} h(s)\,ds.
\end{aligned}
\tag{25}
$$

We have thus shown that, under the stated hypotheses on g and h, the series solution (17), Section 6.1, converges to (17) of this section, and therefore does constitute a solution of the problem. It is a routine task to verify directly that (17) satisfies the differential equation, the boundary conditions, and the prescribed initial conditions.

• EXERCISES

1. Show that the coefficients C_n and D_n in (23) and (22) match those in the preceding section.

2. Show by direct verification that (17) satisfies the prescribed initial conditions, boundary conditions, and equation (1).

3. Discuss the *uniqueness* of these solutions.

4. Assume that $f(x) = 0$ except in the interval $-1 \leqq x \leqq 1$, and $f(x) = 1$ in this interval. Plot $[f(x+t) + f(x-t)]/2$ for $t = 0$, $t = \frac{1}{2}$, and $t = 3$.

5. For the function defined in Exercise 4, plot $\int_{x-t}^{x+t} f(s)\,ds$ for $t = 0$, $t = \frac{1}{2}$, and $t = 3$.

6. The initial displacement of an infinite string is $g(x) = \sin x + \sin\sqrt{2}x$. Its initial velocity is zero.

(a) Find the displacement $y(x,t)$ thereafter.

(b) Would you call this mode of oscillation a standing wave?

6.3 • THE WEIGHTED STRING

Let us now imagine that particles, each having mass m, are suspended from the string at equal intervals d (Figure 6.4). We assume that

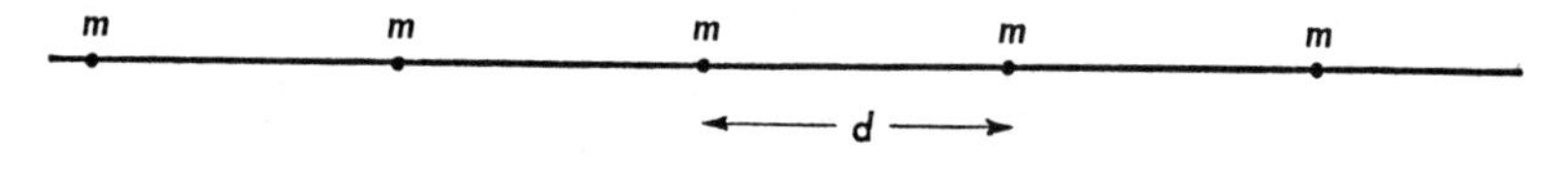

Figure 6.4

the tension T in the string is sufficiently great that we can neglect any initial displacement due to the weight of the particles. We also assume that the mass of the string joining successive weights is negligible when compared with the weight of the particles.

To begin with, we will assume the string is infinitely long, and let the particles be labeled by indices $i = 0, \pm 1, \pm 2, \cdots$. The resultant force on the ith particle is easily seen from Figure 6.5 to be

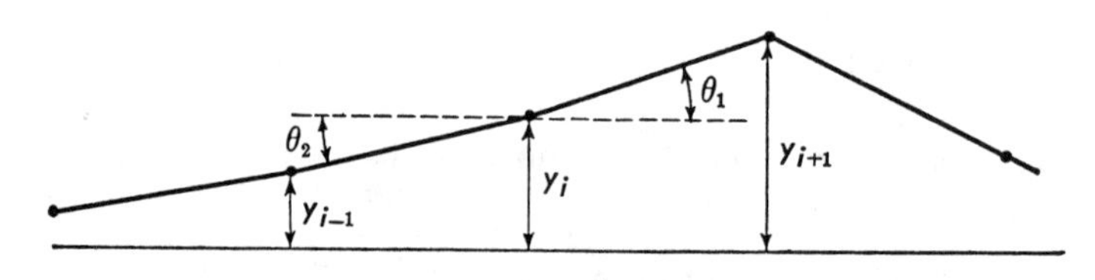

Figure 6.5

$T \sin \theta_1 - T \sin \theta_2$ which, if we assume the angles to be sufficiently small, is approximately

$$\begin{aligned} T \tan \theta_1 - T \tan \theta_2 &= T(Y_{i+1} - y_i)/d - T(y_i - y_{i-1})/d \\ &= T(y_{i+1} - 2y_i + y_{i-1})/d. \end{aligned}$$

By Newton's second law, the differential equation in this case is

$$\frac{T(y_{i+1} - 2y_i + y_{i-1})}{d} = m\,\frac{d^2 y_i}{dt^2}. \tag{1}$$

This may be viewed as a system of infinitely many simultaneous differential equations in the infinitely many functions $y_i(t)$. Or we may view $y_i(t)$ as a single function of two variables, a continuous variable t and a discrete variable i, and (1) as a combined difference and differential equation.

We find particular solutions of (1) by the method of separation of variables. We assume $y_i(t) = u_i f(t)$, where the u_i do not depend on the time t, and substituting into (1) we obtain

$$\frac{T}{d}(u_{i+1} - 2u_i + u_{i-1})f(t) = mu_i f''(t). \tag{2}$$

Dividing by $u_i f(t)$ we obtain

$$\frac{T}{d}\frac{(u_{i+1} - 2u_i + u_{i-1})}{u_i} = m\frac{f''(t)}{f(t)} \tag{3}$$

provided $u_i \neq 0$ and $f(t) \neq 0$. Since the right side is a function of t alone, and the left side depends only on the discrete variable i, we conclude both sides are constants, and obtain the differential equation

$$f''(t) = -k^2 f(t) \tag{4}$$

where we have arbitrarily set the separation constant equal to $-mk^2$, so that we will obtain sinusoidal solutions

$$f(t) = A \sin kt + B \cos kt. \tag{5}$$

We also obtain a difference equation

$$u_{i+1} - 2u_i + u_{i-1} = \left(\frac{mk^2 d}{T}\right) u_i, \tag{6}$$

which we rewrite in the form

$$u_{i+1} - \left(2 + \frac{mk^2 d}{T}\right) u_i + u_{i-1} = 0. \tag{7}$$

A general discussion of difference equations is beyond the scope of this book. It is obvious however that the set of all solutions of (7) constitutes a two-dimensional linear space, since if we fix the values of (say) u_0 and u_1 then from (7) we can determine the values of u_i for all other i (positive or negative). We will, therefore, have the most general possible solution of (7) if we can find two linearly independent solutions. As a guess, we might think of taking $u_i = \sin ip$, by analogy with the continuous string where the solutions were sinusoidal. We have

$$u_{i+1} = \sin (i + 1)p = \sin ip \cos p + \cos ip \sin p, \tag{8}$$

$$u_{i-1} = \sin (i - 1)p = \sin ip \cos p - \cos ip \sin p. \tag{9}$$

Adding these two, we obtain

$$u_{i+1} + u_{i-1} = 2 \sin ip \cos p = 2u_i \cos p, \tag{10}$$

which we rewrite in the form

$$u_{i+1} - (2 \cos p)u_i + u_{i-1} = 0. \tag{11}$$

This is the difference equation satisfied by the function $u_i = \sin ip$. It is easy to see that it is also satisfied by $u_i = \cos ip$. Moreover, it becomes identical with (7) if we take $2 \cos p = 2 + mk^2d/T$. Therefore the general solution of (7) is

$$u_i = C \sin ip + D \cos ip \tag{12}$$

where the separation constant k and the number p must be related by the formula

$$k^2 = \frac{2T}{md}(1 - \cos p). \tag{13}$$

Combining (5) and (12), we obtain

$$y_i(t) = u_i f(t) = (C \sin ip + D \cos ip)(A \sin kt + B \cos kt), \quad (i = 0, \pm 1, \pm 2, \cdots). \tag{14}$$

By direct substitution, we see that (14) does indeed satisfy (1), provided of course that (13) is satisfied.

A solution of the form (14) represents a standing wave in the string. This solution can be rewritten as a sum of traveling waves, just as we did in the preceding section. There is a fundamental difference, however: the velocity of these traveling waves depends on their frequency. Moreover, there is an upper limit to the possible frequencies. This can be seen directly from (13). The maximum possible value of k^2 is obtained when $\cos p = -1$, and this maximum value is $4T/md$. It is interesting to notice that when k^2 has this maximum possible value, p must be an odd multiple of π so the term $\sin ip$ vanishes in (14) and the solution is of the form

$$y_i(t) = \cos i\pi\,(A \sin 2\sqrt{T/md}\, t + B \cos 2\sqrt{T/md}\, t). \tag{15}$$

In this mode of oscillation, each particle is oscillating out of phase with its two neighboring particles. That this mode should be the one with the greatest frequency is physically reasonable, since it is the only mode in which all the string segments are stretched their maximum amount simultaneously.

As before, we can consider a finite string to be a special case of the infinite string. To fix ideas, consider the special case in which we have three particles evenly spaced on the string, which is fixed at its endpoints (Figure 6.6). We consider this to be a special case of the infinite string, with $y_0 = 0$, and $y_4 = 0$. These boundary conditions reduce the number of possible values of p in (14). We must take $D = 0$ since $y_0 = 0$, and since $y_4 = 0$, it follows that $\sin 4p = 0$ and therefore p is an integral multiple of $\pi/4$. Every solution of the form (14) is therefore of the form

$$y_i^{(1)}(t) = \sin(i\pi/4)\cdot (A_1 \sin \sqrt{T(2-\sqrt{2})/d}\, t + B_1 \cos \sqrt{T(2-\sqrt{2})/d}\, t) \tag{16}$$

or

$$y_i^{(2)}(t) = \sin(i\pi/4)\cdot (A_2 \sin \sqrt{T/d}\, t + B_2 \cos \sqrt{T/d}\, t) \tag{17}$$

or

$$y_i^{(3)}(t) = \sin(3i\pi/4)\cdot (A_3 \sin \sqrt{T(2+\sqrt{2})/d}\, t + B_3 \cos \sqrt{T(2+\sqrt{2})/d}\, t). \tag{18}$$

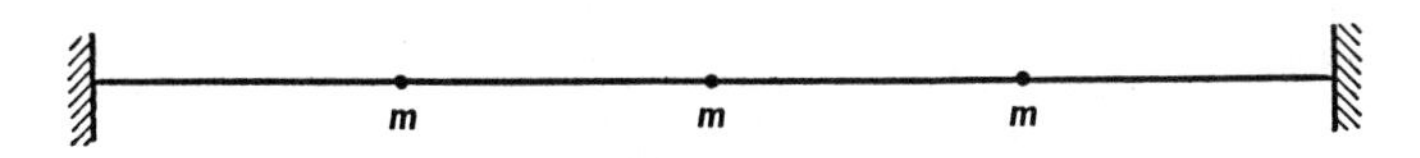

Figure 6.6

These normal modes of oscillation are shown in Figure 2.1, Section 2.2. The general solution is a sum of solutions of this form. If we introduce the three vectors

$$C^{(1)} = (C_1^{(1)}, C_2^{(1)}, C_3^{(1)}) = (\sqrt{2}/2, 1, \sqrt{2}/2) \tag{19}$$

$$C^{(2)} = (1, 0, -1) \tag{20}$$

$$C^{(3)} = (\sqrt{2}/2, -1, \sqrt{2}/2) \tag{21}$$

and let

$k_1 = \sqrt{T(2-\sqrt{2})/d}$, $k_2 = \sqrt{T/d}$, and $k_3 = \sqrt{T(2+\sqrt{2})/d}$,

we can write explicit formulas for the constants A_i, B_i $(i = 1, 2, 3)$ in order that $\sum_{j=1}^{3} y_i^{(j)}(t)$ be a solution having initial displacement D and initial velocity V. We state that the formulas are

$$B_j = \frac{1}{2} \sum_{i=1}^{3} D_i C_i^{(j)} \tag{22}$$

$$A_j = \frac{1}{2k_j} \sum_{i=1}^{3} V_i C_i^{(j)} \tag{23}$$

and leave the easy derivation to the reader.

• **EXERCISES**

1. Discuss solutions of (7) valid when $k^2 = 0$.
2. Write a formula for the velocity of traveling waves in the weighted string, showing that the velocity is a function of the frequency.
3. Find a difference equation analogous to (11) satisfied by the functions $u_i = \sinh ip$ and $u_i = \cosh ip$, $i = 1, 2, 3, \cdots$.
4. Show that there exist solutions of (1) corresponding to values of k^2 greater than $4T/md$, and discuss their physical significance.
5. Derive formulas similar to (22) and (23) for the finite weighted string having four weights instead of three.
6. Obtain a differential equation from (1) by letting m and d tend to zero in such a way that $m/d \to \rho$.

6.4 • STRING WITH VARIABLE TENSION AND DENSITY

Since an entire chapter could be written on this topic alone, we shall necessarily restrict our attention to special cases. We recall that the equation of the vibrating string, derived in Section 6.1, is

$$\frac{\partial}{\partial x}\left(T\frac{\partial y}{\partial x}\right) = \rho\frac{\partial^2 y}{\partial t^2}. \tag{1}$$

The time dependence can be separated out by assuming solutions of the form

$$y(x,t) = u(x)f(t), \tag{2}$$

and in the usual manner we obtain

$$f(t) = A\cos kt + B\sin kt \tag{3}$$

for the time-dependent factor, and

$$\frac{d}{dx}\left(T\frac{du}{dx}\right) + k^2\rho u = 0 \tag{4}$$

as the differential equation which must be satisfied by the corresponding function u.

This is of the form discussed briefly in Section 2.4, with $p(x) = T(x)$, $q(x) = 0$, and $r(x) = \rho(x)$. If these functions satisfy the conditions given in that section, we can carry over the discussion without change. In particular, if the string is stretched between two supports, so that $u(0) = 0$ and $u(L) = 0$, we obtain nontrivial solutions of (4) for certain values of k, and distinct solutions will be orthogonal over the interval $(0,L)$ with respect to the weighting factor $\rho(x)$.

As a first example, let us consider the lateral vibrations of a heavy flexible chain hanging freely from one end. We consider the mass density ρ to be a constant. Let x be distance measured from the bottom of the chain to a point along the chain, and let the length of the chain be denoted L. The tension at any point is due to the weight of the chain and is therefore $T(x) = \rho x g$, where g is the acceleration due to gravity. The equation (1) therefore becomes

$$\frac{\partial}{\partial x}\left(\rho x g \frac{\partial y}{\partial x}\right) = \rho \frac{\partial^2 y}{\partial t^2}. \tag{5}$$

Cancelling the constant factor ρ, (4) then becomes

$$\frac{d}{dx}\left(gx \frac{du}{dx}\right) + k^2 u = 0. \tag{6}$$

If we let $z = 2k\sqrt{x/g}$, so that $x = gz^2/4k^2$, we have

$$\frac{d}{dx} = \frac{dz}{dx}\frac{d}{dz} = \frac{4k^2}{2gz}\frac{d}{dz} \tag{7}$$

and (6) becomes

$$\frac{4k^2}{2gz}\frac{d}{dz}\left(\frac{g^2z^2}{4k^2} \cdot \frac{4k^2}{2gz}\frac{du}{dz}\right) + k^2 u = 0, \tag{8}$$

which simplifies to

$$zu''(z) + u'(z) + zu(z) = 0, \tag{9}$$

which is equivalent to Bessel's equation (Section 4.6) with $n = 0$. The bounded solutions of (9) are therefore scalar multiples of $J_0(z)$ and those of (6) are of the form

$$u(x) = CJ_0(D_k\sqrt{x}), \qquad (D_k = 2k/\sqrt{g}). \tag{10}$$

The permissible values of D_k are now determined by the requirement that $u(L) = 0$. Notice that $x = L$ corresponds to the *top* end of the chain. The first few values of z for which $J_0(z) = 0$ are

$$\begin{aligned} z_1 &= 2.405, \quad z_2 = 5.520, \quad z_3 = 8.654, \\ z_4 &= 11.79, \quad z_5 = 14.93, \quad \cdots \end{aligned} \tag{11}$$

to within sliderule accuracy, and the larger values can be found (to within sliderule accuracy) by the approximate formula

$$z_n = \frac{4n - 1}{4} \pi. \tag{12}$$

Using these values, one can then determine the permissible values of k by

$$k_n = \frac{1}{2} \sqrt{\frac{g}{L}}\, z_n. \tag{13}$$

The normal modes of oscillation of the vertical chain are therefore of the form

$$J_0(2k_n\sqrt{x/g})(A_n \cos k_n t + B_n \sin k_n t) \tag{14}$$

where the frequencies of the harmonics are not integer multiples of a fundamental frequency, as with a string having constant tension, but are spaced according to the zeros of the Bessel function J_0.

As a second example, we consider a uniform flexible chain of length L attached at one end to a vertical rod that rotates at a constant angular velocity α. We assume that gravitational effects are negligible compared to the centrifugal effects, so that the chain whirls around in a horizontal plane.

We recall that the centrifugal force due to a particle of mass m rotating with angular velocity α in a circle of radius x is $m\alpha^2 x$. It follows that the tension in the chain is given by

$$\int_x^L \rho\alpha^2 x \, dx = \frac{\rho\alpha^2}{2} (L^2 - x^2) \tag{15}$$

if we measure x from the point of support. In this case (4) becomes, after cancelling the constant ρ,

$$\frac{\alpha^2}{2} \frac{d}{dx}\left[(L^2 - x^2) \frac{du}{dx}\right] + k^2 u = 0. \tag{16}$$

The boundary conditions are $u(0) = 0$, and we require $u(L)$ to be finite.

Taking $z = x/L$, we easily rewrite (16) in the form

$$\frac{d}{dz}\left[(1 - z^2) \frac{du}{dz}\right] + \frac{2Lk^2}{\alpha^2} u = 0, \tag{17}$$

which we compare with Legendre's equation

$$[(1 - z^2)u']' + n(n + 1)u = 0.$$

We have nontrivial bounded solutions when k satisfies the relation

(18) $$2Lk^2/\alpha^2 = n(n+1), \qquad (n = 0, 1, 2, \cdots)$$

but the corresponding solutions $P_n(z)$ will not satisfy the requirements $P_n(0) = 0$ unless n is *odd*. Summarizing, we obtain in this case normal modes of the form

(19) $$y(x,t) = P_n(x/L)(A_n \cos k_n t + B_n \sin k_n t)$$

where

(20) $$k_n = \alpha \sqrt{\frac{n(n+1)}{2L}}, \qquad (n = 1, 3, 5, \cdots).$$

• EXERCISES

1. Do the functions $p(x) = T(x)$ and $r(x) = \rho(x)$ satisfy the conditions of Section 2.4?
2. Investigate the accuracy of (12) when $n = 1$, 2, and 5, using a sliderule.
3. Show how (12) was derived.
4. Given $y(x,t)$ and $(\partial y/\partial t)(x,t)$ at $t = 0$, show how to find a series solution for $y(x,t)$ in the case of the vertical chain.
5. Let $f(x)$ be a continuous function defined for $0 \leqq x \leqq 1$, and assume $f(0) = 0$. Show how to expand $f(x)$ in a series of Legendre polynomials of odd order:
$$f(x) = C_1P_1(x) + C_3P_3(x) + C_5P_5(x) + \cdots.$$
Do you expect this series to converge to $f(x)$? [Hint: the procedure is similar to expanding a function in a Fourier sine series.]
6. Expand the function $f(x) = x(1 - x)$ in a series of Legendre polynomials of odd order, convergent to $f(x)$ when $0 \leqq x < 1$.
7. In the problem of the rotating chain, let the initial displacement of the chain be $y(x,0) = x(1 - x)$, and let the initial $\partial y/\partial t$ be identically zero. Determine $y(x,t)$.

6.5 • VIBRATING MEMBRANES

In this section, we consider the vibrations of a membrane which, when at rest, is in the xy plane. We assume that the membrane consists of a stretched film of uniform thickness, having a mass per unit area ρ, which we take to be a constant. We assume that the membrane resists further stretching but does not resist bending, and that the vibrations we consider are so small that each particle moves in a direction perpendicular to the xy plane. At any instant of time t we let $w(x,y,t)$ denote the displacement from equilibrium, i.e. the displacement in the z direction, of the particle whose *equilibrium* position is the point (x,y) in the plane.

We must explain what we mean by the term *tension*. We assume that the membrane is stretched even when it is in its equilibrium position. Consider a line segment in the xy plane. With the above assumptions, the force which the particles on one side of the line segment exert on the particles to the other side is in a direction perpendicular to the line segment. The magnitude of this force per unit length (length measured along the line segment) is called the tension and is denoted T. We assume that the deflections are sufficiently small that we can consider T to be a constant.

We now derive the differential equation governing vibrations of the membrane by a procedure somewhat similar to that we used for the vibrating string. In particular, we assume the angles involved to be so small we can replace their sines by their tangents. Consider a simple closed curve C in the xy plane (there is no loss in generality here if we take C to be a circle). Let F denote the scalar component, in the positive z direction, of the resultant force acting on that portion of the membrane enclosed by C. One easily sees that this is given by

$$F = \int_C T \frac{\partial w}{\partial n} \, ds \tag{1}$$

where ds is the element of arc length on C and $\partial w/\partial n$ is the directional derivative of $w(x,y,t)$ in a direction normal to C (outward from the interior enclosed by C). Indeed, $T\,ds$ is the element of force acting across ds, and $\partial w/\partial n$ is the tangent of the angle between this force and the xy plane. Of course, F is a function of t.

We now make use of the two-dimensional analog of (9), Section 5.4, which is called Green's theorem in the plane. This is readily derived (Exercise 9) by applying (9), Section 5.4, to a cylindrical surface, assuming the functions depend only on x and y, and states that

$$\int_C \frac{\partial g}{\partial n} \, ds = \iint_A \nabla^2 g \, dx \, dy$$

where $\nabla^2 g$ is the two-dimensional Laplacian $\partial^2 g/\partial x^2 + \partial^2 g/\partial y^2$ and the double integral is over the area A enclosed by C. Using this theorem, we write (1) in the form

$$F = \iint_A T\nabla^2 w \, dx \, dy. \tag{2}$$

We now apply Newton's second law, setting this force F equal to the acceleration, in the z direction, of the center of mass of the portion of the membrane enclosed by C, times the mass of this portion. That is,

$$F = \left[\iint_A \rho \, dx \, dy\right] \frac{\partial^2 w}{\partial t^2}. \tag{3}$$

As a consequence of (2) and (3), we have

$$\iint_A \left[T\nabla^2 w - \rho \frac{\partial^2 w}{\partial t^2}\right] dx \, dy = 0 \tag{4}$$

where $\partial^2 w/\partial t^2$ is evaluated at the center of mass of A and is regarded as a constant during the integration. Since this must be valid for arbitrary choices of C it follows that the integrand must be identically zero, and we deduce the desired differential equation

$$T\nabla^2 w = \rho \frac{\partial^2 w}{\partial t^2}. \tag{5}$$

Written out in rectangular coordinates, this is

$$T\left[\frac{\partial^2 w}{\partial x^2} + \frac{\partial^2 w}{\partial y^2}\right] = \rho \frac{\partial^2 w}{\partial t^2}. \tag{6}$$

Letting $v^2 = T/\rho$, we write this in the form

$$\frac{\partial^2 w}{\partial x^2} + \frac{\partial^2 w}{\partial y^2} = \frac{1}{v^2}\frac{\partial^2 w}{\partial t^2}. \tag{7}$$

Now let us consider the following problem. We assume that the membrane is rigidly fastened along a piecewise smooth closed curve C which forms the boundary of a connected region D in the xy plane. Then the function $w(x,y,t)$ must satisfy the boundary conditions $w(x,y,t) = 0$ at every point (x,y) along C. We assume we are given the *initial displacement*

$$w(x,y,0) = g(x,y) \tag{8}$$

and the *initial velocity*

$$\frac{\partial w}{\partial t}(x,y,0) = h(x,y) \tag{9}$$

at points (x,y) in D. We desire to determine $w(x,y,t)$ for (x,y) in D and $t > 0$.

In general, it is not possible to solve this problem *completely* by the method of separation of variables. However, we can separate out the time variation in the usual manner, by assuming solutions of the form

$$w(x,y,t) = u(x,y)q(t) \tag{10}$$

and in the usual manner we obtain the differential equations for the functions u and q,

$$q''(t) + k^2v^2q(t) = 0, \tag{11}$$

$$\frac{\partial^2 u}{\partial x^2} + \frac{\partial^2 u}{\partial y^2} = -k^2 u. \tag{12}$$

The solutions of (11) are of the form

$$q(t) = A \cos kvt + B \sin kvt \tag{13}$$

and we seek solutions of (12) satisfying the boundary conditions $u(x,y) = 0$ along C.

We have already noted that the only solution of the equation $\nabla^2 u = pu$ for $p > 0$, taking zero values on the bounding surface of a region in space, is the function identically zero, but that there are many nontrivial solutions if we allow p to be negative. The same situation exists in the two-dimensional case. There will, in general, be an infinite sequence of values of k,

$$k_1 \leqq k_2 \leqq k_3 \leqq \cdots \qquad (k_n > 0 \text{ for all } n) \tag{14}$$

for which we can find nontrivial solutions of (12). It is obvious that, for each k_n, the solutions of (12) for $k = k_n$ will constitute a linear class of functions. It is an easy exercise to show that solutions of (12), corresponding to distinct values of k, which are zero on C, are orthogonal; i.e., that

$$\iint_D u_n(x,y)u_m(x,y)\,dx\,dy = 0 \tag{15}$$

if u_n and u_m are solutions of (12) for k_n and k_m, when $k_n \neq k_m$.

It can be shown that, for each fixed k, the linear space of solutions of (12) (satisfying the boundary condition $u = 0$ on C) is finite-dimensional. By the Gram-Schmidt process, we can select an orthogonal basis in each of these linear spaces. It follows, therefore, that if we permit the sequence (14) to contain each k as many times as the dimension of the corresponding linear space, we can place this sequence in one-to-one correspondence with a sequence of functions

$$u_1, u_2, u_3, \cdots, \tag{16}$$

satisfying (15) whenever $n \neq m$. It can be shown by methods beyond the scope of this book that such a sequence exists and provides an approximating basis for the class V of all functions defined and continuous (on D and C together) and equal to zero on C. Here, we can interpret "approximating basis" in either the mean-square sense or in the sense of uniform convergence; with the former sense we do not even need to require that the functions of class V vanish on C.

The solution to the problem is now obtained by expanding the given functions g and h in terms of the functions (16),

$$g(x,y) = \sum_{n=1}^{\infty} C_n u_n(x,y), \tag{17}$$

$$h(x,y) = \sum_{n=1}^{\infty} D_n u_n(x,y) \tag{18}$$

where the coefficients are determined in the usual manner (Exercise 2). The desired solution is then

$$w(x,y,t) = \sum_{n=1}^{\infty} u_n(x,y)[C_n \cos k_n vt + (D_n/k_n v) \sin k_n vt], \tag{19}$$

in complete analogy with the problem of the vibrating string.

Unfortunately, the problem of finding the sequence (16) is prohibitively difficult except in very special cases. We consider two such cases here, the rectangular membrane and the circular membrane.

Rectangular Membranes

In this case, we take the region D to be the rectangle $0 \leqq x \leqq a$, $0 \leqq y \leqq b$. We separate variables in (12), assuming solutions of the form

$$u(x,y) = X(x)Y(y) \tag{20}$$

where X and Y are required to satisfy the boundary conditions

$$X(0) = 0, \quad X(a) = 0, \quad Y(0) = 0, \quad Y(b) = 0. \tag{21}$$

We shall omit the details, since the procedure is essentially identical to that of Section 6.1. We obtain, for arbitrary positive integers n and m, the following doubly-infinite sequence of functions:

$$u_{nm}(x,y) = \sin\frac{n\pi x}{a}\sin\frac{m\pi y}{b} \tag{22}$$

where n and m are related to k^2 by

$$k^2 = \pi^2\left(\frac{n^2}{a^2} + \frac{m^2}{b^2}\right) \tag{23}$$

and, leaving the routine details to the reader, the solution is of the form

$$w(x,y,t) = \sum_{n=1}^{\infty}\sum_{m=1}^{\infty} \sin\frac{n\pi x}{a}\sin\frac{m\pi y}{b}\,[C_{nm}\cos k_{nm}vt + (D_{nm}/k_{nm}v)\sin k_{nm}vt] \tag{24}$$

where

$$k_{nm} = \pi\sqrt{\frac{n^2}{a^2} + \frac{m^2}{b^2}}. \tag{25}$$

It will be noticed that the functions of the form (22) are eigenfunctions of the two-dimensional Laplacian corresponding to the eigenvalues $-k_{nm}^2$. If only one pair of integers n and m gives this eigenvalue, then to within a scalar multiple (22) is the only eigenfunction which satisfies the prescribed boundary conditions. However, we see from (25) that if the membrane is square, or even under other exceptional circumstances, distinct pairs of positive integers n and m may correspond to the same eigenvalue. In this case one can have distinctly different modes of vibration having exactly the same frequency of vibration.

For example, if $a = b = \pi$, we obtain two solutions of (22) which are eigenfunctions having the same eigenvalue -13, namely $\sin 2x \sin 3y$ and $\sin 3x \sin 2y$. One might argue that these are not really "different" because one can be obtained from the other by rotating the membrane through 90°. But one sees, by the linearity, that any linear combination of these two eigenfunctions is again an eigenfunction with the same eigenvalue, and some of these linear combinations are distinctly different.

To see this, let us look at the *nodal curves* of the vibrations. These are the curves along which the eigenfunction is zero. In the case of the eigenfunction $\sin 2x \sin 3y$ these curves are straight lines, as shown in Figure 6.7. The nodal curves for $\sin 3x \sin 2y$ are obtained by rotating this figure through 90°. On the other hand, the nodal curves for $\sin 2x \sin 3y + \sin 3x \sin 2y$ are quite different; they are shown in Figure 6.8.

As with the vibrating string, any vibration to which we can

ascribe a definite frequency is called a *normal mode* of oscillation, and can be thought of as a "standing wave" in the membrane. Unlike the vibrating string, however, it is possible for distinctly different normal modes to have the same frequency. The most general possible mode of oscillation is given by (24) as a superposition of *normal* modes.

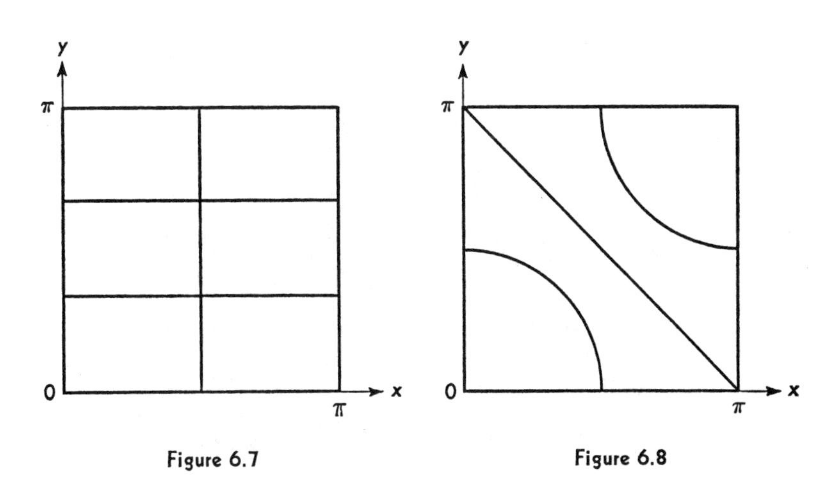

Figure 6.7

Figure 6.8

Circular Membranes

As with the rectangular membrane, an arbitrary vibration of a circular membrane can be written as a superposition of normal modes of oscillation. Again we emphasize that a *normal* mode of vibration is one in which each point of the membrane is oscillating sinusoidally about its equilibrium position, and all points of the membrane are oscillating with the same frequency.

Everyone knows that the sound produced by a drum is less "pure" than that produced by a violin. This can be ascribed to the fact that the frequencies of the normal modes of oscillation of a drum are not integral multiples of a single frequency. Let the lowest possible frequency of vibration of a given drum be denoted f_0; this is called the *fundamental frequency*.

In Figure 6.9 we show the nodal curves for six modes of vibration of a circular membrane. In each instance we give the corresponding frequency in terms of the fundamental frequency. In (a) the drum is vibrating with its fundamental frequency. All points are moving in phase with each other, so the only nodal curve is the edge of the membrane where it is fastened down. In (b) the upper half of the drum is out of phase with the lower half, and the frequency is (approximately) $1.593f_0$. The other figures should be self-explanatory. The shaded regions are those in which the points are moving out of phase with the points in the unshaded regions.

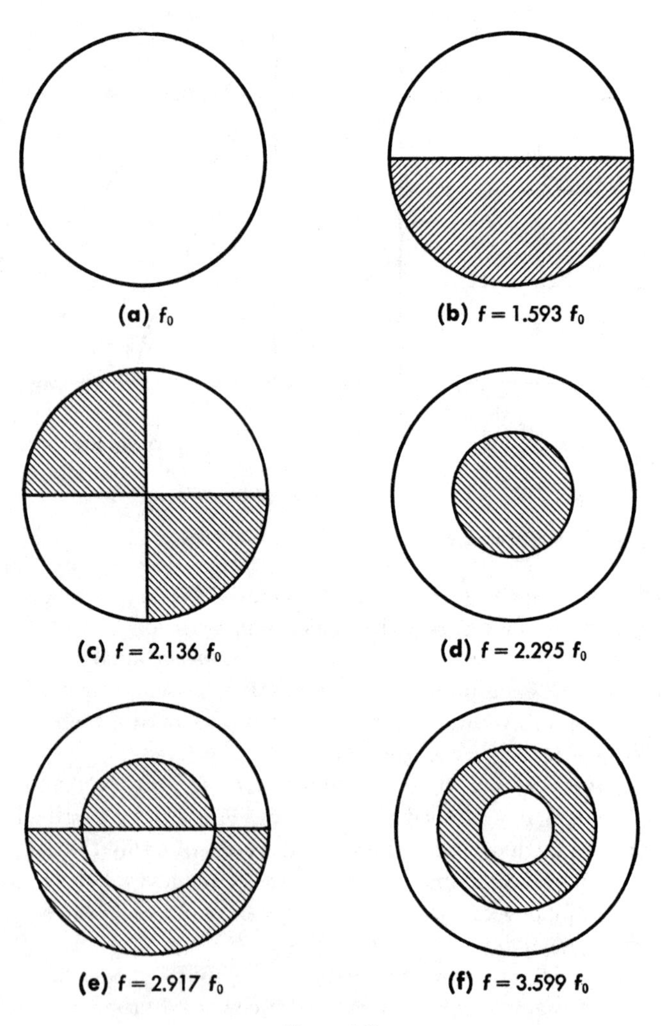

Figure 6.9

These figures appear in many elementary physics textbooks, without any explanation of how the frequencies are obtained. We shall now derive expressions for the normal modes, and show how these frequencies are related to the zeros of the Bessel functions. Let the radius of the membrane be a, so D is the region $x^2 + y^2 \leqq a^2$, or (in polar coordinates) $r \leqq a$.

In polar coordinates, (12) becomes

$$\frac{\partial^2 u}{\partial r^2} + \frac{1}{r}\frac{\partial u}{\partial r} + \frac{1}{r^2}\frac{\partial^2 u}{\partial \theta^2} = -k^2 u. \tag{26}$$

Each solution $u(r,\theta)$ of this equation satisfying the boundary condition $u(a,\theta) = 0$ leads to a mode of oscillation $u(r,\theta)q(t)$ where $q(t)$ is given by (13). The frequency of this mode of oscillation is $kv/2\pi$, where $v = \sqrt{T/\rho}$.

We obtain the eigenfunctions for this boundary-value problem by the method of separation of variables. Assuming a product solution of the form

$$u(r,\theta) = R(r)\Theta(\theta), \tag{27}$$

the usual process of separation leads to the differential equations

$$\Theta'' + \alpha^2\Theta = 0 \tag{28}$$

$$r^2R'' + rR' + (k^2r^2 - \alpha^2)R = 0. \tag{29}$$

The solutions to (28) are

$$\Theta = C \sin \alpha\theta + D \cos \alpha\theta \tag{30}$$

and for these to be physically meaningful, α must be an integer. Taking $\alpha = n$, where $n = 0, 1, 2, \cdots$, every bounded solution of (29) is a scalar multiple of

$$R(r) = J_n(kr). \tag{31}$$

We require that $R(r)$ vanish at $r = a$, so we must choose k so that $J_n(ka) = 0$. Let k_{nm} denote the mth *positive* value of k for which $J_n(k_{nm}a) = 0$. That is, $k_{n1} < k_{n2} < k_{n3} < \cdots$ and $k_{nm} \neq 0$ for all n and m. Then the desired normal modes of vibration are of the form

$$w_{nm}(r,\theta,t) = J_n(k_{nm}r)[C_n \sin n\theta + D_n \cos n\theta] \cdot [A_n \sin k_{nm}vt + B_n \cos k_{nm}vt]. \tag{32}$$

We can now easily interpret n and m in (32) geometrically, by means of the nodal curves. When $n = 0$, the factor involving θ is a constant, so the nodal curves are circles concentric with the origin. In particular, the fundamental mode of vibration is

$$(33) \qquad w_{01}(r,\theta,t) = J_0(k_{01}r)[A_0 \sin k_{01}vt + B_0 \cos k_{01}vt].$$

We recall from Section 4.6 that the first zero of the Bessel function J_0 comes at $z_1 = 2.405$ (approximately), so $k_{01}a = 2.405$ and the fundamental frequency is

$$f_0 = k_{01}v/2\pi = \frac{2.405}{2\pi a}\sqrt{\frac{T}{\rho}}.$$

In the fundamental mode, the only nodal curve is along the edge, where the membrane is fastened down. If $n = 0$ and $m > 1$, we will obtain $m - 1$ circular nodes in the interior of the membrane. Thus case (d) in Figure 6.9 corresponds to $n = 0$ and $m = 2$. Recalling that the second zero of the Bessel function J_0 comes at $z_2 = 5.520$, the frequency in this case is $f = k_{02}v/2\pi = (5.520/2\pi a)\sqrt{T/\rho} = 2.295f_0$. The number 2.295 was found simply by dividing 5.520 by 2.405. Case (f) in the figure corresponds to $n = 0$ and $m = 3$, and the corresponding frequency is

$$f = k_{03}v/2\pi = z_3 f_0/z_1 = (8.654/2.405)f_0 = 3.599f_0.$$

If n is not zero in (32), then we obtain additional nodal curves, which in this case are straight lines corresponding to fixed values of θ. For instance, (e) in Figure 6.9 corresponds to the normal mode

$$(34) \qquad w_{12}(r,\theta,t) = J_1(k_{12}r)(\sin\theta)(A_1 \sin k_{12}vt + B_1 \sin k_{12}vt).$$

To obtain the corresponding frequency, one must know the zeros of the Bessel function $J_1(z)$. We see from the short table of zeros given below that the second positive zero of J_1 is 7.016. Reasoning as before, the frequency must be $(7.016/2.405)f_0 = 2.917f_0$.

We list here the zeros of the Bessel functions J_n, $n = 0, 1, 2, 3$, and 4. The values z_j are those for which $J_n(z_j) = 0$.

j	$n = 0$	$n = 1$	$n = 2$	$n = 3$	$n = 4$
1	2.405	3.832	5.135	6.379	7.586
2	5.520	7.016	8.417	9.760	11.064
3	8.654	10.173	11.620	13.017	14.373
4	11.792	13.323	14.796	16.224	17.616
5	14.931	16.470	17.960	19.410	20.827

For larger values of j, the zeros are given (Exercise 8) approximately by

$$x_j = (4j - 1 + 2n)\frac{\pi}{4} \qquad (j > 5). \tag{35}$$

This formula is correct to within sliderule accuracy for $n = 0$ and $n = 1$, and somewhat less accurate for higher n; the error for $n = 4$ is almost 2 percent. The accuracy improves with increasing j, so no matter what n may be, the formula is very accurate for sufficiently large values of j.

• EXERCISES

1. Derive the more general equation for the vibrations of a membrane that is applicable if T is not assumed to be constant.

2. Write down explicit expressions for C_n and D_n in (17) and (18) in terms of g, h, and u_n.

3. Show how to determine the coefficients C_{nm} and D_{nm} in (24), in terms of the initial displacement and initial velocity of the membrane.

4. Draw a diagram similar to those in Figure 6.8 showing the nodal curves for $\sin x \sin 4y + \sin 4x \sin y$.

5. Complete the discussion of the circular membrane by showing how to determine $w(r,\theta,t)$ given $w(r,\theta,0)$ and $(\partial w/\partial t)(r,\theta,t)$.

6. Derive a formula for the fundamental frequency f_0 of a rectangular membrane, and derive a formula for the frequency f of an arbitrary normal mode in terms of f_0.

7. Show how $f = 2.136f_0$ was determined in (c) of Figure 6.9.

8. Derive (35).

9. Derive Green's theorem in the plane, in the manner suggested.

10. Derive (15).

6.6 • WAVES IN TWO AND THREE DIMENSIONS

We have seen that a standing wave in the vibrating string can be considered to be the superposition of two traveling waves $p(x - vt)$

and $q(x + vt)$ (Section 6.2). A similar situation exists in the vibrating membrane. The situation is more complicated, however. In a vibrating string, there are only two possible directions for traveling waves, whereas in a vibrating membrane there are infinitely many possible directions. Although it is true that a standing wave in a membrane can be considered to be a superposition of traveling waves, the superposition may involve summing an infinite number of such waves, or even integrating with respect to some parameter.

We have not yet discussed superposition by integration to any extent; an introduction will be given in the next section. A general discussion is beyond the scope of this book.

It is interesting to note, however, that there is a formula for waves in two dimensions that is similar to the d'Alembert formula for waves in one dimension [see (17), Section 6.2]. We assume the membrane is infinite in extent, so there are no reflections from boundaries to create standing waves. If the initial displacement is $g(x,y)$ and the initial velocity is $h(x,y)$, the formula is

(1)
$$w(x',y',t) = \frac{1}{2\pi v}\frac{\partial}{\partial t}\left\{\iint_A \frac{g(x,y)\,dx\,dy}{(v^2t^2 - p^2)^{1/2}}\right\} + \frac{1}{2\pi v}\iint_A \frac{h(x,y)\,dx\,dy}{(v^2t^2 - p^2)^{1/2}}.$$

This is called *Poisson's formula.* It gives the displacement at a point (x',y'), at $t > 0$, in terms of integrals of $g(x,y)$ and $h(x,y)$ over the interior (and boundary) of a circle of radius vt centered at (x',y'). Here, $p^2 = (x - x')^2 + (y - y')^2$ and A is therefore the region for which $p \leqq vt$.

Physically, this means that the value of the displacement at (x',y'), at some time $t > 0$, depends on the values of $g(x,y)$ and $h(x,y)$ *throughout* a circular region centered at (x',y'). This suggests, among other things, that we cannot transmit sharply defined signals by creating ripples on the surface of a pond or a tight membrane. Any disturbance set up at $t = 0$ at a point (x,y) affects *all points* in the interior of a circle with radius vt and center at (x,y), so the disturbance created thereby will have a sharp wavefront but a diffused back.

A similar phenomenon occurs in the one-dimensional case. Only in the very special case that $h(x)$ is identically zero can a wave pass by without leaving a permanent trace, as we have already noted.

On the other hand, this diffusive phenomenon does not occur for waves satisfying the *three-dimensional wave equation*

(2)
$$\frac{\partial^2 w}{\partial x^2} + \frac{\partial^2 w}{\partial y^2} + \frac{\partial^2 w}{\partial z^2} = \frac{1}{v^2}\frac{\partial^2 w}{\partial t^2}.$$

It can be shown that, if $w(x,y,t)$ satisfies the initial conditions $w(x,y,0) = g(x,y,z)$ and $\dfrac{\partial w}{\partial t}(x,y,0) = h(x,y,z)$, then for $t > 0$ we have

$$(3) \quad w(x',y',z',t) = \frac{1}{4\pi v^2}\frac{\partial}{\partial t}\left\{\frac{1}{t}\iint_S g(x,y,z)\,dS\right\} + \frac{1}{4\pi v^2 t}\iint_S h(x,y,z)\,dS$$

where S is the *surface* of a sphere centered at (x',y',z') having radius vt.

This means that a disturbance set up at $t = 0$ in the vicinity of a point (x,y,z) affects only points on the surface of a sphere with radius vt and center at (x,y,z). The resulting waves leave no trace after them. Therefore high-fidelity signal transmission is possible in three dimensions, as is well known.

Some readers will recognize (3) as the equation which provides the basis for *Huygen's principle.*

• EXERCISES

1. A solution of (2) of the form $h(r)f(r - vt)$, where $r^2 = x^2 + y^2 + z^2$, is called a *spherical wave.* The function f is called the *wave form* and h is the *attenuating factor.* Show that $h(r)$ must be a constant multiple of $1/r$.

2. A solution of (2) of the form $h(r)f(r - vt)$ where $r^2 = x^2 + y^2$ is called a *cylindrical wave.* The function f is called the *wave form* and h is the *attenuating factor.* Determine $h(r)$. (Compare Exercise 1.)

3. In cylindrical coordinates, if w depends only on time and on the distance r from the z axis, (2) becomes

$$(4) \qquad \frac{1}{r}\frac{\partial}{\partial r}\left(r\frac{\partial w}{\partial r}\right) = \frac{1}{v^2}\frac{\partial^2 w}{\partial t^2}.$$

By the method of separation of variables, find solutions $w(r,t) = R(r)T(t)$ of this equation.

***4.** In spherical coordinates, if w depends only on time and on the distance r from the origin, (2) becomes

$$(5) \qquad \frac{1}{r^2}\frac{\partial}{\partial r}\left(r^2\frac{\partial w}{\partial r}\right) = \frac{1}{v^2}\frac{\partial^2 w}{\partial t^2}.$$

If $u(r,t) = R(r)T(t)$, find differential equations satisfied by R and T, and solve them.

*5. (For more advanced students.) Generalize (2) to n dimensions, and show that "high-fidelity signal transmission by radiation from a central source is possible only in a three-dimensional space." (Warning: a similar argument was given many years ago to show that high-fidelity signal transmission by frequency modulation is impossible! The analysis failed to consider the possibility that the receiver could reverse the distorting process.)

6.7 • THE FOURIER INTEGRAL

The concept of the Fourier integral is fundamental in the harmonic analysis of waves with a "continuous spectrum." We begin with an heuristic explanation of what we mean by the "spectrum" of a function.

Everyone is familiar with the spectrum of colors produced when a beam of light passes through a prism. The velocity of light passing through the prism is a function of its wavelength, and as a consequence beams of different color are "bent" by differing amounts. The result is the familiar optical spectrum of colors ranging from violet through indigo, blue, green, yellow, orange, and red. If the light is more or less random in frequency, as when produced by an incandescent lamp, one obtains a continuous spread of colors. If the source is, on the other hand, a mercury-vapor lamp, one obtains quite distinct spectral lines, characteristic of the frequencies emitted by the mercury atom. We distinguish the latter from the former by speaking of a *discrete* spectrum rather than a continuous spectrum.

A similar phenomenon occurs in acoustics. The babble of a crowd will have a more or less continuous spectrum, but a single violin string will produce nearly a discrete spectrum of frequencies, consisting of a fundamental frequency together with various overtones.

Whether we choose to distinguish between the various components of a spectrum by means of their frequencies or their wavelengths is a matter of arbitrary choice. If we like, we can use other parameters, such as the "angular frequency" or the "wave number." This is also true of the mathematical definition of the spectrum of a function, so the reader must keep in mind that what various authors choose to call the *spectrum of a function* will not necessarily agree.

Nor should he confuse the spectrum of a function with the spectrum of an *operator* which is not discussed in this book.

If a function g is a sum (either a finite sum or the sum of an infinite series) of functions of the form $A_k \cos kx + B_k \sin kx$,

$$g(x) = \sum_{n=0}^{\infty} A_{k_n} \cos k_n x + B_{k_n} \sin k_n x, \tag{1}$$

then those values of k_n for which either A_{k_n} or B_{k_n} are not zero are said to be in the *discrete spectrum* of the function g.

If a function h is obtained by an *integral* superposition of functions of the above kind,

$$h(x) = \int_0^{\infty} [C(k) \cos kx + D(k) \sin kx] \, dk \tag{2}$$

where $C(k)$ and $D(k)$ are continuous functions, then any value of k for which either $C(k) \neq 0$ or $D(k) \neq 0$ is said to be in the *continuous spectrum* of the function h.

By adding two functions like g and h above, one can obtain a function having both a point spectrum and a continuous spectrum.

In the theory of Fourier series, one is concerned with functions having a discrete spectrum. On the other hand, the Fourier integral is relevant to functions having a continuous spectrum.

It should be noted that the theory of Fourier series is relevant to functions of the form (1) only when the members k_n are integral multiples of a single number k. In that event, the functions $\cos k_n x$ and $\sin k_n x$ will be mutually orthogonal over any interval of length $2\pi/k_n$. If the numbers k_n are not integral multiples of a single number k, then the resulting function g will not be periodic, and the coefficients will be related to the given function in a more complicated way (see Exercise 1).

In this section we are concerned entirely with functions having a continuous spectrum. The first part of the section is theoretical. Some readers may wish to scan the second part first, in order to understand some of the motivations for this theory.

Part I. Convergence of Integrals

In this subsection, we shall be concerned with the convergence of *infinite integrals*. The reader may find it instructive to compare this with the discussion of *infinite series* given in Section 3.7. The concepts of uniform convergence and absolute convergence of infinite series have their counterparts in the theory of infinite integrals. It should be noted that the word *infinite* does not refer to the value of the integral, or the function being integrated, but to the length of the interval over which we are integrating. Such integrals are sometimes called "improper integrals of the first kind" to distinguish them from integrals of unbounded functions (which are called "improper integrals of the second kind").

Let us consider a function f which is integrable over (a,x) for all values of $x \geqq a$. [Recall that we are using the elementary Riemann definition of the integral, so this implies the function is bounded in (a,x).] We then define the integral

$$\int_a^\infty f(x)\,dx = \lim_{x\to\infty} \int_a^x f(t)\,dt. \tag{1}$$

If this limit exists, the integral on the left is said to converge and is assigned the value of the limit, in the same way that we assign a number (the "sum") to an infinite series if its sequence of partial sums converges.

If the limit indicated in (1) does not converge, then $\int_a^\infty f(x)\,dx$ is said to be *divergent*. This does *not* imply that $\int_a^x f(t)\,dt$ increases with increasing x; the limit may fail to exist for other reasons. For instance, $\int_0^\infty \cos x\,dx$ is a divergent integral, because $\sin x$ does not tend to a limit as $x \to \infty$, although the "partial integrals" $\int_0^x \cos u\,du$ are bounded.

If the integral

$$\int_0^\infty |f(x)|\,dx = \lim_{x\to\infty} \int_a^x |f(u)|\,du \tag{2}$$

converges, we say that integral (1) is *absolutely convergent*. If f satisfies the conditions stated in the second paragraph of this subsection, then the absolute convergence of (1) implies the convergence of (1). It is, however, possible for an integral to be convergent but not absolutely convergent.

An important example is the integral $\int_0^\infty (\sin x)/x\,dx$. This integral is convergent, but it is not absolutely convergent.

To see this, consider the graph of $(\sin x)/x$. This graph consists of arches alternately above and below the x axis, and these arches together with the axis enclose areas that decrease monotonically and tend to zero with increasing x. Let the areas of these arches be denoted $A_1, A_2, A_3, \cdots$. The alternating series $A_1 - A_2 + A_3 - A_4 + \cdots$ is therefore convergent, and for any M the value of the integral $\int_0^M (\sin x)/x\,dx$ is either equal to a partial sum of the alternating series, or is between successive partial sums of this series:

(3)

$$A_1 - A_2 + \cdots (\pm) A_n \leqq \int_0^M \frac{\sin x}{x} dx \leqq A_1 - A_2 + \cdots (\pm) A_{n+1}$$

where $n \to \infty$ and $M \to \infty$ together. It follows that $\int_0^\infty (\sin x)/x\, dx$ is a convergent integral. It is not absolutely convergent, however, since $\int_0^\infty |(\sin x)/x|\, dx$ is divergent (Exercise 2).

An infinite integral $\int_0^\infty f(x)\, dx$ is the analog, for integrals, of an infinite *numerical* series $\sum_{n=0}^{\infty} c_n$. An infinite integral $\int_0^\infty f(x,t)\, dx$ is the analog of an infinite series of functions $\sum_{n=0}^{\infty} f_n(t)$. From this it should be clear what we mean by the *uniform convergence* of an integral $\int_0^\infty f(x,t)\, dx$ for t in some prescribed interval. The only difference is that, instead of requiring that a sequence of partial sums be uniformly convergent, we require that the one-parameter family of functions

$$g_M(t) = \int_0^M f(x,t)\, dx \tag{4}$$

converge uniformly to some function $g(t)$ for t in the prescribed interval. Essentially, all this means is that, given any positive ϵ, no matter how small, we have $|g(t) - g_M(t)| < \epsilon$ for all M greater than some number M_ϵ which is independent of t.

We recall that a uniformly convergent series of continuous functions must converge to a continuous function, and such a series can be integrated term-by-term over a finite interval. The corresponding theorem for integrals is this: If $f(x,t)$ is a continuous function of t for $a \leqq t \leqq b$, and if $\int_0^\infty f(x,t)\, dx$ exists and converges uniformly in this interval, the resulting function $g(t) = \int_0^\infty f(x,t)\, dx$ is continuous in the interval and $\int_a^b g(t)\, dt = \int_0^\infty [\int_a^b f(x,t)\, dt]\, dx$.

This can be written in the form

$$\int_a^b \left[\int_0^\infty f(x,t)\, dx\right] dt = \int_0^\infty \left[\int_a^b f(x,t)\, dt\right] dx. \tag{5}$$

The proof, under the stated hypotheses, is not essentially different from the proof of the corresponding theorem for uniformly convergent series.

If we do not require that the function $f(x,t)$ be continuous in $a \leqq t \leqq b$, but only integrable over this interval (for every fixed x) then the only change in the remarks made above is that we cannot conclude that the function $g(t)$ is continuous; nevertheless, $g(t)$ will be integrable and (5) will still be valid.

Similar remarks can be made for an integral of the form $\int_{-\infty}^{0} f(x)\,dx$ or $\int_{-\infty}^{0} f(x,t)\,dx$. It should be noticed that, if we refer to an integral over the entire real line, $\int_{-\infty}^{\infty} f(x,t)\,dx$, as uniformly convergent in this interval, we mean that $\int_{-\infty}^{0} f(x,t)\,dx$ and $\int_{0}^{\infty} f(x,t)\,dx$ are both uniformly convergent; we define $\int_{-\infty}^{\infty} f(x,t)\,dx$ to be the sum of these two integrals. (See Exercise 4 for a caution concerning such integrals.)

We now prove a sequence of lemmas that will be useful to us later.

Lemma 1. *If f is piecewise continuous over (a,b), then*

$$\lim_{A\to\infty} \int_a^b f(x) \sin Ax\,dx = 0,$$

and also $\lim_{A\to\infty} \int f_b^a (x) \cos Ax\,dx = 0.$

Notice that this is a generalization of the Riemann lemma proved in Section 2.3.

Proof: If f has a continuous derivative, this is easily proved; we integrate by parts to obtain

$$\int_a^b f(x) \cos Ax\,dx = \left[f(x)\,\frac{\sin Ax}{A}\right]_a^b - \frac{1}{A}\int_a^b f'(x) \sin Ax\,dx,$$

which tends to zero as $A \to \infty$ since the integral on the right side is bounded.

If f is not differentiable, let p be a continuously differentiable function such that $\int_a^b |f(x) - p(x)|\,dx < \epsilon$. Then

$$\left|\int_a^b [f(x) - p(x)] \cos Ax\,dx\right|$$
$$\leqq \int_a^b |f(x) - p(x)||\cos Ax|\,dx \leqq \int_a^b |f(x) - p(x)|\,dx < \epsilon$$

independently of A, and since $\int_a^b p(x) \cos Ax\,dx \to 0$ by the preceding paragraph, it follows that $\int_a^b f(x) \cos Ax\,dx \to 0$ also.

The proof that $\int_a^b f(x) \sin Ax\,dx \to 0$ is similar.

Lemma 2. $\displaystyle\int_0^\infty \frac{\sin x}{x}\,dx = \frac{\pi}{2}.$

Proof: In Section 3.6 we showed that

$$\int_0^\pi \frac{\sin \dfrac{2n+1}{2} u}{\sin \dfrac{u}{2}}\,du = \pi. \tag{6}$$

Applying Lemma 1 to the function $\frac{2}{u} - \frac{1}{\sin u/2}$ (which is bounded in $0 < u < \pi$), we have

$$\lim_{n\to\infty} \int_0^\pi \left(\sin \frac{2n+1}{2} u\right)\left(\frac{2}{u} - \frac{1}{\sin u/2}\right) du = 0. \tag{7}$$

Summing (6) and (7), we obtain

$$\lim_{n\to\infty} \int_0^\pi \frac{2 \sin \frac{2n+1}{2} u}{u} du = \pi. \tag{8}$$

Changing variables, letting $t = (2n + 1)u/2$, we have

$$\lim_{n\to\infty} \int_0^{(2n+1)\pi/2} \frac{\sin t}{t} dt = \pi/2. \tag{9}$$

Earlier in this section, we proved that $\int_0^M (\sin t)/t\, dt$ tends to a limit as $M \to \infty$ *continuously*, and this limit is obviously the same as the limit as $M \to \infty$ through *discrete values*, which [in view of (9)] proves the lemma.

The next lemma is of fundamental importance in proving the Fourier integral theorem. Recall that, if f is piecewise continuous, $f(0+)$ denotes the limiting value of $f(x)$ as x tends to zero through positive values. If f is also piecewise smooth, $f'(0+)$ must exist, and is the limit as x tends to zero (through positive values) of $[f(x) - f(0+)]/x$.

Lemma 3. *Whenever f is a piecewise smooth function,*

$$\lim_{A\to 0} \int_0^b f(x) \frac{\sin Ax}{x} dx = \frac{\pi}{2} f(0+), \qquad (b > 0).$$

Proof:

$$\int_0^b f(x) \frac{\sin Ax}{x} dx = \int_0^b f(0+) \frac{\sin Ax}{x} dx + \int_0^b \frac{f(x) - f(0+)}{x} \sin Ax\, dx$$

$$= f(0+) \int_0^{Ab} \frac{\sin u}{u} du + \int_0^b \frac{f(x) - f(0+)}{x} \sin Ax\, dx.$$

By Lemma 1, the last integral tends to zero as $A \to \infty$, since the integrand is piecewise smooth in the interval $0 < x < b$. [It remains bounded in this interval because, as x tends to zero, $[f(x) - f(0+)]/x$ tends to $f'(0+)$.] The other integral tends to the desired value, by Lemma 2.

We are now ready to prove the *Fourier integral theorem*. Because of the presence of three "variables" u, x, and t, this theorem may seem slightly complicated at first. Observe, however, that the theorem is valid for each fixed x, so x can be considered a "constant" insofar as the integrations are concerned. The theorem will be discussed in detail in the second part of this section.

Fourier Integral Theorem. *If f is piecewise smooth in every finite interval, and if $\int_{-\infty}^{\infty} |f(t)|\, dt$ is a convergent integral, then*

$$(10) \quad \frac{1}{\pi} \int_0^\infty \left[\int_{-\infty}^\infty f(t) \cos u(x - t)\, dt \right] du = \frac{f(x+) + f(x-)}{2}.$$

Proof: Consider the integral

$$(11) \quad \int_{-\infty}^\infty f(t) \cos u(x - t)\, dt.$$

Since $|\cos u(x - t)| \leqq 1$, the convergence of this integral is ensured by the hypothesis that $\int_{-\infty}^{\infty} |f(t)|\, dt$ converges, and since this is independent of u and x, the convergence is *uniform* for all u. Therefore we can interchange the order of integration in the following integral:

$$(12) \quad \int_0^b \left[\int_{-\infty}^\infty f(t) \cos u(x - t)\, dt \right] du$$

to obtain

$$(13) \quad \int_{-\infty}^\infty \left[\int_0^b f(t) \cos u(x - t)\, du \right] dt = \int_{-\infty}^\infty \left[\frac{\sin b(x - t)}{x - t} \right] f(t)\, dt,$$

which we now split up into four integrals, where we take M so large that the first and the last integrals are less in absolute value than some prescribed $\epsilon > 0$.

$$(14) \quad \left[\int_{-\infty}^{-M} + \int_{-M}^{x} + \int_x^M + \int_M^\infty \right] \frac{\sin b(x - t)}{x - t} f(t)\, dt.$$

By changing variables, taking $u = t - x$, the third integral can be written $\int_0^{M-x} [(\sin bu)/u]\, f(x + u)\, du$, which (by Lemma 3) tends to $\pi f(x+)/2$ as $b \to \infty$. Similarly, the second integral tends to

$\pi f(x-)/2$. Therefore, by taking M sufficiently large, we can show that the left side of (10) differs from the right side by an amount less than 2ϵ. But the left side of (10) is independent of M, so it follows that both sides are equal, completing the proof of the theorem.

Part 2. Fourier Transforms and Applications

In this part, we assume that f not only is piecewise smooth but is continuous, so we can write (10) in the form

$$\frac{1}{\pi}\int_0^\infty \left[\int_{-\infty}^\infty f(t)\cos u(x-t)\,dt\right]du = f(x). \tag{15}$$

This is purely a matter of convenience, to enable us to avoid continually rewriting the awkward right side of (10).

Let us rewrite (15) in an alternative form, which shows more clearly how similar it is to a Fourier series. Writing out $\cos u(x - t)$ as $\cos ux \cos ut + \sin ux \sin ut$ and splitting (15) into two integrals, we obtain

$$\begin{aligned} f(x) &= \frac{1}{\pi}\int_0^\infty \left[\cos ux \int_{-\infty}^\infty f(t)\cos ut\,dt\right]du \\ &\quad + \frac{1}{\pi}\int_0^\infty \left[\sin ux \int_{-\infty}^\infty f(t)\sin ut\,dt\right]du. \end{aligned} \tag{16}$$

Defining two new functions by

$$A(u) = \frac{1}{\pi}\int_{-\infty}^\infty f(t)\cos ut\,dt \tag{17}$$

$$B(u) = \frac{1}{\pi}\int_{-\infty}^\infty f(t)\sin ut\,dt, \tag{18}$$

we can write (16) in the form

$$f(x) = \int_0^\infty A(u)\cos ux\,du + \int_0^\infty B(u)\sin ux\,du \tag{19}$$

where the right side can now be compared with

$$f(x) = \sum_{n=0}^\infty A_n \cos nx + \sum_{n=1}^\infty B_n \sin nx, \tag{20}$$

which is the form of the Fourier series expansion for an altogether different kind of function f. Notice the similarity: we are writing f

in terms of sines and cosines in each instance. In (19) the summations over n are replaced by integrations over u, and the "coefficients" are continuous functions rather than sequences.

We have seen that, if a function is periodic, with period 2π, we can (in some instances) expand it in a series like (20). Such a function could not be expanded in the manner of (19) because the integrals in (17) and (18) would not converge. On the other hand, a function satisfying the conditions of the Fourier integral theorem, if expanded in a Fourier series in some interval, will be represented by the series in that interval alone, whereas the Fourier integral theorem provides a representation valid for the entire real line. Such a function must "tend to zero at infinity" sufficiently rapidly that the integrals in (17) and (18) exist, and therefore cannot possibly be periodic (with the trivial exception of the zero function).

If f is an *even* function, (17) becomes

$$A(u) = \frac{2}{\pi}\int_0^\infty f(t)\cos ut\,dt \tag{21}$$

and in this case $B(u)$ will be identically zero.

If f is an *odd* function, (18) becomes

$$B(u) = \frac{2}{\pi}\int_0^\infty f(t)\sin ut\,dt \tag{22}$$

and in this event, $A(u)$ is identically zero.

Regardless of whether or not f possesses any property of symmetry, the function A defined by (21) is called the *Fourier cosine transform* of the function f, and B is called the *Fourier sine transform.*

If f satisfies the hypotheses of the Fourier integral theorem, and if A is its cosine transform, then $\int_0^\infty A(u)\cos ux\,du$ *alone* represents $f(x)$ for *positive* values of x, and for negative x represents the *even extension* of f. Similarly, we can represent the *odd* extension of f by means of the Fourier cosine transform. (Compare the Fourier sine series and the Fourier cosine series.)

The functions $A(u)$ and $B(u)$ together are needed to comprise the full Fourier transform, just as we need all the A_n's and B_n's in the full Fourier series. It is, however, rather inconvenient to call the pair $A(u)$ and $B(u)$ the Fourier transform. Instead, the complex function

$$F(u) = \frac{1}{2\pi}\int_{-\infty}^\infty f(t)e^{-iut}\,dt \tag{23}$$

is commonly referred to as the Fourier transform of f. Under the stated hypotheses, we can recover f from the transform by

$$f(x) = \int_{-\infty}^{\infty} F(u)e^{iux}\, du. \tag{24}$$

The derivation of these expressions is left as an exercise.

We consider two simple applications of this theory. Perhaps the simplest problem is that in which we desire a continuous function $w(x,y)$ that satisfies Laplace's equation

$$\frac{\partial^2 w}{\partial x^2} + \frac{\partial^2 w}{\partial y^2} = 0 \tag{25}$$

when $y > 0$, and assumes prescribed values $f(x)$ on the x axis,

$$w(x,0) = f(x). \tag{26}$$

To ensure uniqueness, we also require that $w(x,y)$ tend to zero as $y \to \infty$.

By the usual method of separation of variables, we find product solutions which tend to zero as $y \to \infty$:

$$e^{-uy}(A \cos ux + B \sin ux), \tag{27}$$

which we seek to superimpose to obtain the desired solution:

$$w(x,y) = \int_0^{\infty} e^{-uy}[A(u) \cos ux + B(u) \sin ux]\, du. \tag{30}$$

When $y = 0$ this reduces to

$$f(x) = \int_0^{\infty} [A(u) \cos ux + B(u) \sin ux]\, du. \tag{31}$$

The desired solution is therefore obtained by taking $A(u)$ and $B(u)$ in (30) to be the functions defined in (17) and (18). If f satisfies the conditions of the Fourier integral theorem, there is no great difficulty in showing that (30) gives a continuous solution of Laplace's equation in the upper half plane, which tends to $[f(x+) + f(x-)]/2$ as $y \to 0$. Unless f possesses a continuous second derivative, the solution w will not satisfy Laplace's equation on the x axis itself (compare the discussion in Section 5.5).

In our second application we will make use of the complex Fourier transform to derive a formula already obtained earlier by other methods. Let us seek a solution $w(x,t)$ to the one-dimensional wave equation

$$\frac{\partial^2 w}{\partial x^2} = \frac{1}{v^2} \frac{\partial^2 w}{\partial t^2}, \tag{32}$$

satisfying the initial conditions

$$w(x,0) = g(x), \quad \frac{\partial w}{\partial t}(x,0) = 0. \tag{33}$$

For fixed t, $w(x,t)$ is a function of x, and we let $W(u,t)$ denote its Fourier transform:

$$W(u,t) = \frac{1}{2\pi}\int_{-\infty}^{\infty} w(x,t)e^{-iux}\,dx. \tag{34}$$

The Fourier transform of $\partial^2 w/\partial t^2$, considered as a function of x, is therefore

$$\frac{1}{2\pi}\int_{-\infty}^{\infty} \frac{\partial^2 w}{\partial t^2}(x,t)e^{-iux}\,dx = \frac{\partial^2 W}{\partial t^2}(u,t) \tag{35}$$

provided we can justify the interchange in order of differentiation and integration. The Fourier transform of $\partial^2 w/\partial x^2$, considered as a function of x, can be evaluated using integration by parts, provided we assume that $\partial^2 w/\partial x^2$ and $\partial w/\partial x$ are continuous and satisfy the conditions of the Fourier integral theorem. We obtain

$$\begin{aligned}
&\frac{1}{2\pi}\int_{-\infty}^{\infty} \frac{\partial^2 w}{\partial x^2}(x,t)e^{-iux}\,dx \\
&\quad = \frac{1}{2\pi}\left[\frac{\partial w}{\partial x}(x,t)e^{-iux}\right]_{-\infty}^{\infty} + \frac{(iu)}{2\pi}\int_{-\infty}^{\infty} \frac{\partial w}{\partial x}(x,t)e^{-iux}\,dx \\
&\quad = \frac{iu}{2\pi}\int_{-\infty}^{\infty} \frac{\partial w}{\partial x}(x,t)e^{-iux}\,dx = \frac{iu}{2\pi}\left[w(x,t)e^{-iux}\right]_{-\infty}^{\infty} \\
&\qquad + \frac{(iu)^2}{2\pi}\int_{-\infty}^{\infty} w(x,t)e^{-iux}\,dx = -u^2 W(u,t).
\end{aligned} \tag{36}$$

For (32) to be valid, the Fourier transform of $\partial^2 w/\partial x^2$ must equal the Fourier transform of $v^{-2}\partial^2 w/\partial t^2$, and therefore

$$-u^2 W(u,t) = \frac{1}{v^2}\frac{\partial^2 W}{\partial t^2}(u,t) \tag{37}$$

or, equivalently,

$$\frac{\partial^2 W}{\partial t^2}(u,t) = -u^2 v^2 W(u,t). \tag{38}$$

The effect of taking the Fourier transform of both sides of (32) has been to reduce it to an equation which is, for each fixed u, an ordinary differential equation in W. This is the familiar harmonic equation, with solutions, when written in complex form,

$$W(u,t) = A(u)e^{iuvt} + B(u)e^{-iuvt}. \tag{39}$$

By the same argument that led to (35), we have, since $(\partial w/\partial t)(x,0) = 0$,

$$\frac{\partial W}{\partial t}(u,0) = 0. \tag{40}$$

Applying this condition to (39) we see that $A(u) = B(u)$. Letting $G(u)$ denote the Fourier transform of $g(x)$, the initial displacement prescribed in (33), clearly $W(u,0) = G(u)$ and (39) becomes

$$W(u,t) = \tfrac{1}{2}G(u)[e^{iuvt} + e^{-iuvt}]. \tag{41}$$

By (24) we have

$$w(x,t) = \int_{-\infty}^{\infty} W(u,t)e^{iux}\,dx = \tfrac{1}{2}\int_{-\infty}^{\infty} G(u)[e^{iu(x+vt)} + e^{iu(x-vt)}]\,du. \tag{42}$$

Applying (24) to G, we have

$$g(x) = \int_{-\infty}^{\infty} G(u)e^{iux}\,du \tag{43}$$

and hence we recognize that (42) consists of two terms, one identical to (43) with x replaced by $x + vt$ and the other with x replaced by $x - vt$. Therefore

$$w(x,t) = \tfrac{1}{2}g(x + vt) + \tfrac{1}{2}g(x - vt), \tag{44}$$

which coincides with the solution that would be obtained directly by using d'Alembert's formula (Section 6.2).

Of course, this is a rather complicated way to obtain a familiar result. It should be noted, however, that a person making frequent use of Fourier transforms would skip most of these steps. Such a person knows that, if the transform of $f(x)$ is $F(u)$, then

(45) the transform of $f'(x)$ is $(iu)F(u)$,

(46) the transform of $f''(x)$ is $-u^2F(u)$,

(47) the transform of $f(x - x')$ is $e^{-iux'}F(u)$,

and many other similar properties, and he would have a table of Fourier transforms at his disposal.

• EXERCISES

1. **(a)** Show that, if $k_1^2 \neq k_2^2$, then $\sin k_1x$ and $\sin k_2x$ are orthogonal relative to the inner product

$$(f|g) = \lim_{M\to\infty} \frac{1}{2M}\int_{-M}^{M} f(x)g(x)\,dx.$$

(b) Show how one would obtain formally the coefficients A_{k_n} and B_{k_n} in (1), given the function g.

***2.** Show that $\int_0^\infty \left|\frac{\sin x}{x}\right| dx$ is divergent.

3. Is a uniformly convergent integral necessarily absolutely convergent?

4. **(a)** Does $\int_{-M}^{M} x^3\, dx$ tend to a limit as $M \to \infty$?
(b) Does $\int_{-\infty}^{\infty} x^3\, dx$ exist in the sense of the definition given in this section?

5. Does the one-parameter family of functions $g_M(t) = t^3 + e^{-M}$ converge *uniformly* over the entire real line as $M \to \infty$?

6. Is it legitimate to reverse the order of integration in (10)?

7. Solve the problem posed in (25) and (26), obtaining another solution tending to zero as $y \to -\infty$.

8. Derive (23) and (24) from the Fourier integral theorem.

9. Prove the assertions made in (45), (46), and (47).

10. Show that the solution obtained in (30) can be written directly in terms of the function f as follows:

$$w(x,y) = \frac{1}{\pi} \int_0^\infty \left[\int_{-\infty}^{\infty} e^{-uy} f(t) \cos u(t - x)\, dt \right] du. \tag{48}$$

11. **(a)** Show that, if we take $f(x) = 0$ for $x < 0$ and $f(x) = 1$ for $x > 0$ in (48) (Exercise 10), we obtain nonsense, but if we reverse the order of integration, we obtain convergent integrals.
(b) Hence, determine a continuous function satisfying Laplace's equation in the upper half plane that tends to $f(x)$ as $y \to 0$. What happens to $w(x,y)$ as $y \to 0$ along the y axis?

12. Letting $f(x) = 0$ when $x > 0$, $f(x) = 1$ when $0 < x < a$, and $f(x) = 0$ when $x > a$, where a is positive,
(a) determine the Fourier sine integral representation of $f(x)$.
(b) determine the Fourier cosine integral representation of $f(x)$.
(c) write out the (real) Fourier integral representation of $f(x)$. In other words, what does the Fourier integral theorem become for this function?
(d) determine the (complex) Fourier transform of f.

13. A rod of infinite length (along the positive x axis) is wrapped with heavy insulating tape. The initial temperature distribution is given by $f(x)$, and at $t = 0$ the temperature at $x = 0$ is changed to zero and maintained at zero thereafter. Find the temperature $T(x,t)$ by

(a) superimposing product solutions, in a manner similar to that of the first example given in the text;

(b) transforming the one-dimensional heat equation, in a manner similar to that used in the second example given in the text.

***14.** (For more advanced students.) What connection, if any, do you see between the Fourier transform and the Laplace transform? In particular, is assertion (46) the same for the Laplace transform? (Note: Laplace transforms are not discussed in this book.)

***15.** (For more advanced students.)

(a) For what class of functions could you define an inner product by $(f|g) = \int_{-\infty}^{\infty} f(x)\overline{g(x)}\,dx$?

(b) Formally, the Fourier transform is $(f|g)$ where $g(x) = e^{iux}/2\pi$. Is g in the class of functions determined in part (a)?

6.8 • ALGEBRAIC CONCEPTS IN ANALYSIS

This section is intended as a primer of linear algebra from the viewpoint of the analyst. Although it is self-contained, it will make difficult reading for anyone not already familiar with matrices and linear transformations. The purpose is to show some of the relationships between topics discussed in this book and topics included in courses in linear algebra.

Linear spaces have been discussed earlier in this book. It is important to distinguish, in the following discussion, between those that are finite-dimensional and those that are infinite-dimensional. For example, both types arise in connection with linear differential operators of the form

$$L = \frac{d^n}{dx^n} + a_1(x)\frac{d^{n-1}}{dx^{n-1}} + \cdots + a_{n-1}(x)\frac{d}{dx} + a_n(x)\cdot. \tag{1}$$

Assuming the n functions $a_j(x)$, $j = 1, 2, \cdots, n$ are continuous in an interval (a,b), L transforms the linear space $C^{(n)}(a,b)$, consisting of functions possessing n continuous derivatives, onto the linear space $C(a,b)$ of continuous functions. These spaces have been discussed before in this book, and both are infinite-dimensional. On the other hand, the *null space* of L, i.e., those functions f of class $C^{(n)}(a,b)$ for which $Lf = \theta$, is finite-dimensional: precisely n-dimensional, where n is the order of the differential operator L. (Recall that θ denotes the zero function.)

On the other hand, the null space of the two-dimensional Laplacian operator

$$\frac{\partial^2}{\partial x^2} + \frac{\partial^2}{\partial y^2} \tag{2}$$

is infinite-dimensional. Elements of this null space are specified, not by two arbitrary parameters (as would be the case with a second order ordinary differential operator), but by two more or less arbitrary functions (page 193).

In any finite-dimensional linear space, it is possible to introduce an inner product and hence a norm. It is also possible to introduce a norm directly that is not in any way related to an inner product. This is already obvious in the xy plane, where the norm $(x^2 + y^2)^{1/2}$ is related to an inner product but the norm $|x| + |y|$ is not.

Different norms are useful for different purposes, and some seem hardly useful at all. For instance, consider the four-dimensional linear space spanned by the functions $x \sin x$, $x \cos x$, $\sin x$, $\cos x$. If we like, we may take these four functions to be mutually orthogonal and have unit norm, so that the norm of a function $C_1 x \sin x + C_2 x \cos x + C_3 \sin x + C_4 \cos x$ is by definition

$$(C_1^2 + C_2^2 + C_3^2 + C_4^2)^{1/2}.$$

These functions are defined for all values of x, and the *translate* of any function in this linear space is also a function in the space. By this we mean that, if we replace x by $x + a$, each function is transformed to a new function that is still in the linear class of functions. For example, $x \sin x$ becomes the function $(x + a) \sin (x + a) = (\cos a)x \sin x + (\sin a)x \cos x + (a \cos a) \sin x + (a \sin a) \cos x$, which is a linear combination of the four functions we started with, and therefore is in the space spanned by them. For most purposes of analysis, it is desirable that the norm of all translates of a given function be the same, but the norm of $(x + a) \sin (x + a)$ is seen to be $(1 + a^2)^{1/2}$, which is not only different from the norm of $x \sin x$, but increases without bound as we take larger values of a. Such a norm is therefore of very limited usefulness in analysis.

If a space is equipped with an inner product, the bases that are most convenient are the *orthogonal bases*. We recall that, if e_1, $e_2, \cdots, e_n$ is an orthogonal basis in a linear space, i.e., if $(e_i|e_j) = 0$ whenever $i \neq j$, then the components of an arbitrary element of the space, $v = v_1e_1 + v_2e_2 + \cdots + v_ne_n$, are given by the formula

$$v_j = \frac{(v|e_j)}{(e_j|e_j)}. \tag{3}$$

We have already noted the formal similarity between (3) and the formula for the coefficients of a Fourier series; there is no need to belabor the point here. The main distinction arising when a linear space is infinite-dimensional is that elements of the space are not necessarily linear combinations of the (infinitely many) elements of the "approximating basis" and problems of convergence arise that do not occur in the finite-dimensional case.

An operator T is said to be *linear* if it has the property $T(\alpha f + \beta g) = \alpha T(f) + \beta T(g)$. The differential operators discussed in this book are examples. The term *linear transformation* is more commonly used in texts on modern algebra. In analysis, it is rather uncommon for the domain and the range of the operator to be the same. For example, (1) is a linear operator whose domain is $C^{(n)}(a,b)$ and whose range is $C(a,b)$. In some very elementary books on modern algebra, the only linear transformations discussed are those having the same domain and range. One refers to them simply as *linear transformations of the space.* Examples can be concocted from analysis. For example, the ordinary derivative operator defines a linear transformation of the four-dimensional space spanned by $x \sin x$, $x \cos x$, $\sin x$, and $\cos x$.

Linear transformations of a finite-dimensional space can be represented by matrices, relative to a basis in the space. There are two conventions, depending on whether one represents vectors by rows or columns. The usual convention used by analysts and physicists is to take columns. If the basis is $e_1, e_2, \cdots, e_n$, we take the jth *column* of the matrix to be the coefficients of the transform of the jth base vector. For example, the operation of differentiation, in the four-dimensional space just mentioned, is represented relative to the basis $x \sin x$, $x \cos x$, $\sin x$, and $\cos x$ by the matrix

$$\begin{pmatrix} 0 & -1 & 0 & 0 \\ 1 & 0 & 0 & 0 \\ 1 & 0 & 0 & -1 \\ 0 & 1 & 1 & 0 \end{pmatrix}. \tag{4}$$

For instance, the second *column* reflects the fact that the derivative of the second base element, $x \cos x$, is $(-1)x \sin x + 0(x \cos x) + 0(\sin x) + 1(\cos x)$.

In general, if t_{ij} denotes the entry in the ith row and jth column, we have

$$Te_j = t_{1j}e_1 + t_{2j}e_2 + \cdots + t_{nj}e_n \tag{5}$$

and by the linearity of the operator T, we can determine Tv for any element v if we know Te_j for every base element e_j. Indeed, if $w = Tv$, the ith component of w is given by

$$w_i = \sum_{j=1}^{n} t_{ij}v_j. \tag{6}$$

The reader familiar with matrix multiplication will recognize this as the usual rule for multiplying a square matrix (t_{ij}) by a matrix (v_j) having but a single column.

For example, if we wish to differentiate the function $3x \sin x + 2x \cos x - \sin x + 4 \cos x$, we can do so using (4) by forming the matrix product

$$\begin{pmatrix} 0 & -1 & 0 & 0 \\ 1 & 0 & 0 & 0 \\ 1 & 0 & 0 & -1 \\ 0 & 1 & 1 & 0 \end{pmatrix} \begin{pmatrix} 3 \\ 2 \\ -1 \\ 4 \end{pmatrix} = \begin{pmatrix} -2 \\ 3 \\ -1 \\ 1 \end{pmatrix}$$

and we see that the answer is $-2x \sin x + 3x \cos x - \sin x + \cos x$.

The *identity transformation*, denoted here by I, is defined by the requirement $Iv = v$ for every element v in the linear space. If the space is four-dimensional, I is represented by the matrix

$$\begin{pmatrix} 1 & 0 & 0 & 0 \\ 0 & 1 & 0 & 0 \\ 0 & 0 & 1 & 0 \\ 0 & 0 & 0 & 1 \end{pmatrix} \tag{7}$$

no matter what basis we choose. In general, the identity transformation in an n-dimensional linear space is represented by the *identity matrix* (δ_{ij}), where $\delta_{ij} = 0$ if $i \neq j$ and $\delta_{jj} = 1$ $(j = 1, 2, \cdots, n)$. The *zero transformation* transforms every element into the zero element θ, and is represented by a matrix with only zero entries.

If T is a linear transformation and c is a number, cT is the transformation which maps each element v into cTv, and is represented by the matrix (ct_{ij}) obtained by multiplying every entry in the matrix representing T by the number c.

If S and T are linear transformations of the same linear space, their *sum* $U = S + T$ is defined by the requirement $Uv = Sv + Tv$ for every v in the linear space. It is represented related to any fixed basis by the matrix (u_{ij}) obtained by adding the corresponding entries in the matrices represented by S and T. That is, we have $u_{ij} = s_{ij} + t_{ij}$.

If S and T are linear transformations of the same linear space, their *product* $M = ST$ is defined to be the composite of the two transformations, first T and then S. That is, for every v in the linear space, $Mv = S(Tv)$. It is represented by the matrix (m_{ij}) which is the product of the matrices (s_{ij}) and (t_{ij}) in the same order; this is defined by

$$m_{ij} = \sum_{k=1}^{n} s_{ik}t_{kj}. \tag{8}$$

For example, the product of (4) with itself is

$$\begin{pmatrix} 0 & -1 & 0 & 0 \\ 1 & 0 & 0 & 0 \\ 1 & 0 & 0 & -1 \\ 0 & 1 & 1 & 0 \end{pmatrix}\begin{pmatrix} 0 & -1 & 0 & 0 \\ 1 & 0 & 0 & 0 \\ 1 & 0 & 0 & -1 \\ 0 & 1 & 1 & 0 \end{pmatrix} = \begin{pmatrix} -1 & 0 & 0 & 0 \\ 0 & -1 & 0 & 0 \\ 0 & -2 & -1 & 0 \\ 2 & 0 & 0 & -1 \end{pmatrix} \tag{9}$$

and represents the operation of taking the second derivative in the linear space. For example, according to the first column of this matrix, $\dfrac{d^2}{dx^2}(x \sin x) = -x \sin x + 2 \cos x$.

If the linear space is infinite-dimensional, cT, $S + T$, and ST are defined as above, but the matrix picture no longer applies. (One could introduce the concept of an infinite matrix, but this is of limited usefulness.)

In general, ST and TS are not the same. For example, consider the four-dimensional linear space with basis $1, x, x^2$, and x^3. Let S denote the operator $x\, d/dx$ and let T denote d^2/dx^2. Then ST is the operator $x\, d^3/dx^3$ but TS is the operator $x\, d^3/dx^3 + 2\, d^2/dx^2$. Notice that, relative to the given basis, the matrices representing these operators are

$$S: \begin{pmatrix} 0 & 0 & 0 & 0 \\ 0 & 1 & 0 & 0 \\ 0 & 0 & 2 & 0 \\ 0 & 0 & 0 & 3 \end{pmatrix} \quad T: \begin{pmatrix} 0 & 0 & 2 & 0 \\ 0 & 0 & 0 & 6 \\ 0 & 0 & 0 & 0 \\ 0 & 0 & 0 & 0 \end{pmatrix}$$

$$(10) \qquad ST: \begin{pmatrix} 0 & 0 & 0 & 0 \\ 0 & 0 & 0 & 6 \\ 0 & 0 & 0 & 0 \\ 0 & 0 & 0 & 0 \end{pmatrix} \quad TS: \begin{pmatrix} 0 & 0 & 4 & 0 \\ 0 & 0 & 0 & 18 \\ 0 & 0 & 0 & 0 \\ 0 & 0 & 0 & 0 \end{pmatrix}$$

For example, $ST(x^3) = 6x$, but $TS(x^3) = 18x$, as we can see either by differentiating directly or by looking at the fourth columns of the matrices representing ST and TS. It is amusing to notice that in this special case TS is the same as $ST + 2T$, so we can obtain the fourth of these matrices two different ways: by multiplying the matrices representing T and S, or by adding twice the matrix representing T to the matrix representing ST.

It is particularly important to keep in mind that the matrix representing a given linear operator depends on the choice of basis. For instance, in the linear space spanned by $1, x, x^2, x^3$, and x^4, the Legendre operator $(1 - x^2)\dfrac{d^2}{dx^2} - 2x\dfrac{d}{dx}$ is represented by the matrix

$$(11) \qquad \begin{pmatrix} 0 & 0 & 2 & 0 & 0 \\ 0 & -2 & 0 & 6 & 0 \\ 0 & 0 & -6 & 0 & 12 \\ 0 & 0 & 0 & -12 & 0 \\ 0 & 0 & 0 & 0 & -20 \end{pmatrix}$$

if we take the basis to be 1, x, x^2, x^3, x^4. If, however, we take the basis to be the first five Legendre polynomials P_0, P_1, P_2, P_3, P_4, the matrix representing the operator in this same linear space is

$$\begin{pmatrix} 0 & 0 & 0 & 0 & 0 \\ 0 & -2 & 0 & 0 & 0 \\ 0 & 0 & -6 & 0 & 0 \\ 0 & 0 & 0 & -12 & 0 \\ 0 & 0 & 0 & 0 & -20 \end{pmatrix} \tag{12}$$

since
$$(1 - x^2)\frac{d^2P_n}{dx^2} - 2x\frac{dP_n}{dx} = -n(n+1)P_n.$$

If $ST = TS = I$, we say that S is the *inverse* of the linear transformation T, and write $S = T^{-1}$. If these are linear transformations of a finite-dimensional space, $ST = I$ implies $TS = I$, but if the space is infinite dimensional it is possible to have $ST = I$ without having $TS = I$ (Exercise 2). The definition of *inverse of a matrix* should be clear from this; the inverse of a matrix, if it exists, is the unique matrix which, multiplied with the given matrix (in either order) produces the identity matrix. We shall not discuss here the problem of determining the inverse of a given matrix, beyond mentioning that it is possible if and only if the matrix is square and has nonzero determinant. By direct multiplication, the reader can verify that the inverse of matrix (4) is

$$\begin{pmatrix} 0 & 1 & 0 & 0 \\ -1 & 0 & 0 & 0 \\ 1 & 0 & 0 & 1 \\ 0 & 1 & -1 & 0 \end{pmatrix}. \tag{13}$$

Since (4) represents *differentiation* in the space spanned by $x \sin x$, $x \cos x$, $\sin x$, and $\cos x$, (13) represents *integration* in the same space. Thus, the first column reflects the fact that an indefinite integral of $x \sin x$ is $-x \cos x + \sin x$.

This example shows, in rudimentary form, that it is possible to find an inverse for a differential operator, provided the domain of the operator is suitably restricted. In Section 2.5 we did this for a second order differential operator acting on an infinite-dimensional linear space. The space was carefully selected so it would contain no nontrivial functions f for which $Lf = \theta$; otherwise no inverse could possibly be found. The inverse of the differential operator turned out to be an integral operator, with $L^{-1}h$ given by

$$\int_a^b G(x,y)h(y)\,dy. \tag{14}$$

The similarity between (14) and (6) is apparent; we may regard (6) as a finite-dimensional analog of (14). Note, however, that in the discussion of differential operators, we were concerned with transformations from one linear space to another, rather than transformations of a single linear space.

We shall now discuss linear transformations of a finite-dimensional space equipped with an inner product. The best reference for this material is Halmos, *Finite-dimensional Vector Spaces* (Van Nostrand, 1958).

If the basis $e_1, e_2, \cdots, e_n$ in the linear space is *orthonormal*, i.e., $(e_i|e_j) = \delta_{ij}$, then the entries in the matrix representing a linear transformation T are given by the inner product

$$t_{ij} = (Te_j|e_i), \tag{15}$$

as we see by taking the inner product of both sides of (5) with the base vector e_i. In the following discussion we assume that *all matrices are relative to an orthonormal basis.*

If an operator T^* is related to T by the relation

$$(Tu|v) = (u|T^*v) \tag{16}$$

for every pair of elements u and v in the linear space, we call T^* the *adjoint* of T. The matrices representing T and T^* are seen from (15) to be related to each other in a simple manner:

$$t^*_{ij} = \overline{t_{ji}}. \tag{17}$$

In words, the entry in the ith row and jth column of the matrix representing T^* is the complex conjugate of the entry in the jth row and ith column of the matrix representing T. More briefly, we say that one matrix is the *conjugate* of the *transpose* of the other matrix.

It is not difficult to see that (17) implies (16), provided that the basis is orthonormal, and therefore every operator T has a unique adjoint T^*. (This assertion is not valid for some infinite-dimensional spaces.)

We shall discuss three classes of operators. An operator is said to be *self-adjoint* if $T = T^*$. An operator is said to be *unitary* if $T^*T = TT^* = I$. An operator is called a *projection* if it is self-adjoint and idempotent, i.e., if $T = T^*$ and $T^2 = I$. We shall take up these three types in turn.

If T is self-adjoint, we have

$$(Tu|v) = (u|Tv) \tag{18}$$

for all pairs u, v in the linear space. In this case (17) becomes

$$t_{ij} = \overline{t_{ji}} \tag{19}$$

i.e., reflecting the matrix across its principal diagonal $i = j$ has the same effect as taking the complex conjugate. If the linear space is *real*, we can ignore the conjugation in (19) and the matrix is *symmetric*. The theory of self-adjoint operators therefore includes, as a special case, the study of symmetric matrices.

We have already mentioned self-adjoint differential operators, and the reader will notice the obvious analogy between (19) and the property $G(x,y) = \overline{G(y,x)}$ for Green's functions. Self-adjoint operators on a finite-dimensional space arise most notably in connection with the vibrations of a linear system having n degrees of freedom, where their eigenvectors are related to the normal modes of vibration in the same way that the eigenfunctions of a self-adjoint differential operator are related to the normal modes of a vibrating string or membrane. For further details, see Slater and Frank, *Mechanics* (McGraw-Hill, 1947), pages 127-135.

As a rather trivial example of such an operator, consider the two-dimensional complex linear space spanned by the functions $\sin x$ and $\cos x$. Letting the inner product be $(f|g) = \int_0^{2\pi} f(x)\overline{g(x)}\, dx$, the functions $(1/\sqrt{\pi}) \sin x$ and $(1/\sqrt{\pi}) \cos x$ provide an orthonormal basis for the space, and the operator $i(d/dx)$ is self-adjoint. It is represented, relative to this basis, by the matrix

$$\begin{pmatrix} 0 & -i \\ i & 0 \end{pmatrix} \tag{20}$$

which clearly possesses property (19).

A fundamental theorem concerning such operators is the following.

Theorem 1. *If T is a self-adjoint linear transformation in a finite-dimensional linear space, there exists an orthonormal basis for the space consisting entirely of eigenvectors for T, and the corresponding eigenvalues are necessarily real numbers.*

This theorem is equally valid for both real linear spaces and complex linear spaces. An *eigenvector* is, of course, a nonzero element u with the property $Tu = cu$ for some scalar c.

If we take the basis in the space to be the basis described by this theorem, the matrix representing T will be *diagonal*, i.e. we will have $t_{ij} = 0$ whenever $i \neq j$.

In the example given above, it is clear that the functions $\sin x$ and $\cos x$ are not eigenvectors for $i(d/dx)$, so (20) is not diagonal. However, the functions $e^{ix} = \cos x + i \sin x$ and $e^{-ix} = \cos x - i \sin x$ are eigenvectors for this operator, and when suitably normalized they provide an orthonormal basis. Relative to this basis, the matrix representing $i(d/dx)$ is

$$\begin{pmatrix} -1 & 0 \\ 0 & 1 \end{pmatrix}. \tag{21}$$

As another example, which is almost as trivial but is nevertheless relevant to the theory of Fourier series, we consider the $(n + 1)$-dimensional linear space spanned by the functions 1, $\cos x$, $\cos 2x, \cdots, \cos nx$. Let f be an even function, $f(x) = f(-x)$, having period 2π, which is integrable over the interval $(0,2\pi)$. For any g in the linear space, define Tg to be the function

$$\int_0^{2\pi} f(x - y)g(y)\, dy. \tag{22}$$

In the notation of Section 2.6, this is simply $f * g$, the convolution of f with g. Although f need not be an element of this linear space, $f * g$ is such an element, and therefore T is a linear transformation of the space.

Suitably normalizing the $n + 1$ functions, we obtain an orthonormal basis, and the matrix representing T is then of the form

$$\begin{pmatrix} A_0 & 0 & 0 & \cdots & 0 \\ 0 & A_1 & 0 & \cdots & 0 \\ 0 & 0 & A_2 & \cdots & 0 \\ \cdots & & & & \end{pmatrix} \tag{23}$$

where the entires A_j are, except for a constant factor, the first $n + 1$ coefficients in the Fourier cosine series representing f.

Next, we consider unitary transformations, characterized by the property

$$(Tu|Tv) = (u|v). \tag{24}$$

Indeed, if $T^*T = TT^* = I$, we have $(u|v) = (u|Iv) = (u|T^*Tv) = (Tu|Tv)$, where the last step follows directly from the definition of the adjoint. Because of (24), $\|Tv\| = \|v\|$ for every v in the linear space, and therefore unitary operators are isometric. That is, they preserve distances, and in particular if v is an eigenvector for T, $Tv = cv$, it must follow that $|c| = 1$. Every eigenvalue of a unitary operator must be of the form $c = e^{i\vartheta}$, a complex number having unit modulus.

Theorem 2. *If T is a unitary linear transformation of a finite-dimensional complex linear space, there exists an orthonormal basis for the space consisting entirely of eigenfunctions of T. The corresponding eigenvalues are on the unit circle in the complex plane.*

This theorem is not valid for real linear spaces. For example, rotation about the origin through an angle of 30° in the Euclidean plane is a unitary transformation, but it admits no eigenvectors at all.

Because of (24), the vectors $Te_1, Te_2, \cdots, Te_n$ must be orthonormal, and therefore the columns of the matrix representing T must be mutually orthogonal unit vectors in the space C^n of complex n-tuples. Indeed, the effect of a unitary transformation is to transform an orthonormal basis $e_1, e_2, \cdots, e_n$ into an orthonormal basis $Te_1, Te_2, \cdots, Te_n$. Conversely, given any two orthonormal bases in the space, one can define a unitary transformation which transforms the first basis into the second, and the inverse (which equals the adjoint in this case) transforms the second into the first.

In particular, if S is a self-adjoint operator, and we take the second basis to be that guaranteed by Theorem 1, then relative to the first basis the matrix representing $T^{-1}ST$ will be diagonal. It is in this form that Theorem 1 is sometimes stated: given any self-adjoint matrix S, there is a unitary matrix T such that $T^{-1}ST$ is a diagonal matrix with real entries.

A fundamental notion in analysis is that of *translation invariance.* A linear class of functions is said to be *translation invariant* if every translate of a function of the class is also of the class. If we let T_a denote the translation operator $(T_af)(x) = f(x - a)$, then for every real number a, T_a is a linear transformation of the linear space, provided the linear space is translation invariant. In most practical cases, T_a is a unitary operator.

For example, let the space be spanned by $\sin x$ and $\cos x$. Since $T_a(\sin x) = \sin (x - a) = \cos a \sin x - \sin a \cos x$, and $T_a(\cos x) =$

$\cos(x - a) = \cos a \cos x + \sin a \sin x$, it is clear that this space is translation invariant. Suitably normalizing $\sin x$ and $\cos x$ relative to the usual inner product, we obtain an orthonormal basis, and the matrix representing T_a is

$$\begin{pmatrix} \cos a & \sin a \\ -\sin a & \cos a \end{pmatrix}. \tag{25}$$

If we permit complex scalars, so that this is a two-dimensional complex linear space, another basis for the space is e^{ix} and e^{-ix} and relative to this basis the matrix representing T_a is diagonal:

$$\begin{pmatrix} e^{-ia} & 0 \\ 0 & e^{ia} \end{pmatrix}. \tag{26}$$

Let us take another look at Theorems 1 and 2. Each of these theorems asserts the existence of "plenty" of eigenvectors for the given operator. By "plenty" in this case, we mean there are sufficiently many eigenvectors that any arbitrary vector v can be written as a linear combination of eigenvectors. Theorem 1 can be considered as a finite-dimensional analog of the Sturm-Liouville theorem (for differential operators) or the Hilbert-Schmidt theorem (for integral operators), since each of these theorems asserts essentially that there are plenty of eigenfunctions for an operator that satisfies certain conditions. In this case, "plenty" means that an arbitrary function of the prescribed class can be *approximated* by linear combinations of eigenfunctions, as discussed earlier.

Further insight into these matters is obtained by considering *projections*. The defining property of a projection P is that it is self-adjoint and idempotent. Since P is self-adjoint, Theorem 1 applies without change. Since $P^2 = P$, it is clear that the only possible eigenvalues are 0 and 1, since if v is an eigenvector, $Pv = cv$, and $P^2v = P(cv) = cPv = c^2v$, from which it follows that $c = c^2$ and hence $c = 0$ or $c = 1$. Hence the entries along the main diagonal of a diagonal matrix representing P must be either 0 or 1.

For example, if $e_1, e_2, \cdots, e_n$ is a basis in the linear space, and (notation as in Section 2.2) we define P by $Pv = \text{proj}\ (v{:}e_1)$, the matrix representing P relative to this basis will contain a single 1 in the upper left corner, and all other entries will be zero.

Theorem 3. *If P is a projection in a linear space, every element v in the linear space can be written as the sum of two mutually orthogonal elements u and w, so that $Pu = u$ and $Pw = \theta$.*

Proof: Let $u = Pv$ and let $w = v - u$. Then it is clear that $v = u + w$, $Pu = P^2v = Pv = u$, and $Pw = P(v - u) = Pv - Pu = u - u = \theta$.

More explicitly, P acts like the identity operator in a subspace, and like the zero operator in another subspace, and the two subspaces are mutually orthogonal.

To fix ideas in the following discussion, let us suppose that T is a self-adjoint operator represented, relative to a suitably chosen basis, by the diagonal matrix

$$\begin{pmatrix} 3 & 0 & 0 & 0 \\ 0 & 2 & 0 & 0 \\ 0 & 0 & 3 & 0 \\ 0 & 0 & 0 & -4 \end{pmatrix}.$$

We can write this matrix as a linear combination of matrices representing projections:

$$3\begin{pmatrix} 1 & 0 & 0 & 0 \\ 0 & 0 & 0 & 0 \\ 0 & 0 & 1 & 0 \\ 0 & 0 & 0 & 0 \end{pmatrix} + 2\begin{pmatrix} 0 & 0 & 0 & 0 \\ 0 & 1 & 0 & 0 \\ 0 & 0 & 0 & 0 \\ 0 & 0 & 0 & 0 \end{pmatrix} - 4\begin{pmatrix} 0 & 0 & 0 & 0 \\ 0 & 0 & 0 & 0 \\ 0 & 0 & 0 & 0 \\ 0 & 0 & 0 & 1 \end{pmatrix}.$$

If the basis is e_1, e_2, e_3, e_4, the first of these matrices represents proj $(v{:}e_1,e_3)$, the second represents proj $(v{:}e_2)$, and the third proj $(v{:}e_4)$. The sum of the three matrices is the identity matrix. Letting these three projections be denoted P_1, P_2, and P_3, we have $T = 3P_1 + 2P_2 - 4P_3$, and $I = P_1 + P_2 + P_3$.

Theorem 4. *Any self-adjoint or unitary operator T on a finite-dimensional complex linear space can be written as a linear combination of projections: $T = c_1P_1 + c_2P_2 + \cdots + c_kP_k$, where the projections have the property $P_1 + P_2 + \cdots + P_k = I$ and P_iP_j is the zero operator whenever $i \neq j$.*

The system of projections $P_1, P_2, \cdots, P_k$ is called *a spectral resolution of the identity* relative to the operator T.

In an amusing article written some time ago, Sir Arthur Eddington likened spectral resolutions to the process of sorting out the various kinds of beasts in a zoo [see *The World of Mathematics*, by J. R. Newman (Simon and Schuster, 1956), page 1564]. We will not spoil the reader's enjoyment of this article by discussing it here. It will be noted that each of the P_j's projects all vectors v into the subspace of vectors u having the property $Tu = c_j u$. Thus if v is an arbitrary element of the space,

$$v = Iv = (P_1 + P_2 + \cdots + P_k)v = P_1 v + P_2 v + \cdots + P_k v$$

and we have written v as a sum of vectors each behaving in a particularly nice manner insofar as the operator T is concerned. The "sorting" analogy given by Eddington is not to be taken literally, however; a beast in a zoo *cannot* be partly lion and partly tiger, whereas a vector *can* have nonzero components in more than one of the subspaces described above.

Now let us draw the connection between this discussion and the discussion of Fourier series. Let us consider the linear space consisting of all continuous real-valued functions of period 2π. Let us consider the formal differential operator d^2/dx^2. Clearly this operator cannot be applied to all functions of the prescribed class. There is, however, a subspace of functions for which this operator does make sense. In particular, then, let us define Tf, for functions having period 2π which are twice continuously differentiable, to be d^2f/dx^2. Integrating by parts twice, we readily verify that

$$\int_0^{2\pi} f''(x)g(x)\,dx = \int_0^{2\pi} f(x)g''(x)\,dx \tag{27}$$

if f and g are in this subspace. This may be written as $(Tf|g) = (f|Tg)$, so in some sense T is self-adjoint, but not in the sense usually required in books on modern algebra since Tf does not make sense for every element f in the linear space. Nevertheless, T possesses eigenfunctions. Every nontrivial function in the subspace S_n (Section 2.6) is an eigenfunction corresponding to the eigenvalue $-n^2$. If we let P_n be defined by

$$(P_nf)(x) = \frac{1}{\pi}\int_0^{2} f(t)\cos n(x - t)\,dt \qquad (n > 0) \tag{28}$$

then, as shown in Section 2.6, P_n is the projection onto S_n. Every function $g = P_nf$ has the property $Tg = c_ng$ where $c_n = -n^2$. By analogy with the foregoing, we would expect to have a spectral resolution of the identity

$$I = P_0 + P_1 + P_2 + \cdots \tag{29}$$

containing an infinite number of projections. The main point of the discussion of convergence of Fourier series (Section 3.8) was to show that, if convergence is defined properly, (29) is in some sense valid. In particular, we showed that the projection

$$E_n = P_0 + P_1 + \cdots + P_n \tag{30}$$

is given by an integral operator involving the Dirichlet kernel, and that

$$\|f - E_n f\| \to 0 \tag{31}$$

with increasing n, for any element f in this class (and indeed, in even a larger class than this). The formula corresponding to the one given in Theorem 4,

$$T = -P_1 - 4P_2 - 9P_3 - 16P_4 - \cdots \tag{32}$$

should be viewed with some caution, however; only in very special cases can one differentiate twice, term-by-term, a Fourier series. This is not surprising, of course, since Tf is only defined for certain elements f in the linear space.

• EXERCISES

1. Draw an analogy between the family of all solutions of a non-homogeneous linear differential equation and a plane in Euclidean space that does not pass through the origin.

2. In Hilbert coordinate space, let

$$T(x_1, x_2, x_3, \cdots) = (0, x_1, x_2, x_3, \cdots).$$

Show that there is an operator S such that $ST = I$ but there is no operator S for which $TS = I$.

3. **(a)** Show that, if $TS = ST$ and u is an eigenvector for T, then Su is also an eigenvector for T provided that $Su \neq \theta$.

(b) Hence generalize Theorem 1, to show that if S and T are both self-adjoint and $ST = TS$, a single basis can be found relative to which both S and T are represented by diagonal matrices.

4. Show in detail that the entries A_j in (23) are coefficients in the Fourier cosine series representing f, except for a factor. What is the factor?

*5. Show that, if a linear space of functions is translation invariant, and if every function f in the space is differentiable, then the derivative of the function is also in the space.

6. By using the statement of Exercise 5 and the theory of ordinary differential equations with constant coefficients, describe the possible structure of finite-dimensional real linear spaces of functions that are translation invariant.

7. What can you say a bout the constants c_j in Theorem 4 if T is **(a)** self-adjoint? **(b)** unitary? **(c)** both unitary and self-adjoint?

8. If T is self-adjoint and $T^4 = I$, what can you say about the possible eigenvalues of T?

9. Under what circumstances is (32) valid, and in what sense?

SUPPLEMENTARY EXERCISES

1. Consider the four functions

$$f_1(x) = 1, f_2(x) = x, f_3(x) = e^x, f_4(x) = xe^x.$$

(a) Is this a linearly independent set of functions?
(b) If $g(x)$ is a function in the linear space spanned by these functions, is the function $g(x - 1)$ also in this linear space? How about $g(x - a)$ for an arbitrary (constant) a?

2. What is the inner product $(\sin^2 x|\cos^2 x)$ relative to the interval $0 \leqq x \leqq 2\pi$?

3. Find a function $G(x,t)$ such that the integral equation $f(x) = \int_0^1 G(x,t)p(t)f(t)\,dt$ is equivalent to the boundary-value problem $f''(x) = p(x)f(x), f(0) = f(1) = 0$.

4. Discuss the question of existence of nontrivial solutions of the boundary-value problem in Exercise 3,
(a) if $p(x) > 0$ for $0 \leqq x \leqq 1$,
(b) if $p(x) < 0$ for $0 \leqq x \leqq 1$,
(c) if $p(x)$ is a constant.

5. The Laguerre functions ϕ_n are defined by

$$\phi_n(x) = \frac{1}{n!} e^{-x/2} L_n(x), \qquad (n = 0, 1, 2, \cdots)$$

where $L_n(x) = e^x \dfrac{d^n}{dx^n}(x^n e^{-x})$ is the Laguerre polynomial of degree n. Show that the Laguerre functions are mutually orthogonal relative to the inner product $(f|g) = \int_0^\infty f(x)g(x)\,dx$.

6. Show that the Laguerre polynomials satisfy the equation

$$xy'' + (1 - x)y' + ny = 0.$$

7. The Hermite polynomials H_n are defined by

$$H_n(x) = (-1)^n e^{x^2} \frac{d^n}{dx^n} e^{-x^2}.$$

Show that they satisfy $y'' - 2xy' + 2ny = 0$.

8. The Hermite functions ψ_n are defined by

$$\psi_n(x) = (2^n n! \sqrt{\pi})^{-1/2} H_n(x) e^{-x^2/2}, \qquad (n = 0, 1, 2, \cdots).$$

(See Exercise 7.) Show that these functions are orthogonal relative to the inner product

$$(f|g) = \int_{-\infty}^{\infty} f(x)g(x)\,dx.$$

9. For what values of α does Tchebycheff's equation

$$(1 - x^2)y'' - xy' + \alpha^2 y = 0$$

have polynomial solutions?

10. In some books on differential equations [see, for example, Spiegel, *Applied Differential Equations* (Prentice-Hall, 1958), pages 107-109] the equation of the catenary is discussed. Compare the physical assumptions made in deriving an equation of the catenary with those made in the study of the vibrating string problem. What, for example, is the main difference?

11. Explain how you would find a cubic polynomial $g(x)$ providing the best least-squares approximation to a function $f(x)$ in the interval $1 \leqq x \leqq 3$.

12. **(a)** Find the first four nonvanishing terms in the expansion of the function $f(x) = 0$ when $-1 < x < 0$, $f(x) = x$ when $0 \leqq x < 1$, in a series of Legendre polynomials.

(b) Does the series converge to $f(x)$ for all x in the interval $-1 < x < 1$?

13. Consider a nonhomogeneous bar with ends at $x = 1$ and $x = -1$, and assume a thermal conductivity proportional to $1 - x^2$. Assume the density of the material and the specific heat are constant, and that the lateral surface of the bar and its two ends are insulated. At $t = 0$ the temperature is $T(x,0) = x^3$. Find the temperature $T(x,t)$ thereafter.

14. If $f(x,y,z)$ is harmonic in a closed bounded region not containing the origin, show that

$$\frac{1}{r} f\left(\frac{x}{r^2}, \frac{y}{r^2}, \frac{z}{r^2}\right), \quad (r^2 = x^2 + y^2 + z^2)$$

is also.

15. Find the electrostatic potential $f(r,\theta)$ in a hollow right circular cylinder of inner radius a and outer radius b if $f(a,\theta) = 0, f(b,\theta) = 1$ when $0 < \theta < \pi$, and $f(b,\theta) = -1$ when $\pi < \theta < 2\pi$. (Here, $r^2 = x^2 + y^2$.)

16. Solve the integral equation

$$e^{-x} = \int_0^\infty \cos xt \cdot f(t)\, dt$$

for the function f.

17. An integrable function f has period 2π. State what conclusions can be drawn concerning the Fourier coefficients of f from each of the following properties:

(a) $f(-x) = f(x)$.
(b) $f(-x) = -f(x)$.
(c) $f(\pi - x) = f(x)$.
(d) $f(\pi - x) = -f(x)$.
(e) $f(\pi + x) = f(x)$.
(f) $f\left(\frac{\pi}{2} - x\right) = f(x)$.
(g) $f\left(\frac{\pi}{2} + x\right) = f(x)$.
(h) $f(-x) = f(x) = f\left(\frac{\pi}{2} - x\right)$.
(i) $f(x) = f(2x)$.

18. Show that $\frac{d}{dx}\,[x^n J_n(\alpha x)] = \alpha x^n J_{n-1}(\alpha x)$.

19. Show that $\frac{d}{dx}\,[x^{-n} J_n(\alpha x)] = -\alpha x^{-n} J_{n+1}(\alpha x)$.

20. Find $\frac{d}{dx}\,[x^3 J_0(x^2)]$.

21. Find $\frac{d}{dx}\,[x^5 J_3(4x)]$.

22. Is this formula valid?

$$\frac{d}{dx}\,[x^4 J_1(x^2)] = 2x^5 J_0(x^2) + 3x^3 J_1(x^2).$$

23. Write $\int J_3(x)\,dx$ in terms of J_2 and J_1.

24. Find $\int x^4 J_1(x)\,dx$.

25. Find $\int J_0(\sqrt{x})\,dx$.

26. Verify:
(a) $\int J_0(x)\cos x\,dx = xJ_0(x)\cos x + xJ_1(x)\sin x + C.$
(b) $\int J_0(x)\sin x\,dx = xJ_0(x)\sin x - xJ_1(x)\cos x + C.$
(c) $\int J_1(x)\cos x\,dx = xJ_1(x)\cos x - J_0(x)(x\sin x + \cos x) + C.$
(d) $\int J_1(x)\sin x\,dx = xJ_1(x)\sin x + J_0(x)(x\cos x - \sin x) + C.$

27. Find $\int xJ_1(x)\sin x\,dx$.

28. Find $\int xJ_0(x)\cos x\,dx$.

29. Show that $\int_0^L xJ_0(k_n x)\,dx = \dfrac{L}{k_n} J_1(k_n L)$, provided that

$J_0(k_n L) = 0$.

30. Show, at least formally, that (using the notation of Exercise 29),

$$1 = \sum_{n=1}^{\infty} \frac{2}{k_n L} \frac{J_0(k_n x)}{J_1(k_n L)}, \qquad (-L < x < L).$$

31. Express $J_4(x)$ in terms of $J_0(x)$ and $J_1(x)$.

32. Given that $J_0(4) = -0.397$ and $J_1(4) = -0.066$ (approximately), find $J_2(4)$ and $J_3(4)$.

33. Expand $f(x) = x^2$ over the interval $0 < x < 3$ in a series of Bessel functions $J_0(k_n x)$ where the k_n's are roots of $J_0'(3k) = 0$.

34. Expand $f(x) = x^2$ over the interval $0 < x < 3$ in a series of functions $J_2(k_n x)$ where $J_2(3k_n) = 0, n = 1, 2, 3, \cdots$.

35. Find the steady-state temperature distribution in the solid consisting of half of a right circular cylinder of height h, radius b, whose lower base, curved surface, and vertical plane face are maintained at zero degrees temperature while the upper base has temperature $f(r,\theta)$. In other words, we require a harmonic function whose values, in cylindrical coordinates, are $T(r,\theta,0) = T(r,0,z) = T(r,\pi,z) = T(b,\theta,z) = 0$, $T(r,\theta,h) = f(r,\theta)$.

APPENDIX: FUNCTIONS ON GROUPS

This appendix is intended exclusively for readers having some prior knowledge of modern algebra, including group theory and the theory of rings and ideals. Most of the necessary background material can be found in Birkhoff and MacLane, *A Survey of Modern Algebra* (Macmillan, 1953), referred to as [BM] in the following discussion.

Let G be any finite group, with elements $\alpha_1, \alpha_2, \cdots, \alpha_n$. If the group operation is written in multiplicative form, for any i, j we have $\alpha_i\alpha_j = \alpha_k$ for some k. We tentatively define the *group algebra* of G in the following manner. The elements of the group algebra are formal linear combinations $c_1\alpha_1 + c_2\alpha_2 + \cdots + c_n\alpha_n$, where the c_j's are complex numbers ($j = 1, 2, \cdots, n$). The product of two elements of the group algebra is defined by formally multiplying them and simplifying using the group operation. That is,

$$(1)\qquad (x_1\alpha_1 + \cdots + x_n\alpha_n)(y_1\alpha_1 + \cdots + y_n\alpha_n) = \sum_{i=1}^{n}\sum_{j=1}^{n}(x_iy_j)(\alpha_i\alpha_j).$$

(In [BM, page 240], a more general definition is given, which permits the scalars to be in an arbitrary field; most of the following discussion is not valid, however, unless the scalars are complex numbers.)

Since $\alpha_i\alpha_j = \alpha_k$ for some k, we can write the right side of (1) in the form $\sum_{k=1}^{n} z_k\alpha_k$ where z_k is the sum of those products x_iy_j for which $\alpha_i\alpha_j = \alpha_k$. Our first object is to show that, when rewritten in a suitable form, the product in (1) is essentially the convolution product of two functions.

Let f be a complex-valued function on G with values given by $f(\alpha_j) = x_j$. Similarly, let g be defined by $g(\alpha_j) = y_j$. Then we can write (1) in the form

$$\left(\sum_{i=1}^{n} f(\alpha_i)\alpha_i\right)\left(\sum_{j=1}^{n} g(\alpha_j)\alpha_j\right) = \sum_{k=1}^{n} h(\alpha_k)\alpha_k \tag{2}$$

where $h(\alpha_k)$ is the following sum

$$h(\alpha_k) = \sum f(\alpha_i)g(\alpha_j) \tag{3}$$

taken over those pairs i and j for which $\alpha_i\alpha_j = \alpha_k$. Since $\alpha_i = \alpha_k\alpha_j^{-1}$ this can be written

$$h(\alpha_k) = \sum_{j=1}^{n} f(\alpha_k\alpha_j^{-1})g(\alpha_j). \tag{4}$$

This function h is called the *convolution product* of the functions f and g, in that order, and is denoted $f * g$.

It is easy to see that (4) can also be written in the form

$$h(\alpha_k) = \sum_{j=1}^{n} f(\alpha_j)g(\alpha_j^{-1}\alpha_k). \tag{5}$$

We see from this discussion that there is a natural one-to-one correspondence between elements of the group algebra $x_1\alpha_1 + x_2\alpha_2 + \cdots + x_n\alpha_n$ and functions $f(\alpha_j) = x_j$ $(j = 1, 2, \cdots, n)$, whereby the product of two elements corresponds to the convolution product of the corresponding functions. (This correspondence is also a linear isomorphism.) It is more convenient for our purposes to work with functions, so we shall now drop the tentative definition previously given for the group algebra, and use instead the definition given in the following paragraph. Also, we shall consider only commutative groups and, since it is more suggestive of Fourier series, *we shall change to additive notation*, so that (4) can be written in the form

$$(f * g)(x) = \sum_{y} f(x - y)g(y). \tag{6}$$

At this point the reader should ignore all of the preceding material in this section, which was intended to serve simply as an introduction for the following.

Let G be any finite commutative group, with the group operation written in additive form. [Thus, we write $-y$ instead of y^{-1} and $x - y$ instead of xy^{-1}.] Let $L(G)$ denote the collection of all complex-valued functions with domain G. The sum of two functions and the product of a function by a complex scalar are defined in the usual way, so that $L(G)$ is an n-dimensional complex linear space where n is the order (i.e., number of elements) of G. Define a multiplication $f * g$ in $L(G)$ by (6), where the sum is over all elements y in G. This multiplication can be shown to be associative, i.e., $(f * g) * p = f * (g * p)$ whenever f, g, and p are elements of $L(G)$. Indeed, $L(G)$ satisfies all the axioms of a linear associative algebra [BM, page 239] over the complex field. We call $L(G)$ the *group algebra* of G.

The group algebra $L(G)$ contains a unity element ϵ, which is the function defined by $\epsilon(0) = 1$, $\epsilon(x) = 0$ if $x \neq 0$. (Here we are using 0 as the identity element of G, as well as the number zero.) That $\epsilon * f = f * \epsilon = f$ for every f in $L(G)$ is easily shown by substituting into (6).

Since G is commutative (the term "Abelian" is used instead of "commutative" in many books), it is easy to show that $f * g = g * f$ for every pair of elements f and g in $L(G)$.

EXAMPLE: Let G be the additive group of integers modulo 4 [BM, pages 27 and 131]. The group operation table is

+	0	1	2	3
0	0	1	2	3
1	1	2	3	0
2	2	3	0	1
3	3	0	1	2

Notice that $3 + 2 = 1$, by this addition table, and therefore $1 - 2 = 3$; similarly, $1 - 3 = 2$.

Now suppose that f and g are defined as shown in the following table. It will follow that $f * g$ is the function shown in the last line of the table.

$x =$	0	1	2	3
$f(x) =$	8	-4	3	$2i$
$g(x) =$	3	7	i	$-i$
$(f * g)(x) =$	$24 + 21i$	$42 - 3i$	$-17 + 8i$	$21 - 6i$

For example, if we take $x = 1$ in (6) we obtain

$$
\begin{aligned}
(f * g)(1) &= f(1-0)g(0) + f(1-1)g(1) + f(1-2)g(2) + f(1-3)g(3) \\
&= f(1)g(0) + f(0)g(1) + f(3)g(2) + f(2)g(3) \\
&= -12 + 56 - 2 - 3i \\
&= 42 - 3i.
\end{aligned}
$$

Returning to the general theory, a subspace $L(G)$ is called a *subalgebra* if $f * g$ is in the subspace whenever f and g are in the subspace. For example, the collection of all functions f with the property $f(2) = f(3) = f(4) = 0$ is a one-dimensional subalgebra, as the interested reader can verify for himself.

We are mainly interested in those subspaces which are *ideals*. A subspace of $L(G)$ is an ideal if $f * g$ is in the subspace whenever g is in the subspace, regardless of whether or not f is in the subspace. (Since this algebra has a commutative multiplication, there is no need to distinguish between left and right ideals, as is done in [BM, pages 384–385].

Clearly every ideal is a subalgebra, but not every subalgebra is an ideal. For instance, it is easy to verify that the subalgebra described above is not an ideal.

We are also interested in the subspaces that are *translation invariant*. By this, we mean that for every a in G, f_a is in the subspace whenever f is in the subspace, where f_a is the "translate" of f defined by

$$f_a(x) = f(x - a). \tag{7}$$

Theorem 1. *A subspace of $L(G)$ is an ideal if and only if it is translation invariant.*

Proof: Let ϵ be the unity element of $L(G)$, so that $\epsilon_a(x) = 1$ when $x = a$ and $\epsilon_a(x) = 0$ when $x \neq a$. Then

$$\epsilon_a * g = g_a \tag{8}$$

since $(\epsilon_a * g)(x) = (g * \epsilon_a)(x) = \sum_y g(x - y)\epsilon_a(y) = g(x - a) = g_a(x)$. If a subspace is an ideal containing g, then by the definition of "ideal" it must contain $\epsilon_a * g = g_a$ for every a, and therefore it is translation invariant. Conversely, if the subspace is translation invariant and contains g, then it must contain $g_a = \epsilon_a * g$ for every a and therefore by linearity must contain $f * g$ whenever f is a linear combination of ϵ_a's. However, it is obvious that the n ϵ_a's provide a basis for $L(G)$, and therefore the subspace contains $f * g$ for *every* f in $L(G)$, which proves the subspace is an ideal.

Next we turn to the analysis of special functions on G. The most important of these are the ϵ_a's mentioned above and the *characters* we are about to introduce. We recall from elementary calculus that the function e^x, or more generally any function of the form $f(x) = e^{kx}$, has the properties

$$f(x + y) = f(x)f(y), \qquad \text{and} \qquad f(x) \neq 0 \tag{9}$$

for every x and y. If k is pure imaginary, so $f(x) = e^{ipx}$ for some real p, then for every real number x we have

$$|f(x)| = 1. \tag{10}$$

Conversely, every continuous complex-valued function of a real variable satisfying (9) and (10) must be of the form e^{ipx} for some real p. We see from (9) and (10) that these functions are homomorphisms of the additive group of real numbers into the *circle group*, i.e., the multiplicative group of complex numbers of unit modulus.

If now we interpret x and y as elements of G, we call any complex-valued function on G a *character* of G if it satisfies (9) and (10). Actually, since G has only finitely many elements, (10) is a consequence of (9) in this case. To see this, note that for every fixed x in G, there is a positive integer k such that x added to itself k times gives 0, and from (9) it follows that $f(0) = f(x + \cdots + x) = [f(x)]^k$. Also, since $f(0 + 0) = f(0)f(0)$ and $f(0) \neq 0$ it follows that $f(0) = 1$, and hence $[f(x)]^k = 1$, from which it follows that $|f(x)| = 1$.

Theorem 2. *Every one-dimensional translation-invariant subspace of $L(G)$, or equivalently every one-dimensional ideal in $L(G)$, is spanned by a character of G.*

In other words, such a subspace consists precisely of functions which are scalar multiples of a character.

Proof: Let V be a one-dimensional translation-invariant subspace. Let f be a nonzero element of V. Since V is translation invariant, f_a is in V for every a in G, and since V is one-dimensional this must be a scalar multiple of f. Therefore

$$f(x - a) = C_a f(x) \tag{11}$$

for every x; the scalar C_a depends on a but is independent of x. It is impossible that $f(x) = 0$ for any x, for by (11) this would imply $f(x - a) = 0$ for every a, and that f is identically the zero function. Therefore $f(x) \neq 0$, and in particular $f(0) \neq 0$. Replacing f by an appropriate scalar multiple of f, if necessary, we can assume that

$f(0) = 1$. Then letting $x = 0$ in (11) we have $f(-a) = C_a f(0) = C_a$ so (11) can be written $f(x - a) = f(-a)f(x)$. Letting $a = -y$ for arbitrary y in G, we obtain $f(x + y) = f(y)f(x)$, showing that (9) is satisfied, and, as noted previously, (10) follows from (9) since G is a finite group.

Therefore V contains a character, and since V is one-dimensional every element of V is a scalar multiple of this character.

EXAMPLE: As in the preceding example, let G be the additive group of integers modulo 4. There are precisely four characters for G, given in the following table.

x	$=$	0	1	2	3
$e_1(x)$	$=$	1	1	1	1
$e_2(x)$	$=$	1	i	-1	$-i$
$e_3(x)$	$=$	1	-1	1	-1
$e_4(x)$	$=$	1	$-i$	-1	i

An inner product in $L(G)$ can be defined in the usual manner,

$$(f|g) = \sum_x f(x)\overline{g(x)} \tag{12}$$

where the sum is over all x in G. It will be noted that, relative to this inner product, the four characters given in the preceding example are mutually orthogonal. For example,

$$\begin{aligned}(e_2|e_4) &= (1)(\bar{1}) + (i)(\overline{-i}) + (-1)(\overline{-1}) + (-i)(\bar{i}) \\ &= 1 - 1 + 1 - 1 = 0.\end{aligned}$$

This is no accident, as we see from the following:

Theorem 3. *Distinct characters of G are mutually orthogonal. The norm of every character of G is $\sqrt{n}$, where n is the order of G.*

Proof: If f is a character of G, then $|f(x)| = 1$ for every x in G, so $(f|f) = \sum_x f(x)\overline{f(x)} = \sum_x |f(x)|^2 = n$, hence $\|f\| = (f|f)^{1/2} = n^{1/2}$.

If f and g are distinct characters of G, then letting $y = x + a$, we have

$$\begin{aligned}(f|g) &= \sum_x f(x)\overline{g(x)} = \sum_y f(y - a)\overline{g(y - a)} \\ &= \sum_y f(y)f^{-1}(a)\overline{g(y)g^{-1}(a)} \\ &= f^{-1}(a)g(a) \sum_y f(y)\overline{g(y)} \\ &= f^{-1}(a)g(a)(f|g)\end{aligned}$$

where we have made strong use of the relation $\bar{z} = z^{-1}$, valid whenever $|z| = 1$. If $f \neq g$, then $f(a) \neq g(a)$ for some a, and $f^{-1}(a)g(a) \neq 1$, so the above calculation implies that $(f|g) = 0$. [Compare Exercise 6.]

We have not, so far, proved that characters of a finite group G necessarily exist. Obviously there is one character that must exist, for any group G, namely the function identically equal to unity. The following theorem shows that $L(G)$ possesses plenty of characters.

Theorem 4. *If G is a commutative group of order n, there exist n distinct characters of G.*

Proof: As x runs through G, $y = x + a$ (for fixed a in G) also runs through G, so $(f|g) = \sum_x f(x)\overline{g(x)} = \sum_y f(y - a)\overline{g(y - a)} = \sum_y f_a(y)\overline{g_a(y)} = (f_a|g_a)$. For each fixed a, let U_a be the operator on $L(G)$ defined by $U_a f = f_a$. It follows that

$$(U_a f|U_a g) = (f|g), \tag{13}$$

so U_a is a unitary operator. It is easy to verify that $U_a U_b = U_{a+b} = U_{b+a} = U_b U_a$, so the n operators U_a commute with each other. By a standard theorem of linear algebra (see the reference by Halmos referred to earlier) there exists a basis $e_1, e_2, \cdots, e_n$ for $L(G)$ each of whose elements is an eigenvector for every one of the unitary operators U_a. Thus, for every a, $U_a e_j$ is a scalar multiple of e_j, and hence each of these e_j's spans a one-dimensional translation-invariant subspace of $L(G)$. It follows from Theorem 2 that, when suitably normalized [by requiring $e_j(0) = 1$], we obtain n distinct characters of G.

Theorem 5. *Every complex-valued function f with domain G can be expanded in a "Fourier series"*

$$f(x) = C_1 e_1(x) + C_2 e_2(x) + \cdots + C_n e_n(x) \tag{14}$$

where the expansion coefficients are given by

$$C_j = \frac{1}{n} \sum_x f(x)\overline{e_j(x)}. \tag{15}$$

Proof: This follows directly from Theorems 3 and 4; the characters form an orthogonal basis for $L(G)$.

Theorem 6. *The projection of any f in $L(G)$ in the direction of e_j is given by*

$$\text{proj}\,(f:e_j) = C_j e_j = \frac{1}{n} e_j * f. \tag{16}$$

Proof:

$$\left(\frac{1}{n} e_j * f\right)(x) = \frac{1}{n} \sum_y e_j(x - y) f(y) = \frac{1}{n} \sum_y e_j(x)\overline{e_j(y)} f(y)$$

$$= \frac{1}{n} [\sum_y f(y)\overline{e_j(y)}] e_j(x) = C_j e_j(x).$$

If the reader does not, by now, see what we are up to, he should review Section 2.6. We see that we have, for finite commutative groups, a miniature version of the theory of Fourier series. We even have something corresponding to the Fourier integral theorem:

Theorem 7. *For every x in G, and every complex-valued function f defined on G, the following formula is valid:*

$$\frac{1}{n} \sum_j \sum_t f(t) e_j(x - t) = f(x). \tag{17}$$

Proof: Combine (14) and (16) as follows:

$$f(x) = \sum_j C_j e_j(x) = \sum_j \frac{1}{n} (e_j * f)(x) = \frac{1}{n} \sum_j e_j(x - t) f(t).$$

As a simple application of (17), let us take f to be the function ϵ_y, i.e., the function which takes the value unity at y and zero elsewhere. On the left side, every term but one in the sum over t will vanish, and we obtain $(1/n) \sum_j e_j(x - y) = \epsilon_y(x)$. This immediately gives the curious formula

$$\sum_j e_j(x)\overline{e_j(y)} = \begin{cases} n \text{ if } x = y \\ 0 \text{ if } x \neq y. \end{cases} \tag{18}$$

This is an orthogonality relation in reverse. Before investigating it further, let us study an *example.*

As before, let G be the additive group of integers modulo 4. Since G has four elements, it follows from Theorem 4 that it has four characters. A table showing these characters was given on page 352. If we read *down* rather than *across* the table, we can consider the elements x in G to be functions of the characters e_j, rather than the reverse, and we readily see that they are orthogonal. Thus, (18) makes sense from the viewpoint of orthogonality relations if in $e_j(x)$

we take x to be *fixed* and let the e_j *vary* and thus obtain for each fixed x a *complex-valued function whose domain is the set of characters.*

Investigating this matter further, we find a very interesting fact: *there is complete duality between the group G and the set of characters of G.* Indeed, if we look at the table of characters for the additive group of integers modulo 4, taking the product of two characters to be the ordinary (not the convolution) product, we find that they form a group. Obviously, e_1 plays the rôle of the identity in this group, and $e_2e_3 = e_4$, etc. This is a multiplicative group, rather than an additive group, but that is not of fundamental significance. (It happens to be isomorphic to the group G itself, and this is true for every finite group G, but we prefer not to emphasize this because for infinite groups this is not always the case.) The *characters* of this group can be identified with the original group G. In other words, for fixed e_j and variable x, $e_j(x)$ gives a character of G, but for fixed x and variable e_j we obtain a character of the group of characters! With this example in mind, the reader should have no difficulty proving the following theorem.

Theorem 8. *The characters of G, under ordinary multiplication, form a group $\hat{G}$, called the character group of G. The character group of $\hat{G}$ is isomorphic to G in a natural manner: for every fixed x in G, $e_j(x)$ is a function of the characters e_j which is a character of $\hat{G}$, and every character of $\hat{G}$ is related to an element of G in this manner.*

Letting $L(G)$ denote, as before, the group algebra of G, and letting $L(\hat{G})$ denote the n-dimensional complex linear space of functions on $\hat{G}$, made into an algebra by multiplying the function in the ordinary manner (not by convolution, as in $L(G)$), we obtain an isomorphism of $L(G)$ onto $L(\hat{G})$, called the *Fourier transform*, by associating to each f in $L(G)$ the function $\hat{f}$ on $L(\hat{G})$ defined by

$$\hat{f}(e_j) = \sum_x f(x)\overline{e_j(x)}. \tag{19}$$

Given the transform $\hat{f}$, one can recover f by the formula

$$f(x) = \frac{1}{n} \sum_j \hat{f}(e_j)e_j(x). \tag{20}$$

The reader will notice that these formulas would look more like those given in the section on Fourier transforms if the factor $1/n$ appeared in (19) rather than (20). We put it in (20) in order to ensure the validity of the "convolution theorem":

$$h(x) = \sum_y f(x-y)g(y) \quad \text{if and only if} \quad \hat{h}(e_j) = \hat{f}(e_j)\hat{g}(e_j), \tag{21}$$

whose verification is a necessary step in proving the isomorphism between $L(G)$ and $L(\hat{G})$.

We leave computational details to the exercises. The remainder of this section will be devoted to somewhat deeper ideas.

First of all, we note that the isomorphism mentioned above renders the ideal structure of $L(G)$ rather transparent. The minimal ideals (in this case the one-dimensional ideals) are in one-to-one correspondence with the characters e_j, the ideal corresponding to e_j consisting of those functions f for which $\hat{f}$ vanishes except at e_j. Indeed, there is nothing at all difficult about answering any question concerning the algebraic structure of $L(G)$; we answer the question by looking at $L(\hat{G})$ instead, since the multiplication in $L(\hat{G})$ is ordinary pointwise multiplication of the functions. If we include the trivial subspace and the entire algebra as ideals, there are exactly 2^n ideals, since there are 2^n subsets of $\hat{G}$ (including the empty subset and the entire set) on which all functions $\hat{f}$ in the ideal are required to vanish. One can even classify all the subalgebras of $L(G)$ if one has the patience to work them out. A typical subalgebra that is not an ideal would be obtained, say, by taking all f for which $\hat{f}(e_i) = \hat{f}(e_j)$ for two fixed elements e_i and e_j in $\hat{G}$, since this property is obviously preserved under the formation of linear combinations and by ordinary multiplication.

In linear algebra, an important concept is that of the *dual space* of a linear space [BM, page 197]. Insofar as asking questions about linear functionals on $L(G)$, there is hardly any reason for using $L(\hat{G})$. Every linear functional ϕ on $L(G)$ is obtained by $\phi(f) = (f|g)$ for some suitable choice of g, and obviously for any choice of g this defines a linear functional ϕ. However, in studying algebras the *multiplicative* linear functionals are of special importance. These are linear functionals ϕ which have the additional property $\phi(f * g) = \phi(f)\phi(g)$. In other words, they are homomorphisms of the algebra into the algebra of complex numbers. The study of multiplicative linear functionals on $L(G)$ is rendered easier by considering the isomorphic algebra $L(\hat{G})$ instead. The null-space of such a homomorphism is the entire space if $\phi(f) = 0$ for all f, or it is a maximal ideal (in our special case, this means an $(n - 1)$-dimensional ideal and hence corresponds to those f for which $\hat{f}$ vanishes at one point). The interested reader will find it easy to prove the following theorem:

Theorem 9. *The nontrivial multiplicative linear functionals on $L(G)$ are all of the form $\phi_j(f) = (f|e_j)$ for some fixed character e_j. Let M_j*

denote the kernel of ϕ_j. *Then* M_j *is the maximal ideal consisting of those* f *in* $L(G)$ *for which* $\hat{f}(e_j) = 0$. *We may regard* ϕ_j *as the natural homomorphism of* $L(G)$ *into the factor algebra* $L(G)/M_j$; *this factor algebra may be identified in a natural way with the field of complex numbers.*

Compare [BM, page 377, Theorem 7].

This completes the purely algebraic discussion. We conclude by reviewing the connections between these theorems and the analysis in preceding sections.

The connection is afforded by the fact that the system of real numbers, under addition, is a commutative group, and so also is the system of real numbers modulo 2π. Any periodic function of period 2π can be regarded as a function on the latter group.

If the group is either the real line or the reals modulo 2π, it is impossible to develop a satisfactory theory for the set of *all* functions on the group. In this book we have restricted our attention almost entirely to functions integrable over the group in the simplest sense: they are bounded, and both the function $f(x)$ and its absolute value $|f(x)|$ can be integrated over the group. Let $L(G)$ denote this class of functions, when G is either of these two groups.

For pedagogical reasons, we required throughout most of this book that the functions be real-valued, and therefore we did not use the complex form of the Fourier series. The reader of this section will already see that matters are actually somewhat simpler when the complex form is used.

For example, if G is the reals modulo 2π, the minimal ideals in $L(G)$ are the one-dimensional spaces spanned by the continuous characters e^{inx}. [In Section 2.6 we considered only real-valued functions, and in that case the minimal ideals are (with a single exception) two-dimensional.] The analog of Theorem 1 is valid in this case for finite-dimensional subspaces of $L(G)$; they are ideals if and only if they are translation invariant. Theorem 2 is also valid but it becomes almost vacuous unless we take $L(G)$ to be a *complex* linear space, so that we can admit the functions e^{inx}. Theorem 3 is valid also, except that the norm of every character e^{inx} is $\sqrt{2\pi}$ rather than $\sqrt{n}$. Theorem 4 is valid only in the sense that there are infinitely many elements of G and also infinitely many characters e^{inx} (one for each integer n). One of the main objects of the chapter on Fourier series was to show in what sense Theorem 5 applies to functions on the group of reals modulo 2π, so there is no point in discussing it here, except to note that, in the infinite dimensional case, problems of convergence arise. An analog to Theorem 6 was proved in Section 2.6. There is an analog to Theorem 7 for Fourier series that was not discussed earlier (see Exercise 14).

A fundamental distinction between finite groups and infinite groups is that $\hat{G}$ may be quite different from G if G is infinite. The character group of the reals modulo 2π is defined to be the set of continuous characters (there exist discontinuous characters, but they are all so badly behaved that they are not even Lebesgue integrable). We see that, in this case, $\hat{G}$ is isomorphic to the additive group of integers. Theorem 8 is valid in this case also. The "Fourier transform" of a function f is then a function f defined for integers n by $\hat{f}(n) = \int_0^{2\pi} f(x)e^{-inx}\,dx$, and except for a scalar factor this is simply the nth coefficient in the complex Fourier series expansion of f. (This is not what analysts call the Fourier transform, needless to say.) Theorem 9 also has an obvious analog when G is the group of reals modulo 2π.

Summarizing, we can say that the Fourier series expansion of a function represents the function as a "sum" of functions having especially simple translation properties. From an algebraic viewpoint, the theory of Fourier series is essentially the theory of writing the "group algebra" as a direct sum of minimal ideals.

Now let us turn to the case where G is the additive group of reals, i.e., the entire real line. In this case, it is not difficult to prove that the continuous characters are the functions of the form e^{ikx} where k can be an arbitrary real number. Thus the characters can be identified with the numbers k, and the character group is also isomorphic with the additive group of reals. Nevertheless, it is convenient to distinguish between G and $\hat{G}$, even in applications (electrical engineers refer to G as the "time domain" and $\hat{G}$ as the "frequency domain").

We take $L(G)$ to be functions having absolutely convergent integrals over the entire real line. [In more advanced work, using the Lebesgue integral, it is necessary to distinguish between functions that are integrable and those that are "integrable square," but since we consider only bounded functions we can be sure that, if a function is absolutely integrable, it will also be integrable square.] The convolution product in this case is defined by

$$(f * g)(x) = \int_{-\infty}^{\infty} f(x - y)g(y)\,dy. \tag{22}$$

Since *this* convolution has not been discussed in this book, we shall restrict our remarks to matters that are fairly obvious from the preceding discussion.

In this context Theorem 1 is valid. Theorem 2 is valid only in the following sense: every one-dimensional translation-invariant space of continuous functions on G is spanned by a character of G. [The point here has nothing to do with the word "continuous"; the assertion is equally valid if "continuous" is replaced by "piecewise smooth" or even "Lebesgue measurable."] The point is that $L(G)$ contains no one-dimensional translation-invariant subspaces. The functions e^{ikx} are not in $L(G)$ because $\int_{-\infty}^{\infty} |e^{ikx}|\, dx$ is not a convergent integral.

Indeed, $L(G)$ contains no finite-dimensional nontrivial ideals at all. Every ideal, except the trivial ideal consisting only of the zero function, is infinite-dimensional. An example of an ideal in $L(G)$ would be the collection of all f for which the Fourier transform $\hat{f}$ is zero on a prescribed interval.

Theorems 7 and 8 have reasonable analogs in this case, but we wish especially to draw attention to Theorem 9. Although $L(G)$ has no minimal ideals (every nontrivial ideal contains smaller nontrivial ideals), it does contain maximal ideals. A typical maximal ideal consists of all f for which $\hat{f}(k) = 0$ for some k in $\hat{G}$. For any g in $L(G)$, $\hat{g}(k)$ is (if the Fourier transform is suitably normalized) the image of g under the natural homomorphism of $L(G)$ into the factor algebra $L(G)/M$, where M is the maximal ideal associated with the point k in $\hat{G}$. From an algebraic viewpoint (if such a viewpoint can be said to be algebraic) the theory of the complex Fourier integral amounts to decomposing the "group algebra" as a "direct integral" of these factor algebras, each of which is a replica of the complex plane.

The exercises which follow have been found helpful by some advanced students as part of their preparation for reading more advanced books, such as Loomis, *An Introduction to Abstract Harmonic Analysis* (Van Nostrand, 1953).

• EXERCISES

1. Let G be the additive group of integers modulo 4, and let $f(0) = 3$, $f(1) = i$, $f(2) = 1$, $f(3) = -i$. Let $g(0) = 1$, $g(1) = 2 + i$, $g(2) = -1$, $g(3) = 2 - i$.

(a) Compute $f * g$.

(b) Find the "Fourier series" for f and g (i.e., write them as linear combinations of the characters given on page 352).

(c) Find the Fourier transforms $\hat{f}$ and $\hat{g}$.

(d) Determine $(f|g)$ and $(\hat{f}|\hat{g})$, using (12) for the inner product in $L(G)$ and

$$(\hat{f}|\hat{g}) = \frac{1}{n}\sum_j \hat{f}(e_j)\overline{\hat{g}(e_j)} \tag{23}$$

for the inner product in $L(\hat{G})$.

(e) Letting $h = f * g$, verify that $\hat{h} = \hat{f}\hat{g}$.

2. Determine the characters for the multiplicative group consisting of the nonzero elements of the residue classes of integers modulo 5 [BM, page 131].

3. Write out a table of characters for the "four group," i.e., the group of order 4 each of whose elements has order 2 [BM, page 138].

4. Determine all characters of a cyclic group of order n [BM, page 134].

5. Show that the orthogonality of characters actually follows from the proof given for Theorem 4, and need not have been proved separately.

6. Using only the character properties of the functions e^{inx}, prove that $\int_0^{2\pi} e^{inx}e^{-imx}\,dx = 0$ when $n \neq m$ without actually integrating.

7. Show that, if the inner products in $L(G)$ and $L(\hat{G})$ are defined by (12) and (23) respectively, the Fourier transform preserves inner products: $(f|g) = (\hat{f}|\hat{g})$.

8. The Fourier transform formula (19) can be written $\hat{f}(e_j) = (f|e_j)$, where the inner product is given by (12). The Fourier transform of an element of $L(\hat{G})$ can be defined similarly, using (23) instead. Show that, with these definitions, the Fourier transform of $\hat{f}$ is not exactly the same as f. How is it related to f? How many times would you need to take the Fourier transform of f in order to get back to f?

9. Using (19), find the Fourier transform of the "Dirac delta" on G, i.e., the function ϵ which is the identity in $L(G)$.

10. Prove Theorem 8.

11. Find an analog, for finite groups, of (47), Section 6.7. In other words, how is the Fourier transform of a translated function f_a related to the Fourier transform of f?

12. Let G be a multiplicative group, not necessarily commutative, with multiplication in $L(G)$ defined by (4). In general, $f * g$ will not be the same as $g * f$. Describe those functions f which have the property $f * g = g * f$ for every g (these functions constitute the "center" of $L(G)$). (Hint: there is an intimate connection between these functions and conjugate elements in G; see [BM, page 147].)

13. A *state* in $L(G)$ is a linear functional ϕ on $L(G)$ having the property $\phi(\epsilon) = 1$, where ϵ is the unity element of $L(G)$.

(a) Show that the collection of all states is a convex subset of the dual space of $L(G)$. (In other words, if $a + b = 1$, where $a \geqq 0$ and $b \geqq 0$, then $a\phi_1 + b\phi_2$ is a state if ϕ_1 and ϕ_2 are both states.)

(b) Show that the extreme points in this convex set are precisely the *multiplicative* linear functionals. (A state ϕ is an extreme point if it cannot be decomposed as in (a) for distinct states ϕ_1 and ϕ_2.) Note: These extreme points are sometimes called *pure states.*

14. Write out a reasonable analog to Theorem 7 for the case where G is the additive group of reals modulo 2π.

15. How would you define $L(G)$ if G were the additive group of integers?

16. If G is a finite commutative group, additively written, and f is a complex-valued function on G, the *adjoint* of f is the function f^* defined by $f^*(x) = \overline{f(-x)}$. For a fixed f, let T_f be the linear transformation of $L(G)$ defined by $T_f g = f * g$.

(a) Show that the adjoint of the operator T_f is the operator T_{f^*}.

(b) How is the Fourier transform of f^* related to the Fourier transform of f?

(c) What can you say about self-adjoint functions, i.e., functions f for which $f(-x) = \overline{f(x)}$? (What do their Fourier transforms look like?)

(d) Show that every f in $L(G)$ can be written as a linear combination of self-adjoint elements of $L(G)$.

(e) Using (d), show that Theorem 4 can be proved without using the theory of unitary operators, using the theory of commuting self-adjoint operators instead.

17. If G is the additive group of reals modulo 2π, does $L(G)$ contain a unity element? If you were a physicist, not too concerned with rigor, would your answer to this question be the same?

18. Under what circumstances will an element f in $L(G)$ have a multiplicative inverse, i.e., an element g such that $f * g = \epsilon$, where ϵ is the unity element of $L(G)$? (Here we are again referring to finite commutative groups only.)

19. In this section, we noted that the n translates ϵ_a of the unity element of $L(G)$ provide a basis for $L(G)$. Find a necessary and sufficient condition for the n translates of a function to provide a basis for $L(G)$.

20. Prove that every ideal in $L(G)$ is generated by an idempotent. In other words, there is an element f such that $f * f = f$, and for any g the projection of g into the ideal is $f * g$.

21. Under what conditions can one find "square roots" of an element f in $L(G)$? (That is, a function g such that $g * g = f$.) Is the square root unique? If not, how many square roots will a given function have?

22. (Notation as in Exercise 16.) If e is the identity element of G, and f is an element of $L(G)$, show that

(a) $(f^* * f)(e) = \|f\|^2$.

(b) $(f^* * f)(x) \leqq (f^* * f)(e)$ for every x in G.

23. If the elements of G are $\alpha_1, \alpha_2, \cdots, \alpha_n$, and f is a complex-valued function on G, we can define a matrix (a_{ij}) by $\alpha_{ij} = f(\alpha_i - \alpha_j)$. What is the connection between the operator T_f defined in Exercise 16 and this matrix?

24. A function f is said to be *positive definite* if

$$\sum_{i=1}^{n} \sum_{j=1}^{n} f(\alpha_i - \alpha_j) c_i \bar{c}_j \geqq 0$$

no matter what complex numbers $c_1, \cdots, c_n$ we choose (notation as in the preceding exercise). Show that f is positive definite if and only if $\hat{f}$ is real and nonnegative.

25. A linear functional ϕ on $L(G)$ is said to be *positive definite* if $\phi(f^* * f) \geqq 0$ for every f in $L(G)$. Show that every positive definite linear functional on $L(G)$ can be given by $\phi(f) =$

$\sum_j c_j\hat{f}(e_j)$ where the numbers $c_1, c_2, \cdots, c_n$ are real and non-negative.

26. Let f be a positive definite function on G. [Exercise 25.] Define $[g|h]$ for elements g and h of $L(G)$ by

$$[g|h] = \sum_{i=1}^{n} \sum_{j=1}^{n} f(\alpha_i - \alpha_j)g(\alpha_i)\overline{h(\alpha_j)}.$$

(a) Show that $[g|h]$ satisfies all the requirements of an inner product in $L(G)$, except that it may be possible for $[g|g]$ to be zero without g being the zero element of $L(G)$.
(b) Let N be the set of all g for which $[g|g] = 0$. Show that N is a subspace of $L(G)$.
(c) Show that $[g|h]$ induces an inner product in $L(G)/N$.
(d) For fixed α in G, let $T_\alpha g$ be the function $g(x - \alpha)$. Show that T_α determines a unitary transformation in $L(G)/N$, and that the mapping $\alpha \to T_\alpha$ is a homomorphism of G into a group of unitary operators.
(e) Prove the result of part (d) more directly by considering the transformation induced by T_α in the space $L(S)$ of complex-valued functions on a finite set S, taking S to be those characters of G on which $\hat{f}$ takes nonzero values. [Hint: use Exercise 11.]

27. Deduce from Exercise 26 that every positive definite function on G is of the form $f(\alpha) = [T_\alpha g|g]$ where T_α is a representation of G by unitary operators in a complex linear space equipped with an inner product (denoted here by brackets $[\ |\ ]$) and g is a fixed vector in this space.

28. Prove a result similar to that in Exercise 27, valid for finite noncommutative groups.

29. Write a brief essay on the relationships between positive definite functions on G, positive definite linear functionals on $L(G)$, and states on $L(G)$. [Defined in Exercises 13, 24, and 25.]

ANSWERS AND NOTES

SECTION 1.1 (Page 7)

1. **(a)** 0; **(b)** $\frac{1}{4}$; **(c)** 0, 1; **(d)** 0, $\frac{1}{2}$; **(e)** $\frac{4}{3}$.
2. **(a)** All nonzero x; **(b)** all x; **(c)** all x except $x = -1$ and $x = 2$; **(d)** all x except $x = 0$.
3. **(a)** Same as before; **(b)** all x; **(c)** all x; **(d)** all x; **(e)** all x except $\pm\pi$, $\pm 2\pi$, $\pm 3\pi$, $\cdots$.
4. See Example 5 (*not* Exercise 5). The next value is 3.
5. If you can, the author would like to see it.
6. $f + (-1)g$, defined only when f and g have the same domain of definition.
7. The class of all square matrices.
8. No. The functions x and $1/x$ do not have the same domain of definition.

Note: It is not necessary that the domain of definition be a subset of the real number system. This is the point of Exercise 7. In later chapters, the domain of definition will frequently be a surface or a region in space.

SECTION 1.2 (Page 13)

1. $g(4) = 2$.
2. $f(1) = f(2) = f(3) = 2\sqrt{3}$.
3. (1, 3, 7, 1).
4. **(a)** $B = A\cos\phi$, $C = A\sin\phi$; **(b)** $A = (B^2 + C^2)^{1/2}$, $\phi = \tan^{-1}(C/B)$.
5. $\sqrt{2}\sin(x + 7\pi/4)$.

6. The wrong value of $\tan^{-1}(-1)$ was used. This shows the value of drawing a diagram.
7. **(a)** 5; **(b)** 2; **(c)** approximately 307°.
8. Approximately $8.31 \sin (2t + 11\pi/18)$.
9. **(a)** $\theta = kt + \phi$; **(b)** $A \sin (kt + \phi)$.
10. Yes. B and C are not uniquely determined.
11. **(a)** Use the answers to Exercise 4(a). The D and S scales suffice. Thus, to find $7 \sin 16.6°$ put 90° opposite 7 and read the answer opposite 16.6° (approximately 2).
    (b) An example will suffice. (These remarks refer to K & E rules and many others.) Consider $3 \sin t + 4 \cos t$. First use the T and D scales: put 45° opposite 4 and opposite 3 read 36.9°. From a diagram we see that this is ridiculous, so we read the complementary angle $\phi = 53.1°$ (in red) instead. Without moving the hairline, bring 53.1° (in red) under the hairline, and opposite 90° read $A = 5$.
12. (a), (c), and (e).

Note: The sum of two nonzero sinusoidal functions having different frequencies *cannot* be a sinusoidal function. For example,

$$\sin pt + \sin qt = 2 \cos \left(\frac{p - q}{2}\right) t \sin \left(\frac{p + q}{2}\right) t$$

is not sinusoidal if p and q are positive and $p \neq q$.

Let us look at this trigonometric identity more closely. Suppose that p and q are large and very nearly equal. Then it is tempting to think of the right side as $A \sin kt$ where k is the average of the angular frequencies p and q and the "amplitude" A is slowly varying with angular frequency $(p - q)/2$. Although this is not sinusoidal, it appears "almost" sinusoidal in a small interval. (This explains the so-called *beat phenomenon* of acoustics.)

With a change in notation, this identity can be written

$$A \cos rt \sin st = \frac{A}{2} \sin (s + r)t + \frac{A}{2} \sin (s - r)t.$$

This expression shows what happens if we "modulate" a sinusoidal function having a large angular frequency s by another having a lower frequency r. We obtain two sinusoidal functions, of equal amplitude, with frequencies greater and less than s. This is the *sideband* phenomenon and shows why, if intelligence is to be communicated by the amplitude modulation of a radio wave, one must allot to the radio station a reasonable band of frequencies to either side of the carrier frequency.

It is common engineering practice to abbreviate $A \sin (kt + \phi)$ by $A\underline{/\phi}$ (read "A at an angle of ϕ") whenever the value of k is the same throughout a problem. One can also use complex notation $Ae^{i\phi}$ (where $i^2 = -1$); this is consistent with the equations given in this section, since $Ae^{i\phi} = (A \cos \phi) + i(A \sin \phi) = B + iC$ where B and C, the x and y components of the vector, are now interpreted as the real and imaginary parts of a complex number, and since addition and subtraction of complex numbers is the same as that of vectors in the plane.

Note that the *period* of a sinusoidal function is the reciprocal of its *frequency.*

Cartesian n-space is sometimes called *n-dimensional coordinate space.* The set of all complex sequences of length n, when equipped with an inner product (see Section 2.1) is sometimes called *n-dimensional unitary space.*

SECTION 1.3 (Page 17)

1. (a), (c), (f), and (h).
2. The plane, lines passing through the origin, and the origin itself.
3. The entire space, all planes and lines passing through the origin, and the origin itself.
5. Cosh x, sinh x.
6. No, not unless the function is a constant function.
7. 0.
8. Wave traveling in the positive x-direction with velocity v.
10. 2π.
11. π.
12. Compare the expansions of $f(x)$ and $f(-x)$ and use the fact that the coefficients are unique. Or use Taylor's formula, noticing that the odd derivatives of an even function are zero at $x = 0$.
13. (a), (g), and (h) are odd; the others are even.
14. f is "evenly equivalent" to f_e.
15. $2\pi/s$, where s is the greatest common divisor of m and n.
16. **(a)** Yes; **(b)** no.
17. Sin $(\pi x/2)$.
18. No. For example, $\sin^2 x$ and $\cos^2 x$.
19. Let $f(x) = \sin x + \sin \sqrt{2}x$. If f has positive period T, $2f + f''$ also has period T, and since $2f + f'' = \sin x$ in this case, this would imply $\sin x$ has period T, and hence $f(x) - \sin x = \sin\sqrt{2}x$ also has period T. Hence $T = 2\pi n$ and also $\sqrt{2}T = 2\pi m$ for positive integers m and n. Thus $\sqrt{2} = m/n$, a contradiction, since $\sqrt{2}$ is not a rational number.
20. Use Exercise 9. First show that the set of all periods has a least positive period or is a dense set. Show that the latter implies f is either discontinuous or is a constant function.
21. True. Make use of Exercise 20. Note that, if T is the least positive period of a function, every period of the function is an integral multiple of T.
22. $f(x) = [f(0)] + [f(x) - f(0)]$.

23. Hint: $|x_n + y_n|^2 + |x_n - y_n|^2 = 2[|x_n|^2 + |y_n|^2]$; hence $|x_n + y_n|^2 \leqq 2|x_n|^2 + 2|y_n|^2$. Thus, if $\sum |x_n|^2 < \infty$ and also $\sum |y_n|^2 < \infty$, it follows that $\sum |x_n + y_n|^2 < \infty$.

Note: In general, the sum of two periodic functions is an *almost periodic function.* Almost periodic functions are not discussed in this book. In Exercise 19 we have an example of an almost periodic function that is not periodic.

SECTION 1.4 (Page 23)

1. Here is one method. If $A \sin x + B \cos x = \theta$ then (taking derivatives) $A \cos x - B \sin x = \theta$. Evaluate at $x = 0$ to obtain $A = B = 0$.

2. Evaluate successive derivatives at $x = 0$.

3. The Wronskian is the determinant of the matrix of coefficients of $C_1u(x_0) + C_2v(x_0) = 0$, $C_1u'(x_0) + C_2v'(x_0) = 0$. If the Wronskian were nonzero, this would imply (by the theory of determinants) that $C_1 = C_2 = 0$. In part (b), consider $u(x) = x^3$ and $v(x) = |x^3|$, which are linearly independent (but linearly dependent in any connected interval not containing $x = 0$). In part (c), consider the derivative of $v(x)/u(x)$ [or $u(x)/v(x)$, whichever is defined in the interval].

4. $1, x, \cdots, x^{n-1}$.

5. Yes.

6. No; in the interval $0 \leqq x \leqq 1$, we have $x = |x|$.

7. $(x^2 - x)/2$, $1 - x^2$; $(x + x^2)/2$.

8. $\dfrac{(n-1)!}{r!(n-r-1)!}$, $r = 0, 1, 2, \cdots, n-1$.

9. $(1, 1)$.

10. 3.

11. No.

12. No.

13. No.

14. **(a)** $2n + 1$; **(b)** yes; **(c)** $n = 2, 3, 4, \cdots$, since $\sin^2 x = \frac{1}{2} - \frac{1}{2}\cos 2x$; **(d)** $n = 3, 4, 5, \cdots$, since $\cos^3 x = \frac{1}{4}\cos 3x + \frac{3}{4}\cos x$; **(e)** yes, this is a disguised form of $1 + 2\cos x + \cdots + 2\cos nx$. *Caution*: The H_n in this exercise is not the same as the H_n which will occur in Chapter 2.

15. This can be proved by induction on n. If $n = 1$, then the linear combinations are all of the form D_1u_1. Any point x_1 for which $u_1(x_1) \neq 0$ is a D-set, since for any prescribed C_1 the function D_1u_1 has value C_1 at x_1 if we take $D_1 = C_1/u_1(x_1)$. (Such a point must exist; see Exercise 12.) Now suppose we have proved the theorem for $n = k$. Let us prove it for $n = k + 1$. That is, we wish to prove there is a D-set for a linearly independent set $u_1, u_2, \cdots, u_{k+1}$ and we know already there is a D-set

$x_1, x_2, \cdots, x_k$ for $u_1, u_2, \cdots, u_k$. Let $u_{k+1}(x_i) = C_i$ for $i = 1, 2, \cdots, k$. By the inductive hypothesis there exists a unique linear combination of $u_1, u_2, \cdots, u_k$ that equals $-C_i$ at x_i $(i = 1, 2, \cdots, k)$. Let this linear combination be denoted f. Then the function $f + u_{k+1}$ is zero at each of the points $x_1, x_2, \cdots, x_k$, but is not identically zero (why?) and therefore there is some point, which we shall denote x_{k+1}, at which the value of $f + u_{k+1}$ is not zero. Now complete the proof by showing that $x_1, x_2, \cdots, x_k, x_{k+1}$ is a D-set for $u_1, u_2, \cdots, u_{k+1}$.

Note: According to Exercise 15, we can (for each i) find a function e_i which is a linear combination of the n functions $u_1, u_2, \cdots, u_n$ having the properties $e_i(x_j) = 0$ when $i \neq j$ and $e_i(x_i) = 1$. Then, if $f(x_i) = C_i$ and f is a linear combinatlon of the u's, $\sum_{i=1}^{n} C_i e_i$ is a linear combination having the same values at each point of the D-set, and by uniqueness $f = \sum_{i=1}^{n} C_i e_i$. The e's provide a basis for the linear space spanned by the u's and coordinates of any f in this linear space, relative to this basis, are identical to the coordinates $f(x_1), f(x_2), \cdots, f(x_n)$ obtained by using the D-set. This shows that the use of a D-set to introduce coordinates into a finite-dimensional function space is not basically different from using coordinates relative to a basis.

SECTION 1.5 (Page 28)

1. **(a)** $-1 \leqq x < \infty$; **(c)** all values; **(d)** all x in the domain except $0, 1, 2, 3, \cdots$; **(e)** $\frac{n}{2} + \frac{(k-n)|k-n|}{2}$, $n - 1 \leqq x \leqq n + 1$, $n = 0, 2, 4, \cdots$; **(f)** $3 \leqq x \leqq 4$.
2. (a), (c), (d), (e), (g), (i), and (j).
3. (h).
4. (a), (c), (d), (e), (i), (j).
5. (a), (c), (e), (i), (j).
6. (e), (i).
7. (j) only.
8. (a), (c).
9. Yes.
10. No, $f(0+)$ does not exist.
11. **(a)** No, review the definition of linear combination; **(b)** No, an infinite series is not a linear combination.

Note: Students familiar with the Lebesgue integral will note that a function is integrable, in the sense in which the term is used in this book, if and only if it is bounded and continuous except for a set of points having Lebesgue measure equal to zero.

SECTION 2.1 (Page 36)

1. $\cos^{-1}(\frac{6}{15})$.
2. 7.
4. 2, 2, 2, 2.
5. If $(x|y) = 0$ and $x = Cy$, then $(x|x) = (x|Cy) = \overline{C}(x|y) = 0$ and hence $x = \theta$, contradicting the hypothesis $x \neq \theta$.
6. 0.
7. $-\pi$.
8. 0 (not 2π).
9. $-2\pi i$.
10. 0.
11. $(f|\alpha g + \beta h) = \overline{(\alpha g + \beta h|f)} = \overline{\alpha(g|f)} + \overline{\beta(h|f)} = \bar{\alpha}(f|g) + \bar{\beta}(f|h)$.
12. **(a)** $\sqrt{2}$. **(b)** $\|f - g\| = 0$. Notice that f and g differ at only one point. This shows that the inner product defined in Example 4 does not satisfy the second half of (6) in the strict sense. In particular, this gives an example of two functions f and g that are equal "almost everywhere."
13. Work through the proof of the Schwarz inequality backwards, investigating what happens if the inequalities are equalities.

Notes: If we relax the definition of the inner product (and hence the norm) to permit $(f|f) = 0$ (and hence $\|f\| = 0$) even though f is not identically zero, we have what is called a *pseudo-inner product.* The Schwarz inequality is still valid for pseudo-inner products (its proof made no use of the more stringent requirement). Exercise 13 is not valid for pseudo-inner products unless one agrees consistently to understand "equal almost everywhere" whenever one reads "equal."

Students familiar with the Lebesgue theory will observe that "almost everywhere" here is not precisely the same as "almost everywhere" in the Lebesgue sense. However, for the classes of functions considered in this book, the two notions are precisely the same, in the sense that $f = g$ almost everywhere means the same thing according to either definition (because of the extra restrictions we impose on f and g).

SECTION 2.2 (Page 42)

1. $f(x) = \sqrt{2}\sin x + \sin 2x + \sqrt{2}\sin 3x$.
2. Multiply numerator and denominator by $1 - e^{-i\theta}$ and use the identity $2\cos\theta = e^{i\theta} + e^{-i\theta}$.
3. Subtract $\frac{1}{2}$ from both sides of (18) and note that

$$\cos n\theta - \cos (n+1)\theta = 2 \sin \frac{2n+1}{2} \theta \sin \frac{\theta}{2}$$

and $1 - \cos \theta = 2 \sin^2 \frac{\theta}{2}$.

4. Use (18).

5. Use (18).

6. One way is to use the trigonometric identity $\sin \alpha \sin \beta = \frac{1}{2} \cos (\alpha - \beta) - \frac{1}{2} \cos (\alpha + \beta)$ to obtain a sum of cosines, and then use (20) and (21).

7. Begin by equating the imaginary parts of (17), letting $z = e^{i\theta}$.

8. Compute $\sin \theta [\sin \theta + \sin 3\theta + \cdots + \sin (2n-1)\theta]$ term by term using the trigonometric identity given in the answer to Exercise 6. Most of the terms will cancel in pairs.

9. Can be reduced to an earlier exercise using a suitable trigonometric identity.

Notes: The techniques illustrated here can hardly begin to demonstrate the value of using complex-valued functions in order to derive useful identities concerning real-valued functions. For example, suppose we want to write $\cos^3 \theta$ as a linear combination of sinusoidal functions. One can use trigonometric formulas, of course, but the following procedure is quite easy:

$$\cos^3 \theta = \left[\frac{e^{i\theta} + e^{-i\theta}}{2}\right]^3 = \frac{1}{8} [e^{i3\theta} + 3e^{i\theta} + 3e^{-i\theta} + e^{-i3\theta}] = \frac{1}{4} (\cos 3\theta + 3 \cos \theta).$$

Physics students are urged at this point to look at *Mechanics* by Slater and Frank (McGraw-Hill, 1947), especially Chapter 8, Section 3.

SECTION 2.3 (Page 60)

1. $\int_0^{2\pi} \sin mx \sin nx \, dx = 0$ when $n \neq m (n, m = 1, 2, 3, \cdots)$. Note that the trigonometric identity given in the answer to Exercise 6, Section 2.2, makes it easy to find $\int \sin mx \sin nx \, dx$ without the use of tables of integrals.

2. Yes.

3. No; no.

4. Expand $(f + g|f + g) + (f - g|f - g)$ and simplify. (Compare the answer to Exercise 23, Section 1.3.)

5. Interpret $f - g$ and $f + g$ as diagonals of a parallelogram having sides f and g.

6. $\sqrt{3}x/2$.

8. $u_1(x) = 1/\sqrt{2}$, $u_2(x) = \sqrt{3/2}x$, $u_3(x) = \frac{1}{2}\sqrt{\frac{5}{2}}(3x^2 - 1)$, $u_4(x) = \frac{1}{2}\sqrt{\frac{7}{2}}(5x^3 - 3x)$.

9. **(a)** Yes; **(b)** no; **(c)** yes.

10. $P_4(x) = (35x^4 - 30x^2 + 3)/8$, $P_5(x) = (63x^5 - 70x^3 + 15x)/8$.

11. $\left(\frac{33}{4e} - \frac{3e}{4}\right) + \frac{3}{e}x + \left(\frac{15e}{4} - \frac{105}{4e}\right)x^2$.

12. (c) and (d) are true, the others are false. [For example, in part (e), what if $f = \theta$?]

13. Cos x is orthogonal to all of them.

14. By Riemann's lemma, $\lim_{n\to\infty} \sqrt{\frac{2}{\pi}} \int_0^\pi x \sin nx\, dx = 0$.

15. $\int_0^\pi x^2\, dx = \sum_{n=1}^{\infty} \frac{2\pi}{n^2}$.

16. g.

17. $\alpha_k = \frac{2}{\pi} \int_0^\pi f(x) \sin kx\, dx$.

18. Yes.

19. Hint: $|\alpha_n\bar{\beta}_n| \leq \frac{1}{2}[|\alpha_n|^2 + |\beta_n|^2]$.

20. Yes.

Notes: Students thinking completely in terms of real functions may be somewhat distracted by the continual presence of complex conjugates. A good example of a sequence of complex-valued functions, orthogonal over $(0, 2\pi)$, is $e^{inx} (n = 0, \pm 1, \pm 2, \cdots)$. Notice that, if $\phi_n(x) = e^{inx}$, then $\overline{\phi_n(x)} = e^{-inx}$. One easily sees that $\int_0^{2\pi} e^{inx}e^{-imx}\, dx = 0$ whenever n and m are integers and $n \neq m$.

The subsection on the condition of finality should be compared with a similar discussion given in *Partial Differential Equations in Physics* by Sommerfeld (Academic Press, 1949).

Readers familiar with the Lebesgue theory are advised to look up the *Riesz-Fischer theorem* in any book on real variables, in order to appreciate the remarks made in the last paragraph of this section.

SECTION 2.4 (Page 68)

1. **(a)** Divide by $a_0(x)$ and multiply by $p(x) = e^{\int a_0(x)\, dx}$.
(b) $a_0(x) \neq 0$ is the most obvious restriction.

2. Note that $f'g - g'f = f'(g + \alpha g') - g'(f + \alpha f')$.

3. **(a)** If you don't have a table of integrals, use the identity $\sin \alpha \sin \beta = \frac{1}{2} \cos (\alpha - \beta) - \frac{1}{2} \cos (\alpha + \beta)$; **(b)** the eigenfunctions are the functions $\sin k_i x$ where $k_1, k_2, k_3, \cdots$ are roots of $\tan kL = -k$; **(c)** use Lemma 2.

4. **(a)** Cos $(n\pi x/L)$, $n = 0, 1, 2, \cdots$.

   (b) Sin $\dfrac{2n\pi x}{b-a}$, $n = 1, 2, 3, \cdots$, cos $\dfrac{2n\pi x}{b-a}$, $n = 0, 1, 2, \cdots$.

   (c) sin $(nx/2)$, $n = 1, 3, 5, 7, \cdots$.

5. $uLv - vLu = u[(pv')' + qv] - v[(pu')' + qu]$
   $= u(pv')' - v(pu')' = (puv' - pvu')'$.

6. If $Lu = \theta$ and $Lv = \theta$ then $uLv - vLu = \theta$ and by Exercise 5 it follows that $puv' - pvu'$ is a constant (its derivative is identically zero). *Note:* More generally, if $Lu = \lambda u$, and $Lv = \lambda v$, we have, by the same argument, that $p(uv' - vu')$ is a constant. The statements made in the next exercise can be similarly generalized.

7. **(a)** Use Exercise 6 and the fact that $p(x) \neq 0$.

   (b) If $uv' - vu'$ is nonzero, then (Exercise 3, Section 1.4) u and v are linearly independent and therefore span the two-dimensional linear space of all solutions of $Ly = \theta$. Conversely, if they span this space, the two solutions f and g of $Ly = \theta$ satisfying the conditions $f(a) = 1, f'(a) = 0, g(a) = 0, g'(a) = 1$ are linear combinations of u and v, and hence their Wronskian $[f,g]$ is a scalar multiple of $[u,v]$. Since $[f,g](a) = 1 \neq 0$, it follows that $[u,v](a) \neq 0$, and by part (a) that the Wronskian is nonzero throughout the interval.

   (c) This can be shown by direct calculation, recalling the formula

$$\frac{d}{dx}\int_a^x g(x, t)\, dt = \int_a^x \frac{\partial g}{\partial x}(x, t)\, dt + g(x, x).$$

   (Differentiation under the integral sign is valid here.)

8. Yes, it is possible. For example, let $T = d/dx$, $V = C'[a,b]$ and $W = C[a,b]$. (The answer is *no* if the spaces are finite-dimensional.)

Note: Problems having solutions for only certain values of a parameter are sometimes called *quantum problems.* One says that the solutions of equations such as (1) are *quantized* by the imposition of boundary conditions such as (2). Obviously (1) has solutions for every value of λ, but these solutions satisfy (2) only when $\lambda = n^2$, $n = 1, 2, 3, \cdots$.

SECTION 2.5 (Page 72)

1. **(a)** No; it is discontinuous on the diagonal $x = t$; **(b)** $K(x, x) = 0$ for $a \leqq x \leqq b$.
2. The denominator is constant and the numerator vanishes on the diagonal $x = t$.

3. If they were linearly dependent, *each* of the functions u and v would satisfy *both* boundary conditions, and this would contradict the hypothesis that L is nonsingular.
4. Hint: Notice that f is the sum of the function defined in Exercise 7, Section 2.4, and a scalar multiple of $u(x)$.
8. **(a)** $u(x) = 1$, $v(x) = x$. Another possibility: $u(x) = x$, $v(x) = 1 - x$.
   (b) $K(x, t) = x - t$.
   (c) $f(x) = \int_0^x (x - t)h(t)\,dt$. Compare Exercise 7, Section 2.4.
   (d) $G(x, t) = \begin{cases} x(t-1) & x < t \\ t(x-1) & x > t. \end{cases}$
9. **(b)** $\displaystyle\int_0^1 G(x, t)t\,dt = \int_0^x t(x-1)t\,dt + \int_x^1 x(t-1)t\,dt = \frac{x^3}{6} - \frac{x}{6}.$
10. (a) and (b) are nonsingular, the others are singular.
11. Hint: When written in the canonical form, the equation becomes
$$\frac{d}{dx}\left(e^x \frac{dy}{dx}\right) = 0, \text{ so } p(x) = e^x.$$
12. No.
13. Those of the form $C_1 \cos x + C_2 \sin x$.
14. $\alpha = (b^3 - a^3)/3$, corresponding to scalar multiples of x.

Note: The function $G(x,t)$ referred to in Lemma 2 is sometimes called the *Green's function* for the given problem. The Green's function can be considered as the response of the system due to a unit impulse. For further information, see Section 4.4, *Methods of Applied Mathematics*, by F. B. Hildebrand (Prentice-Hall, 1952).

SECTION 2.6 (Page 84)

1. **(a)** No, only (3) is valid; must replace $n \neq m$ by $n^2 \neq m^2$ in (2); **(b)** $\sin^2 nx + \cos^2 nx = 1$, so the two functions have average value $\frac{1}{2}$ over any interval of periodicity.
2. Mr. McSnoyd remarked, "This is a silly sort of problem. Why should one expect a square root to occur in one expression just because it appears in another?" Mr. McSlide said, "Probably this exercise was suggested by a question asked in class."
3. See the answer to Exercise 8, Section 2.2.
4. Note that this involves changing the order of integration in an iterated integral.
5. **(a)** $-\pi \cos nx$; **(b)** $\psi_n * \psi_n = \dfrac{\sin nx}{\pi} * \dfrac{\sin nx}{\pi} = \dfrac{1}{\pi^2}$ [answer to part (a)] $= -\phi_n$.
6. $\pi(\sin 2x + 2 \sin 3x + 36 \cos 7x)$.

7. $\pi \cos mx$ if $n \geqq m$, 0 if $n < m$.
8. f is orthogonal to H_n but is not orthogonal to H_{n+1}. (This does not imply f is in H_{n+1}.)
9. $$V_n(x) = \frac{1}{\alpha_n} \int_0^{2\pi} f(x-t) \cos^{2n} \frac{t}{2} \, dt.$$
10. $$\alpha_n = 2\pi \left[\frac{1}{2} \cdot \frac{3}{4} \cdot \frac{5}{6} \cdot \ldots \cdot \frac{(2n-1)}{2n}\right].$$
11. Let $K = \max_{a \leqq x \leqq b} |f'(x)|$. Then $|f(x_2) - f(x_1)|$
    $= |f'(c)(x_2 - x_1)| \leqq K|x_2 - x_1|$.
12. Yes. Take $K = 1$ (or any larger K).
13. No.
14. Yes, possibly. See Exercise 12.
15. Let K be the maximum of the absolute values of the slopes of the line segments.
16. One rigorous proof uses uniform continuity and Exercise 15.
17. **(a)** No; **(b)** yes.
18. Show that, if f satisfies the conditions of the theorem, $\|f - V_n\|$ can be made as small as we like by taking n large enough. The linear space is that described in Exercise 17(b).
19. Note that $\|g - V_n\| = \|g - f + f - V_n\| = \|g - f\| + \|f - V_n\|$. Use Exercise 16 to show that $\|g - f\|$ can be made as small as we like, and Exercise 18 to do the same for $\|f - V_n\|$.

Notes: For an alternative derivation of (16), see Exercises 2 and 3, Section 2.2.

From this point on, we make increasing use of *dummy variables*. To explain what this means, we notice that the letter x is used in two different ways in the integral $\int_a^x f(x)\,dx$, both as a dummy variable and as the *current variable* (the variable upper limit, which remains fixed during integration). It is better to write $\int_a^x f(t)\,dt$, especially when dealing with an expression like $e^x \int_0^x e^{-t}\,dt$, since we are then free to bring the e^x inside the integral sign $\int_0^x e^{x-t}\,dt$. Sheer disaster would result if we moved e^x inside the integral without distinguishing between the dummy and current variables in this case. Note carefully the use of the dummy variable t and the current variable x in the formula $(d/dx) \int_a^x f(t)\,dt = f(x)$, which is valid whenever f is a continuous function.

In the theory of Laplace transforms, the convolution product is defined by $\int_0^\infty f(t)g(x-t)\,dt$, and it is in this form that many students first meet the convolution product. In the theory of Fourier transforms, the formula becomes $\int_{-\infty}^{\infty} f(t)g(x-t)\,dt$.

When written in complex form, the theory presented in this section takes a more natural form. Letting $\varepsilon_n(x) = e^{inx}/2\pi$, we have $\varepsilon_n * \varepsilon_m = \theta$

whenever $n \neq m$, and $\varepsilon_n * \varepsilon_n = \varepsilon_n (n = 0, \pm 1, \pm 2, \cdots)$. In more advanced work, the complex functions ε_n are always used in preference to the function pairs ϕ_n, ψ_n.

In the Lebesgue theory, two important classes of functions are $L_1[a,b]$, the class of measurable functions that are integrable over $a \leqq x \leqq b$, and $L_2[a,b]$, the class of measurable functions f that are "square integrable," i.e., for which $|f(x)|^2$ is (Lebesgue) integrable over $a \leqq x \leqq b$. The inner product $\int_a^b f(x)\overline{g(x)}\,dx$ can be defined whenever f and g are of class $L_2[a,b]$, but not always when f and g are of class $L_1[a,b]$ (for example, the integral may not exist if $f(x) = g(x) = 1/\sqrt{x}$.) However, if f and g are of period $(b - a)$, the integral $\int_a^b f(x - t)g(t)\,dt$, which obviously exists for every x when f and g are of class $L_2[a,b]$, can also be shown to exist for *almost every* x when f and g are of class $L_1[a,b]$ (this is not obvious) and defines a function $f * g$ also of class $L_1[a,b]$. Therefore convolution products play a more fundamental role in the study of $L_1[a,b]$ than do inner products.

Mathematicians of my generation did not learn the fundamental importance of convolution products until a relatively late stage. For example, it was not pointed out to us that (17) is a convolution product; moreover, (17) was written in the form

$$h_n(x) = \int_0^{2\pi} f(t) \frac{\sin (n + \frac{1}{2})(t - x)}{2\pi \sin \dfrac{t - x}{2}}\,dt,$$

which is valid since the Dirichlet kernel is an even function; but this tends to disguise its convolution form.

SECTION 3.2 (Page 95)

1. $\frac{2}{3}\pi^2 - 4(\cos x + \frac{1}{4}\cos 2x + \frac{1}{9}\cos 3x + \cdots)$.
2. $0 \leqq x \leqq 2\pi$.
3. $\dfrac{e^{2\pi} - 1}{\pi}\left[\dfrac{1}{2} + \sum_{n=1}^{\infty} \dfrac{\cos nx}{n^2 + 1} - \sum_{n=1}^{\infty} \dfrac{n \sin nx}{n^2 + 1}\right]$.
4. $0 < x < 2\pi$.
5. Discontinuous at every integral multiple of π, including the origin.
6. $\frac{1}{4}\pi[\sin x + \frac{1}{3}\sin 3x + \frac{1}{5}\sin 5x + \cdots]$.
7. Yes, except at integral multiples of π, where it converges to zero.
8. $\frac{1}{2}\pi[1 + (\frac{1}{3} - 1)\cos 2x + (\frac{1}{5} - \frac{1}{3})\cos 4x + (\frac{1}{7} - \frac{1}{5})\cos 6x + \cdots]$.
9. See (46), Section 3.3.
10. **(a)** See (48), Section 3.3; **(b)** yes.
11. **(a)** Yes; **(b)** yes; **(c)** no; **(d)** no; **(e)** no to (a) and (d), yes to the others; **(f)** no to (a) and (d), yes to the others. *Note*: When $n = 2$, $f'(0) = 0$ but $f'(0+)$ and $f'(0-)$ do not exist.

12. No.
13. Yes.
14. **(a)** Yes; **(b)** yes.
15. No; $f(0) = f(2\pi)$.

Notes: Expressions (2) and (3) are formally found by writing $f(x) = A_0/2 + \sum_{n=1}^{\infty} (A_n \cos nx + B_n \sin nx)$, multiplying both sides by $\cos nx$ (alternatively, $\sin nx$) and integrating over $(0, 2\pi)$, as suggested in the introduction to Chapter 2. This procedure does not in any way "prove" that the series thus obtained converges to $f(x)$, although in the days of Fourier this would probably have constituted a satisfactory proof. It is therefore surprising to note that Fourier did not, in fact, follow this procedure when he first gave a "proof" for the validity of these expressions. Instead, he replaced the functions involved by their Taylor's series and equated the coefficients of powers of x, obtaining an infinite system of linear equations in an infinity of unknowns, and then engaged in a complicated analysis. Although Fourier later refers to the "proof" by integration, it appears that he was not familiar with Euler's work; Euler had, in a sense, derived these equations using term-by-term integration at an earlier time.

The entire theory takes a simpler form when complex exponentials are used instead of sines and cosines. Thus we have $\sum_{n=-\infty}^{\infty} C_n e^{inx}$ instead of (1). Such a series is called a *complex Fourier series* if the coefficients C_n can be obtained from an integrable function f by $C_n = \frac{1}{2\pi} \int_0^{2\pi} f(x)e^{-inx}\, dx$. Notice that there is no awkward factor of $\frac{1}{2}$ in the case $n = 0$, as there is with real Fourier series.

It is important that the reader recognize that (2) and (3) are special cases of (14), Section 2.3. Thus $(\cos nx | \cos nx) = \int_0^{2\pi} \cos^2 nx\, dx = \pi$, so (2) is nothing more than $(f| \cos nx)/(\cos nx| \cos nx)$.

The theorems in this section are valid if "integrable" is interpreted to mean "Lebesgue integrable," and Theorem 4 is valid if "piecewise continuous" is replaced by "Lebesgue integrable."

SECTION 3.3 (Page 105)

1. $x^2/4 - \pi x/4 + \pi^2/24$.
2. $x^3/12 - \pi x^2/8 + \pi^2 x/24$.
3. $-\frac{1}{2} \int_0^x \ln |2 \sin t|\, dt$.
4. See Exercise 1, Section 3.2.

5. **(a)** $-\frac{\pi^2}{3} + 2\pi \sum_{n=1}^{\infty} \frac{\sin nx}{n} - 4 \sum_{n=1}^{\infty} \frac{\cos nx}{n^2}.$

   (b) No.

   (c) Discontinuous at integral multiples of 2π, including $x = 0$.

6. (25).

7. $\frac{\pi^2}{6} - 4\left(\frac{\cos 2x}{2^2} + \frac{\cos 4x}{4^2} + \frac{\cos 6x}{6^2} + \cdots\right).$

8. Can use (36) and (20); $0 \leqq x < \pi$.

9. **(a)** $A_n = 0$ when n is odd; **(b)** $A_n = 0$ when n is even; **(c)** $A_n = 0$ when n is not a multiple of 3.

10. It is, when $0 < x < 2\pi$; compare (19).

11. **(a)** $S_n = \frac{1}{2} \cot \frac{x}{2} - \frac{1}{2} \frac{\cos (2n + 1)x/2}{\sin (x/2)}.$

    (b) Use (8) to find the sum in closed form.

    (c) $\frac{1}{2} \cot (x/2)$, $0 < x < 2\pi$.

12. **(c)** If $I = 0$, the imaginary part of I is zero.

SECTION 3.4 (Page 112)

1. No.

3. **(a)** At $x = 0$, (46), Section 3.3, converges to $2\pi^2$ according to Theorem 4, Section 3.2; **(b)** evaluate at $x = 0$.

4. The periodic extension of x^2, $0 < x < 2\pi$, is discontinuous at $x = 0$, but the odd periodic extension of x^2, $0 < x < \pi$, is continuous at $x = 0$.

5. Same as (3) and (4) for odd n, but $A_n = B_n = 0$ when n is even.

6. See (48), Section 3.3.

7. Except for the factor $\frac{1}{2}$, the same as (20), Section 3.3.

9. $B_n = \frac{2(1 - \cos k_n\pi)}{k_n(\pi + \cos^2 k_n\pi)} = \frac{2(1 - \cos k_n\pi)}{k_n\pi - \frac{1}{2} \sin 2k_n\pi}.$

10. No.

SECTION 3.5 (Page 118)

1. Consider the x components of n unit vectors.

2. **(a)** Adjust the magnitude and location of the discontinuity in f_2 to be the same as that of f; **(b)** if the function has n discontinuities, introduce n sawtooth functions.

SECTION 3.6 (Page 124)

1. **(a)** $f(x)$; **(b)** $\dfrac{f(x+) + f(x-)}{2}$; **(c)** no.
2. $S_n(x) = 1$, $n = 0, 1, 2, \cdots$.
3. 0, 1, -1.
4. **(a)** No; **(b)** no; consider $f(x) = 1$, $x > 0$; $f(x) = -1$, $x < 0$; $f(x) = 0$ at $x = 0$.
6. Also note the comments on this theorem in Section 3.11.

SECTION 3.7 (Page 139)

1. **(a)** $\|f_{nm} - \theta\| = \|f\| = 1/\sqrt{n} \to 0$; **(b)** $N(f_{nm}) = 1$; **(c)** keep in mind that "divergent" means nothing more than "not convergent." For every x, the sequence $f_{nm}(x)$ contains infinitely many 0's and 1's, hence is divergent.
2. $-x \ln 2$.
3. $\sum_{n=1}^{\infty} 1/n$ diverges, by comparison with
$$1 + \tfrac{1}{2} + \tfrac{1}{4} + \tfrac{1}{4} + \tfrac{1}{8} + \tfrac{1}{8} + \tfrac{1}{8} + \tfrac{1}{8} + \tfrac{1}{16} + \cdots$$
$$= 1 + \tfrac{1}{2} + 2(\tfrac{1}{4}) + 4(\tfrac{1}{8}) + 8(\tfrac{1}{16}) + \cdots.$$
4. **(a)** $f(x) = x$, $0 \leqq x < 1$; $f(x) = 0$, $x = 1$.

   (b) $\dfrac{n}{n+1}(1 + n)^{-1/n}$.

   (c) No.
5. **(a)** Yes; **(b)** yes.
6. **(a)** $\sqrt{n}$; **(b)** $\sqrt{1/n}$; **(d)** yes.
7. No. See Exercise 5.
9. **(a)** Yes; **(b)** not a sphere but a cube; **(c)** yes.

Note: It is not necessary to use the results of Section 2.6 in the proofs of these theorems. The broken-line functions described in the proof of Theorem 11 are piecewise smooth and continuous, and hence by the remarks at the beginning of Section 3.8 must have uniformly convergent Fourier expansions.

SECTION 3.8 (Page 149)

3. Hint: Apply the Schwarz inequality to a typical partial sum to show that the series is uniformly convergent.
5. No.

7. (a) Hint: $|D_n(x)| \leqq \dfrac{n}{\pi} + \dfrac{1}{2\pi}$ for every x.

(b) Hint: $|\sin (n + \frac{1}{2})x| \leqq 1$, so $|D_n(x)| \leqq 1/2x$ in this interval.

(d) Hint: When $n > 2$, $1 + 1/2n < 5/4$.

8. Note Theorem 7, Sectlon 3.7, and the *definition* of a Fourier series.

11. The point here is that one cannot neglect the lower limit of integration since $\cos (2n + 1)x$ does not vanish at $x = 0$.

12. $-\pi^3x^2/64 + \pi x^4/96 - 5\pi^5/1536$.

13. Not if it has a nonzero constant term.

14. Theorem 5 is such a theorem. (Caution: This fact was proved earlier and was used in the proof of Theorem 5.)

17. $\dfrac{1}{2\pi}\displaystyle\int_0^{2\pi} f(x - t)g(t)\, dt = \sum_{n=-\infty}^{\infty} c_n\gamma_n e^{inx}$. (For the definition of complex Fourier series, see the notes to Section 3.2.)

18. (a) Hint: A piecewise continuous function can have at most a finite number of discontinuities in an interval of finite length; (b) Note: This is not valid if "piecewise continuous" is replaced by "integrable."

19. $\int_0^{2\pi} [f(x)]^2\, dx = 2 \int_0^{\pi} x^2\, dx$. A common error is to take $a_0 = \pi/2$ rather than $a_0 = \pi$.

20. (a) See (61), Section 3.3; (b) $\dfrac{1}{\pi}\displaystyle\int_{-\pi}^{\pi} x^4\, dx = \dfrac{2\pi^4}{9} + \sum_{n=1}^{\infty} \dfrac{16}{n^4}$.

21. Both sides equal $4\pi^3$. Note Exercise 3, Section 3.4.

22. (a) No; (b) no.

23. Use (9) and the Weierstrass M-test to prove uniform convergence.

24. If you have trouble, begin with a specific broken-line function, then generalize your result.

Notes: Theorems 4 and 5 in this section are valid for integrable functions. To see this, one needs only to recall the precise definition of the Riemann integral of a bounded function. An integrable function can be approximated in the mean-square sense by piecewise continuous functions (the step functions whose integrals are the upper and lower Riemann sums) and since Theorems 4 and 5 are valid for piecewise continuous functions, they must also be valid for any functions that are integrable in the proper Riemann sense. However, these theorems are not valid when the Lebesgue theory of integration is used, unless one requires not only the Lebesgue integrability of $f(x)$ but also that of $|f(x)|^2$.

Some of the expressions in this section must be modified if the functions are complex-valued. For example, in (2) one must replace a_n^2 by $|a_n|^2$, $[f(x)]^2$ by $|f(x)|^2$, etc. It will be noted that (7) becomes $\sum_{k=1}^{\infty} (|A_k|^2 + |B_k|^2)^{1/2}$.

Similar expressions hold for complex Fourier series (see the notes to Section 3.2). Thus, if $\sum_{n=-\infty}^{\infty} C_n e^{inx}$ is the complex Fourier expansion of a piecewise continuous function of period 2π, then $\sum_{n=-\infty}^{\infty} |C_n|^2$ is convergent, by Bessel's inequality, but if the function is continuous and piecewise smooth one can say even more: that $\sum_{n=-\infty}^{\infty} |C_n|$ is convergent.

It will be noticed that the convergence of (7) implies the convergence of $\sum_{k=1}^{\infty} (|A_k| + |B_k|)$, a fact that will be useful in Section 5.5.

An alternative proof that (8) is uniformly convergent, not making direct use of the Schwarz inequality, is by using the inequalities

$$\left(|a_n| - \frac{1}{n}\right)^2 = |a_n|^2 - 2\frac{|a_n|}{n} + \frac{1}{n^2} \geqq 0$$

$$\left(|b_n| - \frac{1}{n}\right)^2 = |b_n|^2 - 2\frac{|b_n|}{n} + \frac{1}{n^2} \geqq 0,$$

which imply

$$\frac{|a_n|}{n} + \frac{|b_n|}{n} \leqq \frac{1}{2}(|a_n|^2 + |b_n|^2) + \frac{1}{n^2}.$$

Since the right side is the general term of a convergent series, it follows that the Fourier coefficients a_n, b_n of any (bounded) integrable function have the property

$$\sum_{n=1}^{\infty} \left(\frac{|a_n|}{n} + \frac{|b_n|}{n}\right) < \infty.$$

Hence, in the notation of this section, $\sum_{n=1}^{\infty} (|A_n| + |B_n|) < \infty$, and this implies the uniform and absolute convergence of (8). This can also be used as the basis for an alternative proof of Theorem 6, since it implies the convergence of the right side of (27).

SECTION 3.9 (Page 162)

1. $1, \frac{2}{2}, \frac{2}{3}, \frac{2}{4}, \frac{3}{5}, \frac{4}{6}, \frac{4}{7}, \frac{4}{8}, \frac{5}{9}, \frac{6}{10}, \cdots$.
2. $\frac{1}{2}$.
3. $s = 1 + 0 - s$.
4. $s = \frac{1}{2}$. No fundamental difference.
5. $s = \frac{2}{5}$. In Theorem 3, let s be the sum of $\frac{4}{5} - \frac{1}{5} - \frac{1}{5} - \frac{1}{5} - \frac{1}{5} +$ (repeating every five terms) and t the sum of $\frac{1}{5} + \frac{1}{5} + \frac{4}{5} + \frac{1}{5} + \frac{1}{5} +$ (repeating). Then $s = \frac{2}{5} - t$ but the desired sum is $s + t$.
6. $a = 0$. *Note*: The theory of divergent series is essentially a theory of oscillating series.

7. **(a)** Yes; **(b)** no [let $A_n = B_n = \frac{1}{2} + (-1)^n \frac{1}{2}$].
8. **(a)** $S_2 = \frac{11}{4}$, $S_3 = 3$, $S_4 = \frac{11}{4}$; **(b)** $\frac{23}{8}$; **(c)** no.
11. **(b)** No.
13. There is no analog to (16); (18) is not valid.
15. $\dfrac{n-1}{n} \sin x$. (No need to integrate here.)
16. These are the arithmetic means of $0 + 0 + \cos 2x + 0 + 0 + \cdots$.
17. No.
19. $\frac{1}{2} + s = -(s - 1)$.
20. Yes.
27. See (12), Section 3.3.

SECTION 3.10 (Page 176)

1. Apply the theorem mentioned to the differentiated series.
2. The derivative of $\phi_n * f$.
3. **(a)** Integrate by parts; **(b)** integrate by parts as many times as the function is continuously differentiable.
5. Look at the integrals which define the Fourier coefficients of f_r. As r tends to 1, they must tend to those defining the Fourier coefficients of f (why?).
6. (a) $\frac{1}{2}$.
7. $[g(0+) + g(0-)]/2$.
11. This is not so difficult as it appears if one keeps in mind that $(f_r)^{(m)} = (f^{(m)})_r$, when $0 \leqq r < 1$, is valid from both the formal series viewpoint and from the viewpoint of the sums of the series in question.
12. This is trivial from the definition of strong convergence.
13. Consider $g_m^{(3)}$.
14. Not in general.
15. Note Exercise 3(c).
16. Summable to zero when x is not an integer multiple of 2π.
17. $-\pi g''(0)$.
18. $-\pi$ times the third derivative of the function in question.
21. Hint: The value of

$$\pi\delta(x - \pi) - \tfrac{1}{2} + \cos x + \sum_{n=2}^{\infty} (-1)^n \frac{\cos nx}{n^2 - 1}$$

at $x = 0$ is $-\frac{1}{2} + 1 + \frac{1}{2}[1 - \frac{1}{3} + \frac{1}{2} - \frac{1}{4} + \frac{1}{3} - \frac{1}{5} + \cdots] = \frac{3}{4}$.

22. Also note that $f(0) = 0$.
26. This is (49), Section 3.3. In (c), consider the period 2π extension of $\pi e^x/2 \sinh \pi$.

27. **(a)** $\frac{2}{\pi}(\cos x + \cos 3x + \cdots)$.

(b) $\frac{1}{2}$ when $0 < x < \pi$; $-\frac{1}{2}$ when $\pi < x < 2\pi$.

28. Let $x = 0$ in (36).

SECTION 3.11 (Page 188)

1. **(a)** $p_1 = v_1 + v_2 = (y_1 + y_4) + (y_2 + y_5)$.

(b) $q_1 = v_1 - v_2 = (y_1 + y_4) - (y_2 + y_5)$.

(c) By (11), $A_3 = \frac{2}{6} \sum_{k=0}^{5} y_k \cos 3 \cdot \frac{2}{6}k$

$$= (y_0 - y_1 + y_2 - y_3 + y_4 - y_5)/3$$
$$= (w_0 - w_1 + w_2)/3 = (r_0 - s_1)/3.$$

(d) $66 - 6.6 - 2.2 = 57.2$ (approximately).

2. See the solution to 1(c) for the basic idea.

4. Cos $2x + \sin x$ (you will probably get $A_2 = 1.000$ and $B_1 = 1.001$).

6. (a) and (e) have norm $\sqrt{6}$; the others, $\sqrt{3}$.

8. Use elementary calculus.

9. Note that $\phi_j(x_j) = 1$, and $\phi_j(x_k) = 0$ if $j \neq k$. Also, you must show that these products can be written as trigonometric polynomials.

10. No.

11. **(a)** No; **(b)** no.

12. No.

13. Similar to (2).

14. Cancel an n and use (15) and (23), Section 3.3.

SECTION 4.1 (Page 194)

1. **(a)** Linear, second order; **(b)** nonlinear; **(c)** linear, second order; **(d)** linear, third order; **(e)** linear, second order.

2. $f(x, y) = x^3y^2/2 + y^3x/3 + g(x) + h(y)$.

3. $f(x, y) = g(x) + h(y) + e^{xy}/xy$.

4. **(a)** $\partial z/\partial x = \cos(x + vt) + e^{(x-vt)}$.

$\partial z/\partial t = v \cos(x + vt) - ve^{(x-vt)}$.

(b) $\partial z/\partial x = 5(x + vt)^4 + 7(x - vt)^6$

$\partial z/\partial t = 5v(x + vt)^4 - 7v(x - vt)^6$.

(c) $\partial z/\partial x = f'(x + vt) + g'(x - vt)$

$\partial z/\partial t = vf'(x + vt) - vg'(x - vt)$.

6. $f(u) = g(u) = (\sin u)/2$.

8. **(b)** $y = [\sin(x + t)]/2 + [\sin(x - t)]/2$.

9. **(a)** $\dfrac{\partial^2 z}{\partial x^2} = \dfrac{\partial^2 z}{\partial u^2} + 2\dfrac{\partial^2 z}{\partial p \partial u} + \dfrac{\partial^2 z}{\partial p^2}$

$$\frac{\partial^2 z}{\partial t^2} = v^2 \frac{\partial^2 z}{\partial u^2} - 2v^2 \frac{\partial^2 z}{\partial p \partial u} + v^2 \frac{\partial^2 z}{\partial p^2}.$$

10. $z = f(x + iy) + g(x - iy)$.
11. **(a)** No; **(b)** many answers; for example, $y = e^{x-t} - e^{x+t}$.
12. $f(u) = e^{ku}$ for arbitrary k; $f(u) = 0$ is another (trivial) solution.
13. $(x'y'' - y'x'')/(x')^3$.
14. $\dfrac{\partial}{\partial r}\left(r \dfrac{\partial u}{\partial r}\right) + \dfrac{1}{r}\dfrac{\partial^2 u}{\partial \theta^2}$.
15. No.
16. $ad - bc \neq 0$.

SECTION 4.2 (Page 200)

1. If you have trouble here, read the section again. Many such arguments can be given.
2. The relation is generally valid provided the surface is smooth enough for the integral to be defined.
3. No, one would expect the integral to be positive if f is a positive solution of (8), since heat would be flowing into the region bounded by the surface.
4. If f were positive within the region, by (8) $\nabla^2 f$ would be positive there, contradicting (3).
5. Yes.

Note: The rigorous justification for the intuitive statements in this section will be found in Section 5.4.

SECTION 4.3 (Page 208)

4. **(b)** $n = 1, 3, 5, 7, \cdots$.
5. Minimizing $\|x^n + C_1x^{n-1} + C_2x^{n-2} + \cdots + C_n\|$ is the same as minimizing $\|x^n - Q(x)\|$ where $Q(x) = -C_1x^{n-1} - \cdots - C_n$ is in the n-dimensional space spanned by $1, x, x^2, \cdots, x^{n-1}$. According to Section 2.3, the minimum is obtained by taking Q to be the projection of x^n into this n-dimensional space. Then $x^n - Q(x)$ is a polynomial of degree n that is orthogonal to the n-dimensional space, so $x^n - Q(x)$ must be a scalar multiple of P_n, i.e., $x^n + C_1x^{n-1} + \cdots + C_n$ must be a scalar multiple of P_n, as we desired to show. (A computational proof can also be given.)
6. $\frac{1}{4}P_0(x) + \frac{1}{2}P_1(x) + \frac{5}{16}P_2(x) + \cdots$.
7. $\dfrac{(n!)^2 2^{n+1}}{(2n + 1)!}$.

8. $\cos^3 \phi = \frac{2}{5}P_3(\cos \phi) + \frac{3}{5}P_1(\cos \phi)$.

10. **(b)** $A_k = \dfrac{2k+1}{2L} \displaystyle\int_{-L}^{L} f(x)P_k\left(\frac{x}{L}\right) dx.$

11. **(a)** Yes; **(b)** every value of n; **(c)** no; for example, suppose $f(\phi) = \sin \phi$, which is an odd function.

14. Replace n by $n - 1$ in Exercise 13 and juggle two factors.

15. Integrand is an odd function.

16. If $n - m > 1$, say, xP_m is a polynomial of degree less than n, hence is orthogonal to P_n.

17. Apply (10) to $f(x) = xP_n(x)$. If $n = 0$, we have $xP_0 = P_1$.

18. At $x = 1$ we have $2P_n'(1) = n(n+1)P_n(1)$ by (19), i.e., the value of f determines the value of f'.

19. See the preceding exercise.

20. Use (3).

21. Note: Compare Exercise 8, Section 2.5.

22. $\dfrac{1}{n!}\displaystyle\int_{-1}^{x} (x-t)^n(1+t)^n\,dt = \int_{-1}^{x}\int_{-1}^{x}\cdots\int_{-1}^{x}(1+x)^n\,dx\cdots dx.$

25. It is based on the relation given in Exercise 24.

SECTION 4.4 (Page 220)

3. The second two terms are *not* negligible when r is large.

4. $\Theta = A\theta + B$. Not single-valued unless $A = 0$.

5. Hint: $\dfrac{d}{d\phi}\left(-\sin\phi\,\dfrac{d}{dz}\right) = -\cos\phi\,\dfrac{d}{dz} - \sin\phi\,\dfrac{dz}{d\phi}\dfrac{d^2}{dz^2}.$

6. Scalar multiple of P_{-n-1}.

7. $r^2P_2(\cos\phi)$. Note: $r^{-3}P_2(\cos\phi)$ is unbounded at the origin.

8. $\left(\dfrac{r}{3}\right)^3 P_3(\cos\phi)$.

9. $\dfrac{2}{3r}P_0(\cos\phi) + \dfrac{4}{3r^3}P_2(\cos\phi)$.

11. **(c)** Use mathematical induction.

12. Replace r by $1/r$ and multiply both sides by $1/r$.

13. Gives the same result as Exercise 11.

14. According to (22) it is Abel summable to $(2 - 2z)^{-1/2}$.

15. Look at (39) and (40), Section 3.9.

18. Both sides become $F'(y)f'(y)\dfrac{\partial y}{\partial a}\dfrac{\partial y}{\partial x} + F(y)\dfrac{\partial^2 u}{\partial a\partial x}.$

19. Hint: By (32) it holds for $n = 1$. If (34) is valid for a certain n, then

$$\frac{\partial^{n+1}u}{\partial x^{n+1}} = \frac{\partial^n}{\partial a^{n-1}\partial x}\left[g(y)^n \frac{\partial u}{\partial a}\right].$$

But

$$\frac{\partial}{\partial x}\left[g(y)^n \frac{\partial u}{\partial a}\right] = \frac{\partial}{\partial a}\left[g(y)^n \frac{\partial u}{\partial x}\right] = \frac{\partial}{\partial a}\left[g(y)^{n+1} \frac{\partial u}{\partial a}\right].$$

Note: For a derivation of (1), see Davis, *Introduction to Vector Analysis* (Allyn and Bacon, 1961), Section 4.12.

SECTION 4.5 (Page 230)

1. **(a)** 5; **(b)** 4; **(c)** 8; **(d)** 6.
2. **(a)** 21; **(b)** yes.
3. $(x^2 + y^2 + z^2)^2$; $(\alpha\beta + z^2)^2$.
4. **(a)** No; **(b)** no.
5. $\alpha\beta + \beta^2$.
7. Many possibilities. For example, $e^{(c^2/4)\alpha}e^{(k^2/c^2)\beta} \sin kz$.
10. **(a)** The dimension of the domain equals the dimension of the null space plus the dimension of the range. **(b)** The problem is to show that the range of the linear transformation is all of V_{n-2} and not a proper subspace of V_{n-2}.
11. A scalar multiple of $8\alpha z^4 - 12\alpha^2\beta z^2 + \alpha^3\beta^2$.
13. You obtain a function identically equal to zero.
14. $\dfrac{d^{2n}}{dx^{2n}}(z^2 - 1)^n$ is a constant.
15. **(a)** $3 \sin \phi \cos \phi$ or $\frac{3}{2} \sin 2\phi$; **(b)** $-\frac{1}{2} \sin \phi \cos \phi$.
16. Valid if $m = m'$ and $n \neq n'$.
17. Only one term in (16) is needed, that for $m = 1$ and $n = 2$ (compare Exercise 15). The solution is the real part of $\frac{2}{3}r^2e^{i\theta}P_2^{(1)}(\cos \phi)$, i.e., $r^2 \cos \theta \sin 2\phi$.
18. $P_2^{(0)} = (3 \cos^2 \phi - 1)/2$, $P_2^{(1)}(\cos \phi) = 3 \sin \phi \cos \phi$, $P_2^{(2)}(\cos \phi) = 3 \sin^2 \phi$, $P_3^{(0)}(\cos \phi) = (5 \cos^3 \phi - 3 \cos \phi)/2$, $P_3^{(1)}(\cos \phi) = 3 \sin \phi \cdot (5 \cos^2 \phi - 1)/2$, etc.
19. A surface harmonic of degree n is a linear combination of $Y_n^{(-n)}$, $Y_n^{(-n+1)}, \cdots, Y_n^{(n)}$.
20. (e) Yes.
21. Yes.
22. **(a)** $r^2(3 \cos^2 \phi - 1)$; **(b)** $\frac{1}{3} - \frac{1}{3}r^2(\frac{3}{2} \cos^2 \phi - \frac{1}{2}) + \frac{3}{2}r^2 \sin^2 \phi \cos 2\theta$.
23. No (consider $x^2 + y^2 + z^2 - 1$).

SECTION 4.6 (Page 244)

2. Two exist separately for $r < 0$ and $r > 0$.

4. Insofar as (18) is concerned, they do. See Example 6 (*not* Exercise 6) in Section 2.4.

5. To obtain (18) is easy. To obtain the others, such as (30), take $r = \lambda x$ in (4) to obtain Bessel's equation in the form

$$\frac{d}{dx}\left[x\frac{d}{dx}J_n(\lambda x)\right] + \left(\lambda^2 x - \frac{n^2}{x}\right)J_n(\lambda x) = 0.$$

Multiply both sides by $2x\dfrac{d}{dx}J_n(\lambda x)$ to obtain

$$\frac{d}{dx}\left[x\frac{d}{dx}J_n(\lambda x)\right]^2 + (\lambda^2 x^2 - n^2)\frac{d}{dx}[J_n(\lambda x)]^2 = 0,$$

which can be integrated directly. Integrating by parts in the second term and using $rJ_n'(r) = nJ_n(r) - rJ_{n+1}(r)$, we obtain, after some manipulating,

$$2\lambda^2\int_0^c x[J_n(\lambda x)]^2\,dx = \Big[\{nJ_n(\lambda x) - \lambda x J_{n+1}(\lambda x)\}^2 + (\lambda^2x^2 - n^2)\{J_n(\lambda x)\}^2\Big]_0^c.$$

Hence

$$\int_0^c x[J_n(\lambda x)]^2\,dx = \frac{c^2}{2}\{[J_n(\lambda c)]^2 + [J_{n+1}(\lambda c)]^2\} - \frac{nc}{\lambda}J_n(\lambda c)J_{n+1}(\lambda c),$$

from which the others follow.

6. Yes.

7. Use (28).

9. **(b)** This is Bessel's equation.

10. **(a)** This follows from (53).

13. Replace ϕ in (46) by $\phi - \pi/2$ and observe that the integrand has period π.

14. Yes.

15. $\dfrac{\pi}{2}J_0(x)$.

16. Differentiate (49) with respect to ϕ and set $\phi = 0$.

17. Try Riemann's lemma.

18. $B_n = \dfrac{2}{c^2[J_{n+1}(a_n c)]^2}\displaystyle\int_0^c rJ_n(a_n r)f(r)\,dr.$

SECTION 5.1 (Page 250)

$$\int_{-1}^{1}\int_{0}^{2\pi} K\frac{\partial T}{\partial r}\, a\, d\theta\, dz.$$

SECTION 5.2 (Page 256)

2. **(a)** 25°; **(b)** $\dfrac{400}{\pi}\displaystyle\sum_{\text{odd } n}\dfrac{\sinh(n\pi y/L)}{n\sinh n\pi}\sin\dfrac{n\pi x}{L}$; **(c)** if the "fourth" edge is at the top, the temperature is about 12° (see the note below).

3. $(3/\sinh \pi)\sinh y \sin x - (4/\sinh 2\pi)\sinh 2y \sin 2x$.

4. (Sinh $2y/\sinh 2\pi)\sin 2x + (\sinh x/\sinh \pi)\sin y$.

5. $\frac{1}{4}r\sin\theta - \frac{3}{4}r\cos\theta + \frac{5}{256}r^4\sin 4\theta$.

6. $T(r, \theta) = \dfrac{2\pi^2}{3} - 4\displaystyle\sum_{n=1}^{\infty}\dfrac{3^n}{n^2 r^n}\cos n\theta.$

7. $T(r, \theta) = \left(-\dfrac{5}{9}r + \dfrac{20}{r}\right)\cos\theta + \left(\dfrac{20r}{9} - \dfrac{20}{r}\right)\sin\theta.$

8. **(a)** $T(r, \theta) = \dfrac{4}{\pi}\left(r\sin\theta + \dfrac{r^3}{3}\sin 3\theta + \dfrac{r^5}{5}\sin 5\theta + \cdots\right).$

9. $T(r, \theta) = \dfrac{1 - r^2}{2\pi(1 - 2r\cos\theta + r^2)}, \; 0 \leqq r < 1.$

10. $h(\theta) = 3/[2\pi(5 - 4\cos\theta)]$.

11. $3(r/a)^{\pi/\alpha}\sin(\pi\theta/\alpha)$.

12. $C_n(r/a)^{n\pi/2\alpha}\sin(n\pi\theta/2\alpha),\ n = 1, 3, 5, \cdots$.

Note: Even if realizable physically, boundary conditions involving an infinite temperature gradient on a surface or edge need not produce an infinite rate of flow of heat. In the macroscopic theory, this is purely a surface or edge phenomenon; the solution will not be discontinuous in the *interior* of the body. From this viewpoint, Exercise 2 is not so ridiculous as it may appear at first.

SECTION 5.3 (Page 261)

3. Let $m = 0$.

4. **(a)** Constant functions and those of the form C/r; **(b)** $T(r) = 125 - 150/r$; 87.5° (*not* 75°).

6. In (17) change $(r/r_0)^k$ to $(r_0/r)^{k+1}$. No change in (18).

7. **(a)** $A + B\ln r$; **(b)** 81.6°.

SECTION 5.4 (Page 271)

1. For example, $r^3 \cos 2\theta P_3^{(2)}(\cos \phi)$ and $r^{-4} \cos 2\theta P_3^{(2)}(\cos \phi)$ have the same values on the sphere $r = 1$.
2. Since rf must be bounded, f can be constant only if f is zero.
4. At a local maximum interior to the region, we would have $\partial^2 f/\partial x^2 \leqq 0$, $\partial^2 f/\partial y^2 \leqq 0$, $\partial^2 f/\partial z^2 \leqq 0$, hence $\nabla^2 f \leqq 0$. By (16), $f \leqq 0$ at this point, and since the point is a local maximum, $f \leqq 0$ throughout. If f is nonnegative, this implies f is identically zero, and by continuity is zero on the boundary. Hence, if f is positive on the boundary, it cannot have a local maximum at an interior point.
5. By considering $h = f - g$, it suffices to show that if a solution h of (16) is not negative on the boundary, it must also be nonnegative within. If it were negative within, it would have a minimum value within which would also be negative, but at a local minimum $\partial^2 f/\partial x^2 \geqq 0$, $\partial^2 f/\partial y^2 \geqq 0$, and $\partial^2 f/\partial z^2 \geqq 0$, hence by (16), f would be positive, which is a contradiction.
6. **(a)** Use reasoning similar to Exercises 4 and 5 (one might say that f is the temperature distribution for a reaction that is "more endothermic" than that represented by h); **(b)** consider a harmonic positive function h.
7. There are many answers.
8. No.

Notes: Here is the usual definition of *subharmonic function.* Let v be a continuous function defined in a bounded region G and on its boundary. Let K denote a sphere, entirely within G, and let $(v)_K$ denote that continuous function equal to v outside and on the boundary of K and harmonic within K. If $(v)_K \geqq v$ for every sphere K we say v is *subharmonic.*

A more satisfying analysis of all problems involving Laplace's equation and its relative (16) is obtained by introducing an integral operator that directly gives the values of the function in the interior of the region as an integral involving values of the function on its boundary. In two dimensions, this procedure will be illustrated by (10), Section 5.5. For an outline of the procedure in three dimensions and a list of references, see the article by M. M. Schiffer in Chapter 6 of Beckenbach's *Modern Mathematics for the Engineer* (McGraw-Hill, 1956).

SECTION 5.5 (Page 276)

1. One method: Apply (7) to a cylindrical surface, assuming the functions independent of z, to obtain a relationship between a line integral and an integral over a plane area.
3. No, not in general.

Note: To prove the convergence of (8), we notice [compare (7), Section 3.8] that $\sum_{k=1}^{\infty} (|A_k|^2 + |B_k|^2)^{1/2}$ converges (the absolute value symbols are not necessary, of course, in the real case considered in Section 3.8) and since $|A_k| \leq (|A_k|^2 + |B_k|^2)^{1/2}$ it follows that $\sum_{k=1}^{\infty} |A_k|$ converges. Similarly, $\sum_{k=1}^{\infty} |B_k|$ converges. It follows that (8) converges. An alternative proof is given in the notes to Section 3.8.

SECTION 5.6 (Page 279)

1. $T(r, t) = 100 - \sum_{n=1}^{\infty} \dfrac{200}{k_n a} \dfrac{J_0(k_n x)}{J_1(k_n a)} e^{-\alpha^2 k_n^2 t}$ where $J_0(k_n a) = 0$, $n = 1, 2, 3, \cdots$.
2. Modify (43), Section 3.3, by introducing the factors $e^{-n^2\alpha^2\pi^2 t/L^2}$ in each term.

SECTION 6.1 (Page 286)

1. $y(x, t) = 2 \sin (\pi x/L) \cos (\pi vt/L) - (L/\pi v) \sin (2\pi x/L) \cos (2\pi vt/L)$.
2. 9.
4. $y(x, t) = \dfrac{32h}{\pi} \sum_{\text{odd } n} \dfrac{1}{n^3} \sin \dfrac{nx}{2} \cos \dfrac{nvt}{2}$.
7. See Exercise 9.
10. Odd harmonics.
11. Hint: Look at the series for $\partial y/\partial t$.

SECTION 6.2 (Page 293)

6. **(a)** Sin $x \cos vt + \sin \sqrt{2}x \cos \sqrt{2}vt$; **(b)** no (see Exercise 19, Section 1.3).

SECTION 6.3 (Page 298)

1. In this case, $u_{i+1} = u_k + (u_i - u_{i-1})$. Let $a = u_0$ and $b = u_1 - u_0$. Then $u_n = a + nb$.
2. $\dfrac{1}{2\pi} \sqrt{\dfrac{2T}{md}\left(1 - \cos \dfrac{kd}{v}\right)}$ where v is the velocity of the traveling wave.

3. $u_{i+1} - (2\cosh p)u_i + u_{i-1} = 0.$
4. One obtains solutions that fall to zero exponentially. (This is similar to what a low pass filter does to high frequencies.)
6. You should obtain the wave equation for a string with constant density.

SECTION 6.4 (Page 301)

3. It was obtained from (8), Section 4.6.
5. $C_n = (2n+1)\int_0^1 f(x)P_n(x)\,dx.$
6. $\frac{1}{4}P_1(x) - \frac{7}{24}P_3(x) + \frac{11}{192}P_5(x) - \cdots.$
7. $y(x, t) = \frac{1}{4}P_1(x)\cos at - \frac{7}{24}P_3(x)\cos\sqrt{6}at + \frac{11}{192}P_5(x)\cos\sqrt{15}at - \cdots.$

SECTION 6.5 (Page 311)

1. $\dfrac{\partial}{\partial x}\left(T\dfrac{\partial w}{\partial x}\right) + \dfrac{\partial}{\partial y}\left(T\dfrac{\partial w}{\partial y}\right) = \rho\dfrac{\partial^2 w}{\partial x^2}.$
2. $C_n = \dfrac{\iint_D g(x, y)u_n(x, y)\,dx\,dy}{\iint_D [u_n(x, y)]^2\,dx\,dy}.$ Similarly for D_n.
3. $D_{nm} = \dfrac{4v}{ab}\displaystyle\int_0^b\int_0^a \dfrac{\partial w}{\partial t}(x, y, 0)\sin\dfrac{n\pi x}{a}\sin\dfrac{n\pi y}{b}\,dx\,dy.$
4. One nodal curve is straight across the diagonal; the other is elliptical.
7. 5.135/2.405.

SECTION 6.6 (Page 313)

1. Write (2) in spherical coordinates and substitute $w = h(r)f(r - vt)$. For f to be nontrivial, it follows that $h(r) = C/r$.
2. Write (2) in cylindrical coordinates and substitute $w = h(r)f(r - vt)$. The resulting equations are incompatible for an arbitrary waveform f, which shows that relatively undistorted cylindrical waves are in general impossible.
3. $w(r, t) = J_0(kr/v)(A\cos kt + B\sin kt)$ is a possibility.
4. $w(r, t) = \dfrac{1}{r}\left\{A\sin\dfrac{kr}{v} + B\cos\dfrac{kr}{v}\right\}\left\{C\sin kt + D\cos kt\right\}$ (Compare Exercise 1.)

SECTION 6.7 (Page 325)

3. No. For example, $\int_0^\infty \frac{\sin x}{x}\,dx$ is uniformly convergent but not absolutely convergent.
4. **(a)** Yes; **(b)** no; we require that

$$\int_{-\infty}^{\infty} f(x)\,dx = \int_0^\infty f(x)\,dx + \int_{-\infty}^{0} f(x)\,dx,$$

where both integrals on the right-hand side converge.
5. Yes; e^{-M} is independent of t.
6. No.
7. $e^{uy}[A \cos ux + B \sin ux]$.

11. **(b)** $w(x, y) = 1 - \frac{1}{\pi}\tan^{-1}\frac{y}{x} = 1 - \frac{\theta}{\pi}$. Along the y axis, the value is $\frac{1}{2}$.

12. **(a)** $f(x) = \frac{2}{\pi}\int_0^\infty \frac{1 - \cos au}{u} \sin ux\,du,\ 0 < x < \infty.$

(b) $f(x) = \frac{2}{\pi}\int_0^\infty \frac{\sin au \cos ux}{u}\,du,\ 0 < x < \infty.$

(c) $f(x) = \frac{1}{\pi}\int_0^\infty \left(\frac{\sin au \cos xu}{u} + \frac{1 - \cos au}{u}\sin xu\right) du,$

$-\infty < x < \infty.$

(d) $\frac{1 - e^{-iau}}{iu}.$

13. $\frac{1}{2\alpha\sqrt{\pi t}}\int_0^\infty f(u)\left[\exp\left\{-\frac{(u - x)^2}{4\alpha^2 t}\right\} - \exp\left\{-\frac{(u + x)^2}{4\alpha^2 t}\right\}\right] du.$

15. **(a)** Integrable square functions; **(b)** no.

SECTION 6.8 (Page 341)

1. In both instances, elements of the set are of the form $f = f_p + C_1u_1 + C_2u_2 + \cdots + C_nu_n$, where $C_1, C_2, \cdots, C_n$ are scalars.
2. Let $S(x_1, x_2, x_3, \cdots) = (x_2, x_3, x_4, \cdots)$.
3. **(a)** $T(Su) = S(Tu) = S(\lambda u) = \lambda(Su)$, so if $Su \neq \theta$, Su is by definition an eigenvector for T, having the same eigenvalue λ.
   (b) Hint: Let H_c be the subspace of all v such that $Tv = cv$. Then, by part (a), Sv is in H_c whenever v is in H_c, and since S is a self-adjoint operator on H_c (why?) we can find an orthonormal basis of H_c, for each distinct eigenvalue c, consisting of eigenvectors of T. Caution: H_c need not be one-dimensional, and an eigenvector of T is not necessarily an eigenvector of S.

4. $A_j = \int_0^{2\pi} f(y) \cos jy\, dy$, which differs from the Fourier coefficient by a factor of π.
5. Believe it or not, this was once the topic of a master's thesis.
6. Look at the discussion of differential equations with constant coefficients, in any standard textbook on differential equations.
7. **(a)** The C_j are real; **(b)** $|C_j| = 1$ for all j; **(c)** $C_j = 1$ or -1 for all j.
8. If $Tv = cv$, $T^4 v = v$, it follows that $c^4 = 1$, so the possible eigenvalues are $1, i, -1, -i$.
9. See Section 3.10.

SUPPLEMENTARY EXERCISES (Page 343)

1. **(a)** Yes; **(b)** yes; yes.
2. $\pi/4$.
3. $G(x, t) = x - 1$ if $t < x$; $G(x, t) = t - 1$ if $x < t$.
4. **(a)** Does not exist; **(b)** may or may not exist; **(c)** exists if $p = -n^2$, $n = 1, 2, 3, \cdots$.
6. Hint: Let $u(x) = x^n e^{-x}$. Then $xu' = nu - xu$. Differentiate $n + 1$ times to obtain

$$xu^{(n+2)} + (n + 1)u^{(n+1)} = nu^{(n+1)} - xu^{(n+1)} - (n + 1)u^{(n)}.$$

Simplify and substitute $u(x) = e^{-x}L_n(x)$.

7. Hint: Obviously $H_n'(x) = 2xH_n(x) - H_{n+1}(x)$. Also, if $u(x) = e^{-x^2}$, $u'(x) + 2xu(x) = 0$. Differentiate this n times.
8. Hint: $\psi_n''(x) - x^2\psi_n(x) = -(2n + 1)\psi_n(x)$. Can you modify the proof of orthogonality in the discussion of the Sturm-Liouville theorem?
9. Integral values.
10. The weight of the chain or rope is not neglected.
11. $g(x) = A_0 + A_1P_1(x - 2) + A_2P_2(x - 2) + A_3P_3(x - 2)$ where the P_n are Legendre polynomials and $A_n = \dfrac{2n + 1}{2} \displaystyle\int_1^3 f(x)P_n(x - 2)\, dx$.
12. **(a)** $f(x) = \frac{1}{4} + \frac{1}{2}P_1(x) + \frac{5}{16}P_2(x) - \frac{3}{32}P_4(x) + \cdots$.
    (b) Yes.
13. $T(x, t) = \frac{3}{5}P_1(x)e^{-2ct} + \frac{2}{5}P_3(x)e^{-12ct}$ where ^{n} is a constant depending on the material. *Note:* Since the thermal conductivity is zero at $x = \pm 1$, it is not necessary to assume the two ends are insulated.
14. Use the Laplacian in spherical coordinates.
15. Hint: Begin by expanding $f(b, \theta)$ in a Fourier sine series. Compare the discussion of the steady-state temperature distribution in a circular annulus.
16. $f(x) = \dfrac{2}{\pi} \displaystyle\int_0^\infty e^{-t} \cos tx\, dx = \dfrac{2}{\pi} \dfrac{1}{1 + x^2}\cdot$ (Use the Fourier integral theorem.)

17. **(a)** $b_n = 0$; **(b)** $a_n = 0$; **(c)** $a_{2n+1} = 0$, $b_{2n} = 0$;
(d) $a_{2n} = 0$, $b_{2n+1} = 0$; **(e)** $a_{2n+1} = 0$, $b_{2n+1} = 0$;
(f) $a_{4n+2} = 0$, $a_{4n+1} = b_{4n+1}$, $a_{4n+3} = -b_{4n+3}$, $b_{4n} = 0$;
(g) $a_n = 0$ and $b_n = 0$ except perhaps when $n = 0, 4, 8, \cdots, 4k, \cdots$;
(h) $b_n = 0$; also, $a_n = 0$ except perhaps for $n = 0, 4, 8, \cdots, 4k, \cdots$;
(i) All terms vanish except the constant term.

18. Use the series definition. *Note:* This and the result in Exercise 19 are used repeatedly in later exercises. Thus, to find $\frac{d}{dx}[x^3 J_1(x)]$ write

$$\frac{d}{dx}[x^2 \cdot x J_1(x)] = x^2 \cdot x J_0(x) + 2x \cdot x J_1(x)$$

$$= x^3 J_0(x) + 2x^2 J_1(x).$$

20. $3x^2 J_0(x^2) - 2x^4 J_1(x^2)$. Recall that $J_0'(t) = -J_1(t)$.

21. $2x^4 J_3(4x) + 4x^5 J_2(4x)$.

22. No. A common error is $\frac{d}{dx}[xJ_1(x^2)] = xJ_0(x^2)\frac{d}{dx}(x^2)$. The correct answer is $2x^5 J_0(x^2) + 2x^3 J_1(x^2)$.

23. $\int J_3(x)\,dx = \int x^2[x^{-2}J_3(x)]\,dx = \int x^2[-x^{-2}J_2(x)]'\,dx$ by Exercise 19. Now integrate by parts to obtain $-J_2(x) - \frac{2}{x}J_1(x) + C$.

24. $(4x^3 - 16x)J_1(x) - (x^4 - 8x^2)J_0(x) + C$. *Note:* $x^4 J_1(x) = x^2 \frac{d}{dx}[x^2 J_2(x)]$.

25. $2\sqrt{x}J_1(\sqrt{x}) + C$.

26. Differentiate the right sides.

27. $\frac{1}{3}x^2 J_0(x)\cos x + \frac{1}{3}x^2 J_1(x)\sin x - \frac{2}{3}xJ_1(x)\cos x$.

28. $\frac{1}{3}x^2 J_0(x)\cos x + \frac{1}{3}x^2 J_1(x)\sin x + \frac{1}{3}xJ_1(x)\cos x$.

29. Use Exercise (18).

31. $(48/x^3 - 8/x)J_1(x) - (24/x^2 - 1)J_0(x)$.

32. $J_2(r) = 0.364$, $J_3(4) = 0.430$ (approximately).

33. $\frac{9}{2} + \sum_{n=1}^{\infty} \frac{4}{k_n^2 J_0(3k_n)} J_0(k_n x)$.

34. $\sum_{n=1}^{\infty} \frac{-6}{k_n J_1(3k_n)} J_2(k_n x)$.

35. $T(r, \theta, z) = \sum_{n=1}^{\infty}\sum_{m=1}^{\infty} A_{nm} J_n(k_{nm} r) \sin n\theta \sinh k_{nm} z$ where $J_n(k_{nm} b) = 0$

and

$$A_{nm} = \frac{\frac{2}{\pi}\int_0^b \int_0^\pi rf(r, \theta) \sin n\theta J_n(k_{nm} r)\, d\theta\, dr}{\frac{b^2}{2}(\sinh k_{nm} h) J_{n+1}^2(k_{nm} b)}.$$

APPENDIX (Page 359)

1. **(a)** If $h = f * g$, then $h(0) = 4$, $h(1) = 8 + 4i$, $h(2) = -4$, $h(3) = 8 - 4i$.
   (b) $f = e_1 + e_2 + e_3$, $g = e_1 + e_2 - e_3$.
   (c) $\hat{f}(e_1) = 4$, $\hat{f}(e_2) = 4$, $\hat{f}(e_3) = 4$, $\hat{f}(e_4) = 0$.
   $\hat{g}(e_1) = 4$, $\hat{g}(e_2) = 4$, $\hat{g}(e_3) = -4$, $\hat{g}(e_4) = 0$.
   (d) 4.
   (e) $\hat{h}(e_1) = 16$, $\hat{h}(e_2) = 16$, $\hat{h}(e_3) = -16$, $\hat{h}(e_4) = 0$.
2. In the table given in the text, replace 0, 1, 2, 3 by 1, 3, 4, 2 (in that order). (This group is isomorphic to the additive group of integer modulo 4.)
3. 1, 1, 1, 1; 1, 1, -1, -1; 1, -1, 1, -1; 1, -1, -1, 1.
4. $e_k(a^s) = \exp(2\pi iks/n)$, $k,s = 0, 1, 2, \cdots, n - 1$.
5. The basis can be chosen to be orthogonal (see the reference cited).
6. Imitate the proof of Theorem 3 (i.e., replace x by $y - a$ in the integral).
7. $(\hat{f}|\hat{g}) = 1 \sum_j \hat{f}(e_j)\overline{\hat{g}(e_j)} = \frac{1}{n} \sum_j [\sum_x f(x)\overline{e_j(x)}][\sum_y \overline{g(y)}e_j(y)]$. Using (18), this equals $\sum_x f(x)\overline{g(x)}$.
8. $\hat{\hat{f}}(x) = f(-x)$. Four times.
9. $\hat{\epsilon}(e_j) = 1$ for all j.
10. Routine verification.
11. The transform of $f(x - a)$ is $\overline{e_j(a)}\hat{f}(e_j)$.
12. $f(x) = f(a^{-1}xa)$ for every x and every a; i.e., these functions are constant on each conjugate class.
13. **(a)** Trivial; **(b)** $\phi(f) = \sum_{j=1}^{n} C_j\hat{f}(e_j)$ for some choice of constants C_j. Since $\phi(\epsilon) = 1$, we must have $\sum_{j=1}^{n} C_j = 1$. Show that the extreme points correspond to those ϕ for which one of the C_j's is unity and the rest equal zero, and multiplicativity implies the same thing.
14. Corresponding to (17), we would have

$$\frac{1}{2\pi} \sum_{n=-\infty}^{\infty} \int_0^{2\pi} f(t)e^{in(x-t)}\, dt = f(x)$$

valid, for example, if f is continuously differentiable and of period 2π.

15. All functions f for which $\sum_{n=-\infty}^{\infty} |f(n)| < \infty$ is one possibility;

$(f * g)(n) = \sum_{n=-\infty}^{\infty} f(n - m)g(m)$.

16. **(a)** A routine calculation shows $(T_f g|h) = (g|T_{f^*}h)$ for all g and h; **(b)** it is the complex conjugate; **(c)** those whose transforms are real-valued; **(d)** $f = \frac{1}{2}(f + f^*) + \frac{1}{2i}(if - if^*)$; **(e)** a commutative self-adjoint family of operators can be "simultaneously diagonalized."

17. No. A physicist might say that the Dirac delta function is such a unity element.

18. When $\hat{f}(e_j) \neq 0$ for every j.

19. When $\hat{f}(e_j) \neq 0$ for every j.

20. If the ideal consists of all h for which $\hat{h}$ is zero on some subset of $\hat{G}$, let $\hat{f}$ be the function that is zero on this subset and unity elsewhere.

21. Always; let $\hat{g}(e_j)$ be a square root of $\hat{f}(e_j)$. Not unique; if $\hat{f}(e_j) \neq 0$ for every j, there will be 2^n distinct such roots, where n is the order of G (fewer if $\hat{f}$ vanishes for some e_j).

22. Just calculate in (a). In (b) use the Schwarz inequality.

23. This matrix represents the operator relative to a certain basis in $L(G)$.

26. **(a)** Notice that $[g|h] = (T_f g|h)$, and since f is positive definite, $(T_f g|g) \geqq 0$ for all g. (This is a rewording of the definition of *positive definite*.)

(b) Hint: If $[g|g] = 0$ and $[h|h] = 0$ then $[g + h|g + h] = [g|h] + [h|g]$, but $|[g|h]| \leqq [g|g]^{1/2}\,[h|h]^{1/2}$ by the Schwarz inequality. Hence $[g + h|g + h] = 0$.

27. Hint: If $\hat{f}(e_j) \neq 0$ for all j, one can take g to be the unity element of $L(G)$. In any case, one can take g to be identically equal to unity on the set mentioned in Exercise 26(e).

INDEX

A CATALOG OF SELECTED

DOVER BOOKS

IN SCIENCE AND MATHEMATICS

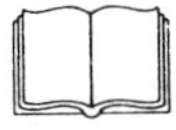

RELATIVITY, THERMODYNAMICS AND COSMOLOGY, Richard C. Tolman. Landmark study extends thermodynamics to special, general relativity; also applications of relativistic mechanics, thermodynamics to cosmological models. 501pp. 5⅜ × 8½. 65383-8 Pa. $12.95

APPLIED ANALYSIS, Cornelius Lanczos. Classic work on analysis and design of finite processes for approximating solution of analytical problems. Algebraic equations, matrices, harmonic analysis, quadrature methods, much more. 559pp. 5⅜ × 8½. 65656-X Pa. $12.95

SPECIAL RELATIVITY FOR PHYSICISTS, G. Stephenson and C.W. Kilmister. Concise elegant account for nonspecialists. Lorentz transformation, optical and dynamical applications, more. Bibliography. 108pp. 5⅜ × 8½. 65519-9 Pa. $4.95

INTRODUCTION TO ANALYSIS, Maxwell Rosenlicht. Unusually clear, accessible coverage of set theory, real number system, metric spaces, continuous functions, Riemann integration, multiple integrals, more. Wide range of problems. Undergraduate level. Bibliography. 254pp. 5⅜ × 8½. 65038-3 Pa. $7.95

INTRODUCTION TO QUANTUM MECHANICS With Applications to Chemistry, Linus Pauling & E. Bright Wilson, Jr. Classic undergraduate text by Nobel Prize winner applies quantum mechanics to chemical and physical problems. Numerous tables and figures enhance the text. Chapter bibliographies. Appendices. Index. 468pp. 5⅜ × 8½. 64871-0 Pa. $11.95

ASYMPTOTIC EXPANSIONS OF INTEGRALS, Norman Bleistein & Richard A. Handelsman. Best introduction to important field with applications in a variety of scientific disciplines. New preface. Problems. Diagrams. Tables. Bibliography. Index. 448pp. 5⅜ × 8½. 65082-0 Pa. $12.95

MATHEMATICS APPLIED TO CONTINUUM MECHANICS, Lee A. Segel. Analyzes models of fluid flow and solid deformation. For upper-level math, science and engineering students. 608pp. 5⅜ × 8½. 65369-2 Pa. $13.95

ELEMENTS OF REAL ANALYSIS, David A. Sprecher. Classic text covers fundamental concepts, real number system, point sets, functions of a real variable, Fourier series, much more. Over 500 exercises. 352pp. 5⅜ × 8½. 65385-4 Pa. $10.95

PHYSICAL PRINCIPLES OF THE QUANTUM THEORY, Werner Heisenberg. Nobel Laureate discusses quantum theory, uncertainty, wave mechanics, work of Dirac, Schroedinger, Compton, Wilson, Einstein, etc. 184pp. 5⅜ × 8½. 60113-7 Pa. $5.95

INTRODUCTORY REAL ANALYSIS, A.N. Kolmogorov, S.V. Fomin. Translated by Richard A. Silverman. Self-contained, evenly paced introduction to real and functional analysis. Some 350 problems. 403pp. 5⅜ × 8½. 61226-0 Pa. $9.95

PROBLEMS AND SOLUTIONS IN QUANTUM CHEMISTRY AND PHYSICS, Charles S. Johnson, Jr. and Lee G. Pedersen. Unusually varied problems, detailed solutions in coverage of quantum mechanics, wave mechanics, angular momentum, molecular spectroscopy, scattering theory, more. 280 problems plus 139 supplementary exercises. 430pp. 6½ × 9¼. 65236-X Pa. $12.95

NUMERICAL METHODS FOR SCIENTISTS AND ENGINEERS, Richard Hamming. Classic text stresses frequency approach in coverage of algorithms, polynomial approximation, Fourier approximation, exponential approximation, other topics. Revised and enlarged 2nd edition. 721pp. 5⅜ × 8½.
65241-6 Pa. $14.95

THEORETICAL SOLID STATE PHYSICS, Vol. I: Perfect Lattices in Equilibrium; Vol. II: Non-Equilibrium and Disorder, William Jones and Norman H. March. Monumental reference work covers fundamental theory of equilibrium properties of perfect crystalline solids, non-equilibrium properties, defects and disordered systems. Appendices. Problems. Preface. Diagrams. Index. Bibliography. Total of 1,301pp. 5⅜ × 8½. Two volumes. Vol. I 65015-4 Pa. $14.95
Vol. II 65016-2 Pa. $14.95

OPTIMIZATION THEORY WITH APPLICATIONS, Donald A. Pierre. Broad-spectrum approach to important topic. Classical theory of minima and maxima, calculus of variations, simplex technique and linear programming, more. Many problems, examples. 640pp. 5⅜ × 8½. 65205-X Pa. $14.95

THE MODERN THEORY OF SOLIDS, Frederick Seitz. First inexpensive edition of classic work on theory of ionic crystals, free-electron theory of metals and semiconductors, molecular binding, much more. 736pp. 5⅜ × 8½.
65482-6 Pa. $15.95

ESSAYS ON THE THEORY OF NUMBERS, Richard Dedekind. Two classic essays by great German mathematician: on the theory of irrational numbers; and on transfinite numbers and properties of natural numbers. 115pp. 5⅜ × 8½.
21010-3 Pa. $4.95

THE FUNCTIONS OF MATHEMATICAL PHYSICS, Harry Hochstadt. Comprehensive treatment of orthogonal polynomials, hypergeometric functions, Hill's equation, much more. Bibliography. Index. 322pp. 5⅜ × 8½. 65214-9 Pa. $9.95

NUMBER THEORY AND ITS HISTORY, Oystein Ore. Unusually clear, accessible introduction covers counting, properties of numbers, prime numbers, much more. Bibliography. 380pp. 5⅜ × 8½. 65620-9 Pa. $9.95

THE VARIATIONAL PRINCIPLES OF MECHANICS, Cornelius Lanczos. Graduate level coverage of calculus of variations, equations of motion, relativistic mechanics, more. First inexpensive paperbound edition of classic treatise. Index. Bibliography. 418pp. 5⅜ × 8½. 65067-7 Pa. $11.95

MATHEMATICAL TABLES AND FORMULAS, Robert D. Carmichael and Edwin R. Smith. Logarithms, sines, tangents, trig functions, powers, roots, reciprocals, exponential and hyperbolic functions, formulas and theorems. 269pp. 5⅜ × 8½. 60111-0 Pa. $6.95

THEORETICAL PHYSICS, Georg Joos, with Ira M. Freeman. Classic overview covers essential math, mechanics, electromagnetic theory, thermodynamics, quantum mechanics, nuclear physics, other topics. First paperback edition. xxiii + 885pp. 5⅜ × 8½. 65227-0 Pa. $19.95